Welding

Skills, Processes and Practices for Entry-Level Welders

Book 3

- Shielded Metal Arc Welding
- Gas Tungsten Arc Welding

Larry Jeffus
Lawrence Bower

Australia • Brazil • Japan • Korea • Mexico • Singapore • Spain • United Kingdom • United States

Welding Skills, Processes and Practices for Entry-Level Welders: Book Three
Larry Jeffus/Larry Bower

Vice President, Editorial: Dave Garza

Director of Learning Solutions: Sandy Clark

Executive Editor: David Boelio

Managing Editor: Larry Main

Senior Product Manager: Sharon Chambliss

Editorial Assistant: Lauren Stone

Vice President, Marketing: Jennifer McAvey

Executive Marketing Manager: Deborah S. Yarnell

Senior Marketing Manager: Jimmy Stephens

Marketing Specialist: Mark Pierro

Production Director: Wendy Troeger

Production Manager: Mark Bernard

Content Project Manager: Cheri Plasse

Art Director: Benj Gleeksman

Technology Project Manager: Christopher Catalina

Production Technology Analyst: Thomas Stover

For product information and technology assistance, contact us at
Professional & Career Group Customer Support, 1-800-648-7450
For permission to use material from this text or product,
submit all requests online at **cengage.com/permissions.**
Further permissions questions can be e-mailed to
permissionrequest@cengage.com.

Library of Congress Control Number: 2009920893

ISBN-13: 978-1-4354-2796-9
ISBN-10: 1-4354-2796-3

Delmar
5 Maxwell Drive
Clifton Park, NY 12065-2919
USA

Cengage Learning products are represented in Canada by Nelson Education, Ltd.

For your lifelong learning solutions, visit **delmar.cengage.com**
Visit our corporate website at **cengage.com.**

Notice to the Reader

Publisher does not warrant or guarantee any of the products described herein or perform any independent analysis in connection with any of the product information contained herein. Publisher does not assume, and expressly disclaims, any obligation to obtain and include information other than that provided to it by the manufacturer. The reader is expressly warned to consider and adopt all safety precautions that might be indicated by the activities described herein and to avoid all potential hazards. By following the instructions contained herein, the reader willingly assumes all risks in connection with such instructions. The publisher makes no representations or warranties of any kind, including but not limited to, the warranties of fitness for particular purpose or merchantability, nor are any such representations implied with respect to the material set forth herein, and the publisher takes no responsibility with respect to such material. The publisher shall not be liable for any special, consequential, or exemplary damages resulting, in whole or part, from the readers' use of, or reliance upon, this material.

1 2 3 4 5 XX 11 10 09

Brief Contents

Contents

Preface

ABOUT THE SERIES

Welding: Skills, Processes and Practices for Entry-Level Welders is an exciting new series that has been designed specifically to support the American Welding Society's (AWS) SENSE EG2.0 training guidelines. Offered in three volumes, these books are carefully crafted learning tools consisting of theory-based texts that are accompanied by companion lab manuals, and extensive instructor support materials. With a logical organization that closely follows the modular structure of the AWS guidelines, the series will guide readers through the process of acquiring and practicing welding knowledge and skills. For schools already in the SENSE program, for those planning to join, or for schools interested in obtaining certifiable outcomes based on nationally recognized industry standards in order to comply with the latest Carl D. Perkins Career and Technical Education Requirements. *Welding: Skills, Processes and Practices for Entry-Level Welders* offers a turnkey solution of high quality teaching and learning aids.

Career and technical education instructors at the high school level are often called upon to be multi-disciplinary educators, teaching welding as only one of as many as five technical disciplines in any given semester. The *Welding: Skills, Processes and Practices for Entry-Level Welders* package provides these educators with a process-based, structured approach and the tools they need to be prepared to deliver high level training on processes and materials with which they may have limited familiarity or experience. Student learning, satisfaction and retention are the target of the logically planned practices, supplements and full color textbook illustrations. While the AWS standards for entry level welders are covered, students are also introduced to the latest in high technology welding equipment such as pulsed gas metal arc welding (GMAW-P). Career pathways and career clusters may be enhanced by the relevant mathematics applied to real world activities as well as oral and written communication skills linked to student interaction and reporting.

Book 1, the core volume, introduces students to the welding concepts covered in AWS SENSE Modules 1, 2, 3, 8 and 9 (Occupational Orientation, Safety and Health of Welders, Drawing and Welding Symbol Interpretation, Thermal Cutting, and Weld Inspection Testing and Codes). Book 1 contains all the material needed for a SENSE program that prepares students for qualification in Thermal Cutting processes. The optional Books 2 and 3 cover other important welding processes and are grouped in logical combinations. Book 2 corresponds to AWS SENSE Modules 5 and 6 (GMAW, FCAW), and Book 3 corresponds to AWS SENSE Modules 4 and 7 (SMAW, GTAW).

The texts feature hundreds of four-color figures, diagrams and tight shots of actual welds to speed beginners to an understanding of the most widely used welding processes.

FEATURES

- Produced in close collaboration with experienced instructors from established SENSE programs to maximize the alignment of the content with SENSE guidelines and to ensure 100% coverage of Level I-Entry Welder Key Indicators.
- Chapter introductions contain general performance objectives, key terms used, and the AWS SENSE EG2.0 Key Indicators addressed in the chapter.
- Coverage of Key Indicators is indicated in the margin by a torch symbol and a numerical reference.
- Contains scores of fully illustrated Practices, which are guided exercises designed to help students master processes and materials. Where applicable, the Practices reproduce and reference actual AWS technical drawings in order to help students create acceptable workmanship samples.
- Each section introduces students to the materials, equipment, setup procedures and critical safety information they need in order to weld successfully.
- Hundreds of four-color figures, diagrams and tight shots of actual welds to speed beginners to an understanding of the most widely used welding processes.
- End of chapter review questions develop critical thinking skills and help students to understand "why" as well as "how."

SUPPLEMENTS

Each book in the Welding Skills series is accompanied by a ***Lab Manual*** that has been designed to provide hands-on practice and reinforce the student's understanding of the concepts presented in the text. Each chapter contains practice exercises to reinforce the primary objectives of the lesson, including creation of workmanship samples (where applicable), and a quiz to test knowledge of the material. Artwork and safety precautions are included throughout the manuals.

Instructor Resources (on CD-ROM), designed to support Books 1–3 and the accompanying Lab Manuals, provide a wealth of time-saving tools, including:

- An Instructor's Guide with answers to end-of-chapter Review Questions in the texts and Lab Manual quizzes.
- Modifiable model Lesson Plans that aid in the design of a course of study that meets local or state standards and also maps to the SENSE guidelines.
- An extensive ExamView Computerized Test Bank that offers assessments in true/false, multiple choice, sentence completion and short answer formats. Test questions have been designed to expose students to the types of questions they'll encounter on the SENSE Level 1 Exams.
- PowerPoint Presentations with selected illustrations that provide a springboard for lectures and reinforce skills and processes covered in the texts. The PowerPoint Presentations can be modified or expanded as instructors desire, and can be augmented with additional illustrations from the Image Library.
- The Image Library contains nearly all (well over 1000!) photographs and line art from the texts, most in four-color.
- A SENSE Correlation Chart that shows the close alignment of the *Welding* series to the SENSE Entry Level 1 training guidelines. Each Key Indicator within each SENSE Module is mapped to the relevant text and lab manual page or pages.

TITLES IN THE SERIES

Welding: Skills, Processes and Practices for Entry-Level Welders: Book 1, Occupational Orientation, Safety and Health of Welders, Drawing and Welding Symbol Interpretation, Thermal Cutting, Weld Inspection Testing and Codes
(Order #: 1-4354-2788-2)
Lab Manual, Book One (Order #: 1-4354-2789-0)

Welding: Skills, Processes and Practices for Entry-Level Welders: Book 2, Gas Metal Arc Welding, Flux Cored Arc Welding (Order #: 1-4354-2790-4)
Lab Manual, Book Two (Order #: 1-4354-2795-5)

Welding: Skills, Processes and Practices for Entry-Level Welders: Book 3, Shielded Metal Arc Welding, Gas Tungsten Arc Welding (Order #: 1-4354-2796-3)
Lab Manual, Book Three (Order #: 1-4354-2797-1)

AWS Acknowledgment

The Authors and Publisher gratefully acknowledge the support provided by the American Welding Society in the development and publication of this textbook series. "American Welding Society," the AWS logo and the SENSE logo are the trade and service marks of the American Welding Society and are used with permission.

For more information on the American Welding Society and the SENSE program, visit **http://www.aws.org/education/sense/** or contact AWS at (800) 443-9353 ext. 455 or by email: **education@aws.org**.

Acknowledgments

The authors and publisher would like to thank the following individuals for their contributions:

Garey Bish, *Gwinnett Technical College, Lawrenceville, GA*
Julius Blair, *Greenup County Area Technology Center, Greenup, KY*
Rick Brandon, *Pemiscot County Career & Technical Center, Hayti, MO*
Stephen Brandow, *University of Alaska Southeast, Ketchikan, Ketchikan, AK*
Francis X Brieden, *Career Technology Center of Lackawanna County, Scranton, PA*
John Cavenaugh, *Community College of Southern Nevada, Las Vegas, NV*
Clay Corey, *Washington-Saratoga BOCES, Fort Edward, NY*
Keith Cusey, *Institute for Construction Education, Decatur, IL*
Craig Donnell, *Whitmer Career Technology Center, Toledo, OH*
Steve Farnsworth, *Iowa Lakes Community College, Emmetsburg, IA*
Ed Harrell, *Traviss Career Center, Lakeland, FL*
Robert Hoting, *Northeast Iowa Community College , Sheldon, IA*
Steve Kistler, *Moberly Area Technical Center, Moberly, MO*
David Lynn, *Lebanon Technology & Career Center, Lebanon, MO*
Frank Miller, *Gadsden State Community College , Gadsden, AL*
Chris Overfelt, *Arnold R Burton Tech Center, Salem, VA*
Kenric Sorenson, *Western Technical College, LaCrosse, WI*
Pete Stracener, *South Plains College, Levelland, TX*
Bill Troutman, *Akron Public Schools, Akron, OH*
Norman Verbeck, *Columbia/Montour AVTS, Bloomsburg, PA*

About The Authors

Larry Jeffus is a dedicated teacher and author with over twenty years experience in the classroom and several Delmar Cengage Learning welding publications to his credit. He has been nominated by several colleges for the Innovator of the Year award for setting up nontraditional technical training programs. He was also selected as the Outstanding Post-Secondary Technical Educator in the State of Texas by the Texas Technical Society. Now retired from teaching, he remains very active in the welding community, especially in the field of education.

Lawrence Bower is a welding instructor at Blackhawk Technical College, an AWS SENSE School, in Janesville, Wisconsin. Mr. Bower is an AWS-certified Welding Inspector and Welding Educator. In helping to create *Welding: Skills, Processes and Practices for Entry-Level Welders,* he has brought to bear an excellent mix of training experience and manufacturing know-how from his work in industry, including fourteen years at United Airlines, and six years in the US Navy as an aerospace welder.

CHAPTER

1 Shielded Metal Arc Equipment, Setup, and Operation

OBJECTIVES

After completing this chapter, the student should be able to

- describe the process of shielded metal arc welding (SMAW)
- list and define the three units used to measure a welding current
- contrast how adding various chemicals to the coverings of the electrodes affects the welding arc
- contrast the three different types of current used for welding and their effects on welds
- contrast constant current (CC) and constant voltage (CV) welding power supplies and which type the shielded metal arc welding process requires
- define open circuit voltage and operating voltage
- identify arc blow and apply three different techniques to control it
- tell what the purpose of a welding transformer is and what kind of change occurs to the voltage and amperage with a step-down transformer
- tell the purpose of a rectifier
- read a welding machine duty cycle chart and explain its significance
- determine the proper welding cable size
- service and repair SMAW electrode holders
- identify three problems that can occur as a result of poor work lead clamping
- describe the factors that should be considered when placing an arc welding machine in a welding area

KEY TERMS

amperage
anode
cathode
duty cycle
electrons
inverter
magnetic flux lines
open circuit voltage
operating voltage
output
rectifier
step-down transformer
voltage
wattage
welding cables
welding leads

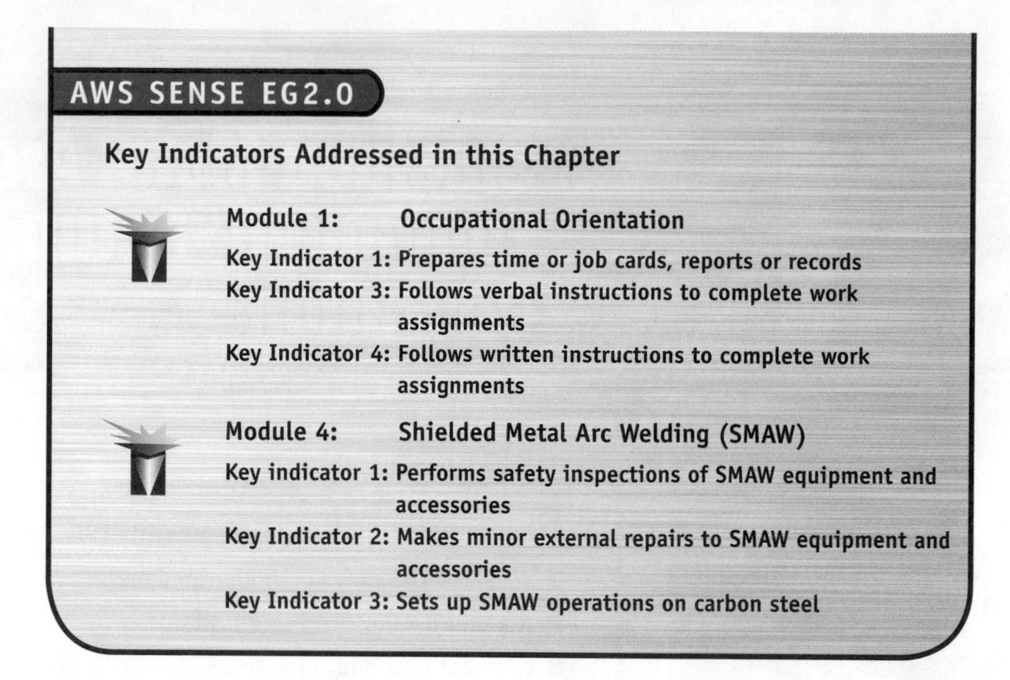

AWS SENSE EG2.0

Key Indicators Addressed in this Chapter

Module 1: Occupational Orientation

Key Indicator 1: Prepares time or job cards, reports or records

Key Indicator 3: Follows verbal instructions to complete work assignments

Key Indicator 4: Follows written instructions to complete work assignments

Module 4: Shielded Metal Arc Welding (SMAW)

Key indicator 1: Performs safety inspections of SMAW equipment and accessories

Key Indicator 2: Makes minor external repairs to SMAW equipment and accessories

Key Indicator 3: Sets up SMAW operations on carbon steel

INTRODUCTION

Shielded metal arc welding (SMAW) is a welding process that uses a flux-covered metal electrode to carry an electrical current, **Figure 1.1.** The current forms an arc across the gap between the end of the electrode and the work. The electric arc creates sufficient heat to melt both the electrode and the work. Molten metal from the electrode travels across the arc to the molten pool on the base metal, where they mix together. The end of the electrode and molten pool of metal is surrounded, purified, and protected by a gaseous cloud and a covering of molten flux is produced as the flux coating of the electrode burns or vaporizes. As the arc moves away, the mixture of molten electrode and base metal solidifies and becomes one piece. At the same time, the molten flux solidifies forming a solid slag. Some electrode types produce heavier slag coverings than others.

SMAW is a widely used welding process because of its low cost, flexibility, portability, and versatility. The machine and the electrodes are low cost. The machine itself can be as simple as a 110-V step-down transformer. The electrodes are available from a large number of manufacturers in packages from 1 lb (0.5 kg) to 50 lb (22 kg).

The SMAW process is very flexible in terms of the metal thicknesses that can be welded and the variety of positions it can be used in. Metal as thin as 1/16 in. (2 mm) thick, or approximately 16 gauge, to several feet thick can be welded using the same machine with different settings and sizes of electrodes. The flexibility of the process also allows metal in this thickness range to be welded in any position.

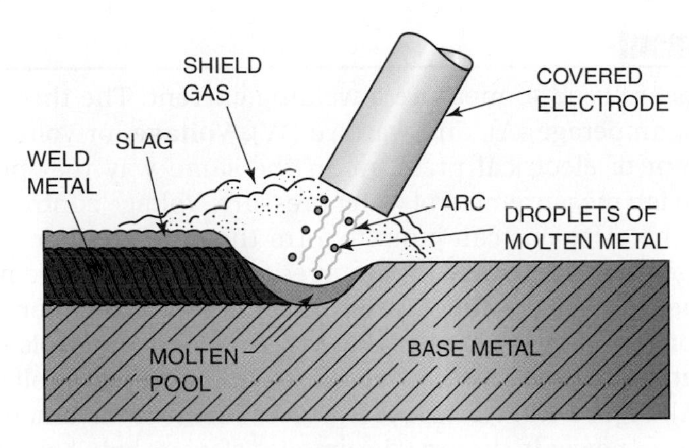

Figure 1.1
Shielded metal arc welding

SMAW is a very portable process because it is easy to move the equipment, and engine-driven generator-type welders are available. Also, the limited amount of equipment required for the process makes moving easy.

The process is versatile, and it is used to weld almost any metal or alloy, including cast iron, aluminum, stainless steel, and nickel.

WELDING CURRENT

The source of heat for arc welding is an electric current. An electric current is the flow of **electrons.** Electrons flow through a conductor from negative (−) to positive (+), **Figure 1.2.** Resistance to the flow of electrons (electricity) produces heat. The greater the resistance, the greater the heat. Air has a high resistance to current flow. As the electrons jump the air gap between the end of the electrode and the work, a great deal of heat is produced. Electrons flowing across an air gap produce an arc.

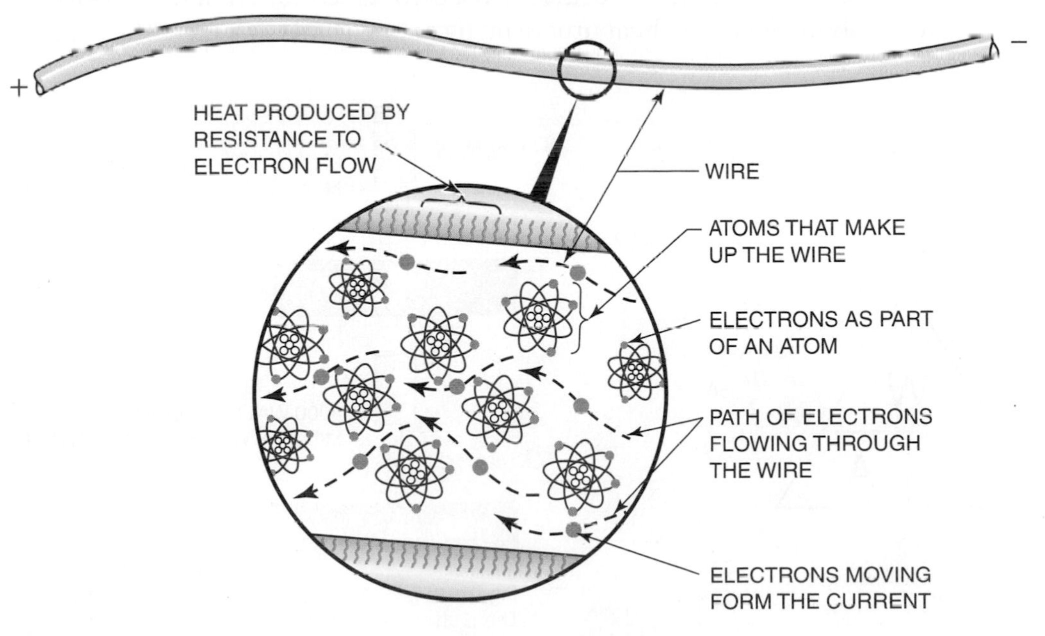

Figure 1.2
Electrons traveling along a conductor

Measurement

Three units are used to measure a welding current. The three units are voltage (V), amperage (A), and wattage (W). **Voltage,** or volts (V), is the measurement of electrical pressure, in the same way that pounds per square inch is a measurement of water pressure. Voltage controls the maximum gap the electrons can jump to form the arc. A higher voltage can jump a larger gap. **Amperage,** or amperes (A), is the measurement of the total number of electrons flowing, in the same way that gallons is a measurement of the amount of water flowing. Amperage controls the size of the arc. **Wattage,** or watts (W), is calculated by multiplying volts (V) times amperes (A), **Figure 1.3.** Wattage is a measurement of the amount of electrical energy or power in the arc. The amount of watts being put into a weld per inch (cm) controls the width and depth of the weld bead, **Figure 1.4.**

Temperature

The temperature of a welding arc exceeds 11,000°F (6000°C). The exact temperature depends on the resistance to the current flow. The resistance is affected by the arc length and the chemical composition of the gases formed as the electrode covering burns and vaporizes. As the arc lengthens, the resistance increases, thus causing a rise in the arc voltage and temperature. The shorter the arc, the lower the arc temperature produced.

Most shielded metal arc welding electrodes have chemicals added to their coverings to stabilize the arc. These arc stabilizers reduce the arc resistance, making it easier to maintain an arc. By lowering the resistance, the arc stabilizers also lower the arc temperature. Other chemicals within the gaseous cloud around the arc may raise or lower the resistance.

The amount of heat produced is determined by the size of the electrode and the amperage setting. Not all of the heat produced by an arc reaches the weld. Some of the heat is radiated away in the form of light and heat waves, **Figure 1.5.** Additional heat is carried away with the hot gases formed by the electrode covering. Heat also is lost through conduction in the work. In total, about 50% of all heat produced by an arc is missing from the weld.

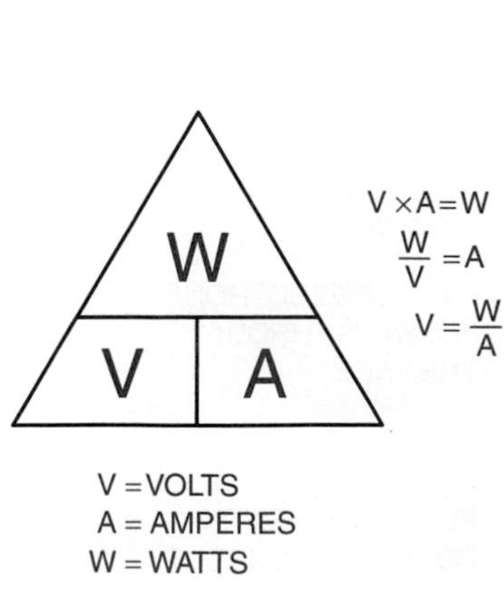

Figure 1.3
Ohm's law

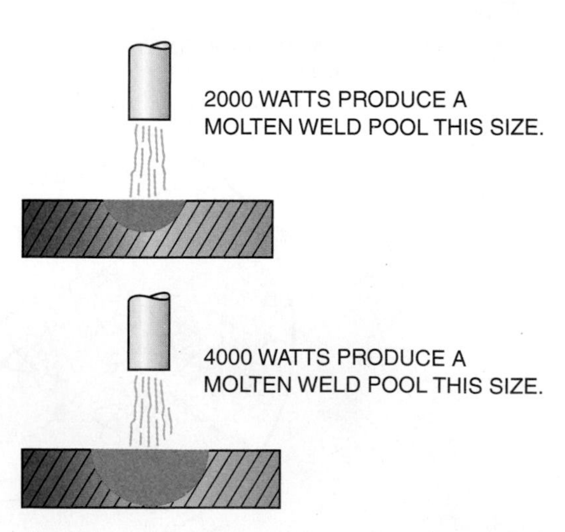

Figure 1.4
The molten weld pool size depends upon the energy (watts), the metal mass, and thermal conductivity

Figure 1.5
Energy is lost from the weld in the forms of radiation and convection

The 50% of the remaining heat the arc produced is not distributed evenly between both ends of the arc. This distribution depends on the composition of the electrode's coating, the type of welding current, and the polarity of the electrode's coating.

Currents

The three different types of current used for welding are alternating current (AC), direct current electrode negative (DCEN), and direct current electrode positive (DCEP). The terms DCEN and DCEP have replaced the former terms *direct current straight polarity (DCSP)* and *direct current reverse polarity (DCRP)*. DCEN and DCSP are the same currents, and DCEP and DCRP are the same currents. Some electrodes can be used with only one type of current. Others can be used with two or more types of current. Each welding current has a different effect on the weld.

DCEN

In direct current electrode negative, the electrode is negative, and the work is positive, **Figure 1.6**. DCEN welding current produces a high electrode melting rate.

DCEP

In direct current electrode positive, the electrode is positive, and the work is negative, **Figure 1.7.** DCEP current produces the deepest penetrating welding arc characteristics.

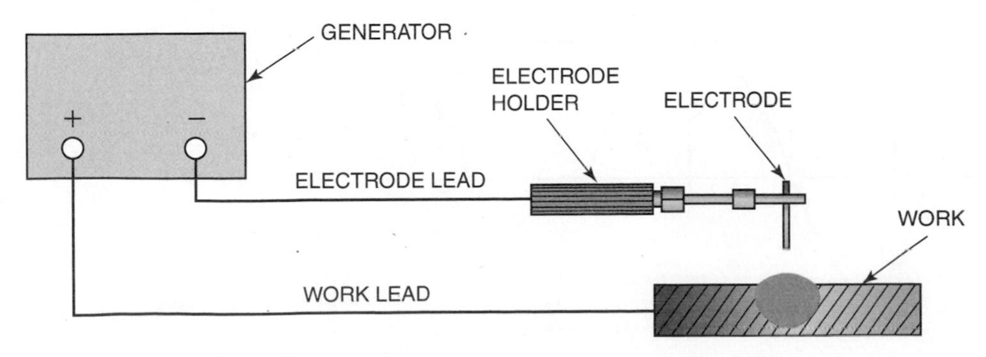

Figure 1.6
Electrode negative (DCEN), straight polarity (DCSP)

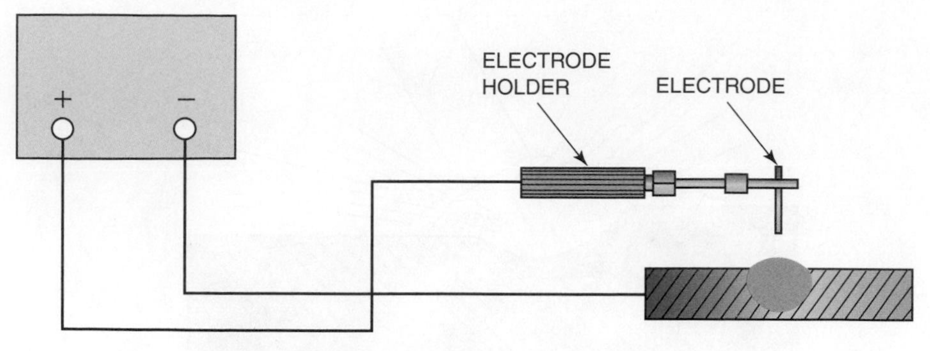

Figure 1.7
Electrode positive (DCEP), reverse polarity (DCRP)

AC

In alternating current, the electrons change direction every 1/120 of a second so that the electrode and work alternate from **anode** to **cathode, Figure 1.8.** The positive side of an electrode arc is called the anode, and the negative side is called the cathode. The rapid reversal of the current flow causes the welding heat to be evenly distributed on both the work and the electrode—that is, half on the work and half on the electrode. The even heating gives the weld bead a balance between penetration and buildup.

TYPES OF WELDING POWER

Welding power can be supplied as

- Constant voltage (CV)—The arc voltage remains constant at the selected setting even if the arc length and amperage increase or decrease.
- Rising-arc voltage (RAV)—The arc voltage increases as the amperage increases.
- Constant current (CC)—The total welding current (watts) remains the same. This type of power is also called drooping-arc voltage (DAV), because the arc voltage decreases as the amperage increases.

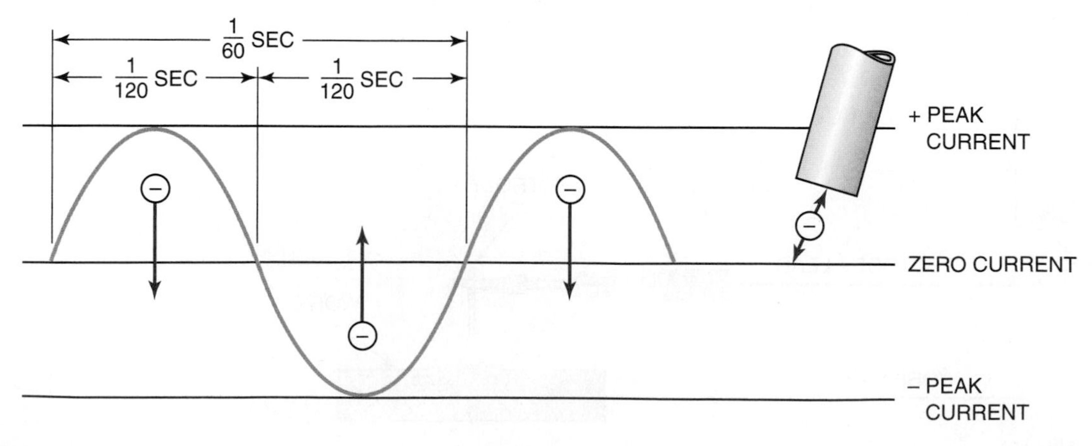

Figure 1.8
Alternating current sine wave (AC)

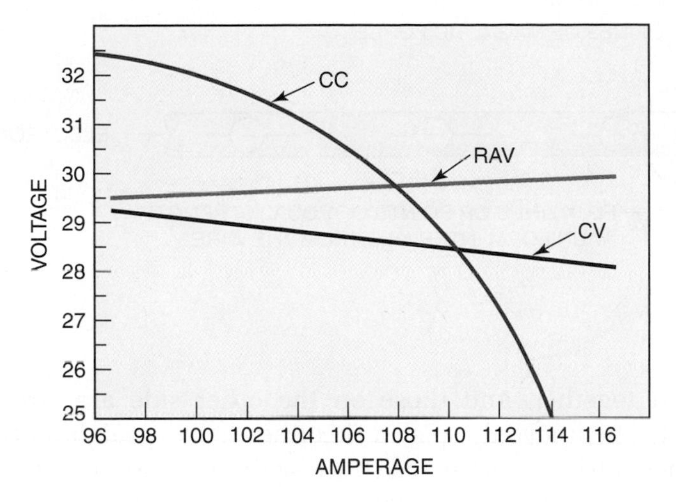

Figure 1.9
Constant voltage (CV), rising arc voltage (RAV), and constant current (CC)

The shielded metal arc welding (SMAW) process requires a constant current arc voltage characteristic, illustrated by the CC line in **Figure 1.9.** The shielded metal arc welding machine's voltage **output** decreases as current increases. This output power supply provides a reasonably high open circuit voltage before the arc is struck. The high open circuit voltage quickly stabilizes the arc. The arc voltage rapidly drops to the lower closed circuit level after the arc is struck. Following this short starting surge, the power (watts) remains almost constant despite the changes in arc length. With a constant voltage output, small changes in arc length would cause the power (watts) to make large swings. The welder would lose control of the weld.

OPEN CIRCUIT VOLTAGE

Open circuit voltage is the voltage at the electrode before striking an arc (with no current being drawn). This voltage is usually between 50 V and 80 V. The higher the open circuit voltage, the easier it is to strike an arc. The higher voltage also increases the chance of electrical shock.

CAUTION

The maximum safe open circuit voltage for welders is 80 V.

OPERATING VOLTAGE

Operating voltage, or closed circuit voltage, is the voltage at the arc during welding. This voltage will vary with arc length, type of electrode being used, type of current, and polarity. The operating voltage will be between 17 V and 40 V.

ARC BLOW

When electrons flow, they create lines of magnetic force that circle around the line of flow, **Figure 1.10.** Lines of magnetic force are referred to as **magnetic flux lines.** These lines space themselves evenly along a current-carrying wire. If the wire is bent, the flux lines on one side are

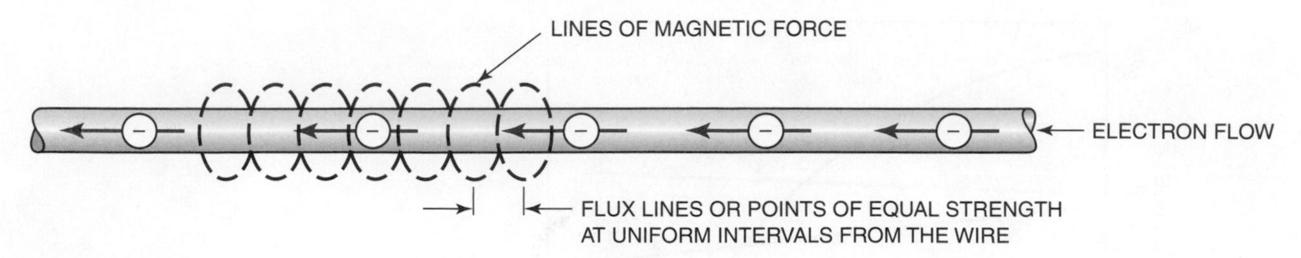

Figure 1.10
Magnetic force around a wire

compressed together, and those on the other side are stretched out, **Figure 1.11.** The unevenly spaced flux lines try to straighten the wire so that the lines can be evenly spaced once again. The force that they place on the wire is usually small. However, when welding with very high amperages, 600 A or more, the force may cause the wire to move.

The welding current flowing through a plate or any residual magnetic fields in the plate will result in uneven flux lines. These uneven flux lines can, in turn, cause an arc to move during a weld. This movement of the arc is called *arc blow.* Arc blow makes the arc drift as a string would drift in the wind. Arc blow is more noticeable in corners, at the ends of plates, and when the work lead is connected to only one side of a plate, **Figure 1.12.** If arc blow is a problem, it can be controlled by connecting the work lead to the end of the weld joint and making the weld in the direction toward the work lead, **Figure 1.13.** Another way of controlling arc blow is to use two work leads, one on each side of the weld. The best way to eliminate arc blow is to use alternating current. AC usually does not allow the flux lines to build long enough to bend the arc before the current changes direction. If it is impossible to move the work connection or to change to AC, a very short arc length can help control arc blow. A small tack weld or a change in the electrode angle can also help control arc blow.

Arc blow may not be a problem as you are learning to weld in the shop, because most welding tables are all steel. If you are using a pipe stand to hold your welding practice plates, arc blow can become a problem. In this case, try reclamping your practice plates or switch to alternating current.

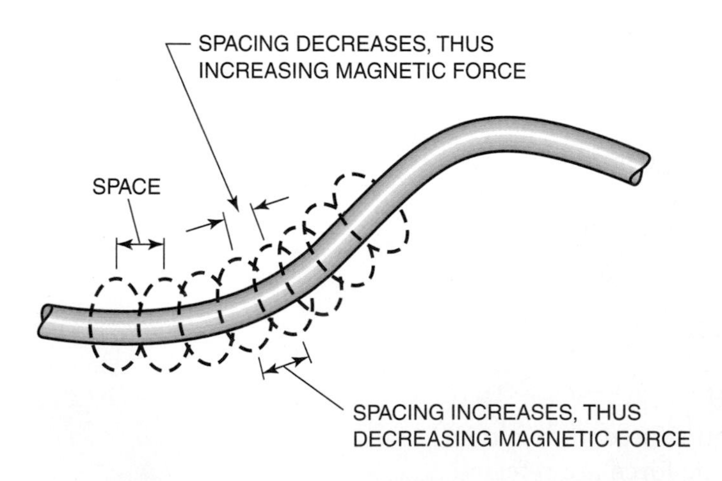

Figure 1.11
Magnetic forces concentrate around bends in wires

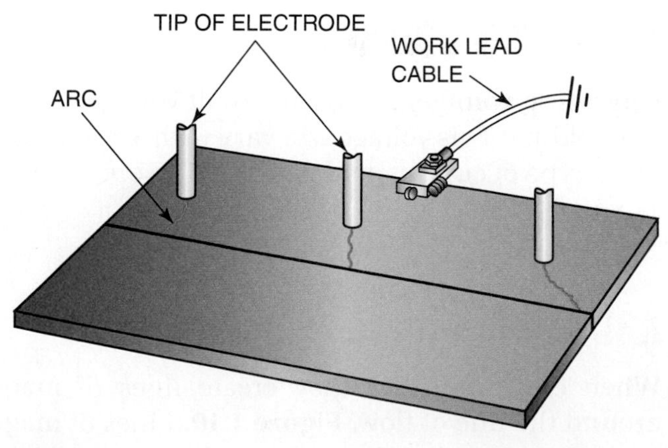

Figure 1.12
Arc blow

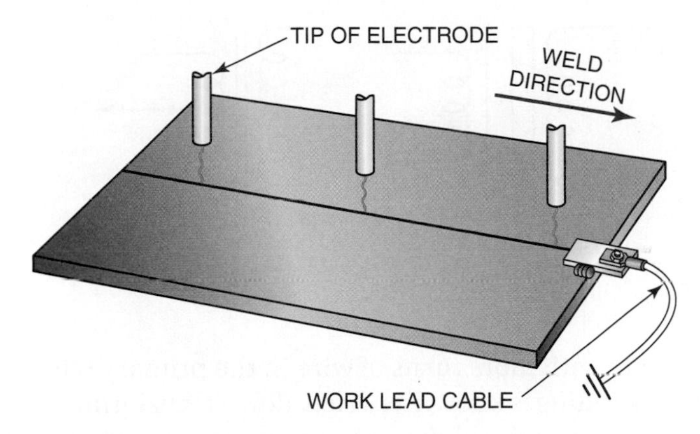

Figure 1.13
Correct current connections to control arc blow

TYPES OF POWER SOURCES

Two types of electrical devices can be used to produce the low-voltage, high-amperage current combination that arc welding requires. One type uses electric motors or internal combustion engines to drive alternators or generators. The other type uses **step-down transformers.** Because transformer-type welding machines are quieter, are more energy efficient, require less maintenance, and are less expensive, they are now the industry standard. However, engine-powered generators are still widely used for portable welding.

Transformers

A welding transformer uses the alternating current (AC) supplied to the welding shop at a high voltage to produce the low-voltage welding power. As electrons flow through a wire they produce a magnetic field around the wire. If the wire is wound into a coil the weak magnetic field of each wire is concentrated to produce a much stronger central magnetic force. Because the current being used is alternating or reversing each 1/120 of a second, the magnetic field is constantly being built and allowed to collapse. By placing a second, or secondary, winding of wire in the magnetic field produced by the first, or primary, winding, a current will be induced in the secondary winding. The placing of an iron core in the center of these coils will increase the concentration of the magnetic field, **Figure 1.14.**

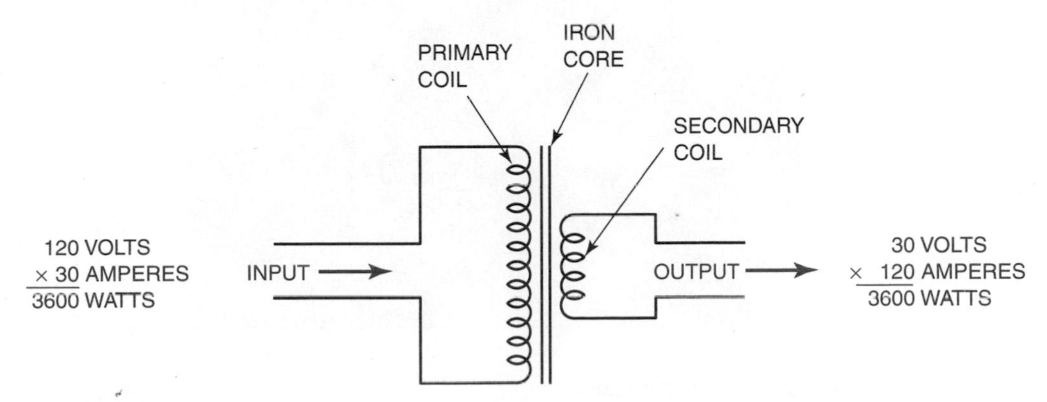

Figure 1.14
Diagram of a step-down transformer

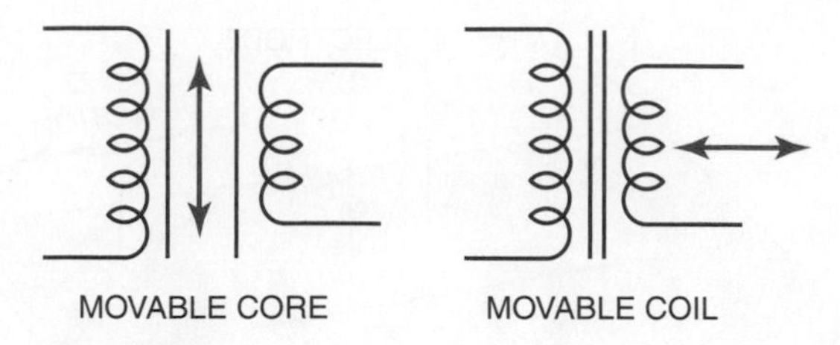

Figure 1.15
Major types of adjustable welding transformers

A transformer with more turns of wire in the primary winding than in the secondary winding is known as a step-down transformer. A step-down transformer takes a high-voltage, low-amperage current and changes it into a low-voltage, high-amperage current. Except for some power lost by heat within a transformer, the power (watts) into a transformer equals the power (watts) out because the volts and amperes are mutually increased and decreased.

A transformer welder is a step-down transformer. It takes the high line voltage (110 V, 220 V, 440 V, etc.) and low-amperage current (30 A, 50 A, 60 A, etc.) and changes it into 17 V to 45 V at 190 A to 590 A.

Welding machines can be classified by the method through which they control or adjust the welding current. The major classifications are multiple-coil, called tap-type, movable-coil or movable-core, **Figure 1.15,** and inverter-type machines.

Multiple-coil Machines

The multiple-coil machine, or tap-type machine, allows the selection of different current settings by tapping into the secondary coil at a different turn value. The greater the number of turns, the higher the amperage is

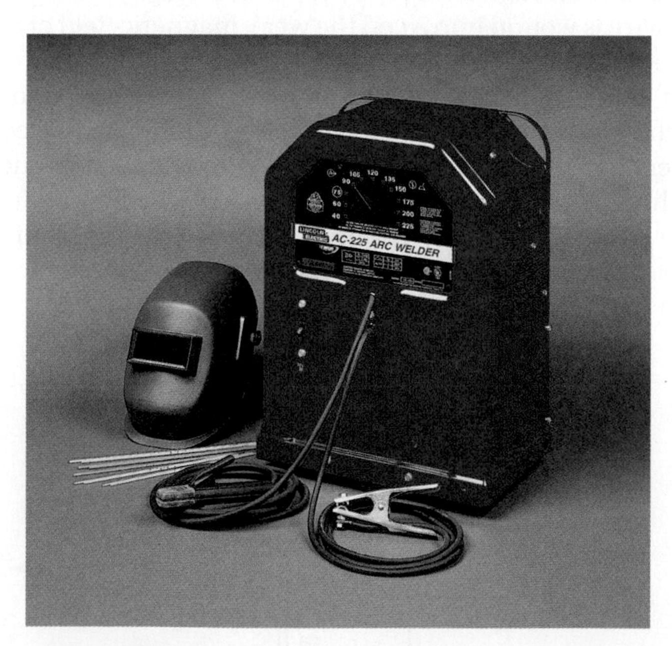

Figure 1.16
Tap-type transformer welding machine
Courtesy of Lincoln Electric Company

induced in the turns. These machines may have a large number of fixed amperes, **Figure 1.16,** or they may have two or more amperages that can be adjusted further with a fine adjusting knob. The fine adjusting knob may be marked in amperes, or it may be marked in tenths, hundredths, or in any other unit.

EXPERIMENT 1-1

Estimating Amperages

Module 1
Key Indicator 1, 3, 4

Module 4
Key Indicator 3

Using a pencil and paper, you will prepare a rough estimate of the amperage setting of a welding machine. **Figure 1.17** shows a welding machine with low, medium, and high tap amperage ranges. A fine adjusting knob is marked with ten equal divisions, and each division is again divided by ten smaller lines.

The machine is set on the medium range, 50 A to 250 A, and the fine adjusting knob is turned until it points to the line marked 5 (halfway between 0 and 10). This means that the amperage is halfway from 50 to 250, or 150 A. If the fine adjusting knob points between 2 and 3, the resulting amperage is one-quarter of the way from 50 to 250, or about 100 A. If the knob points between 7 and 8, the amperage is three-quarters of the way from 50 to 250, or about 200 A. If the knob points at 4, the amperage is more than 100 but a little less than 150, or about 130 A to 140 A. What is the amperage if the knob points at 6?

Since this is a method of estimating only, the amperage value obtained is close enough to allow an arc to be struck. The welder can then finish the fine adjusting to obtain a good weld.

Complete a copy of the "Student Welding Report" listed in Appendix I or provided by your instructor.

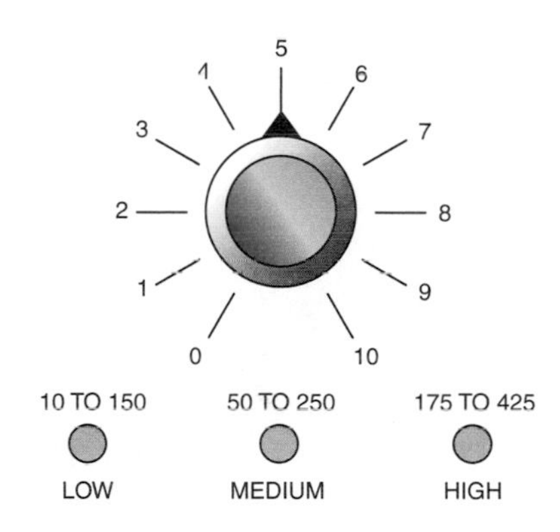

Figure 1.17
Tap-type welder knobs

EXPERIMENT 1-2

Calculating the Amperage Setting

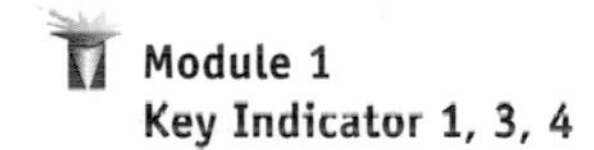
Module 1
Key Indicator 1, 3, 4

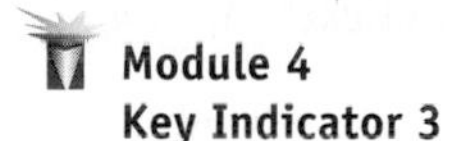
Module 4
Key Indicator 3

Using a pencil and paper or calculator, you will calculate the exact value for each space on the fine adjusting knob of a welding machine.

With the machine set on the medium range, from 50 A to 250 A, first subtract the low amperage from the high amperage to get the amperage spread ($250 - 50 = 200$). Now divide the amperage spread by the number of units shown on the fine adjusting knob ($200 \div 10 = 20$). Each unit is equal to a 20-A increase, **Table 1.1.** When the knob points to 0, the amperage is 50; when the knob points to 1, the amperage is 70; and at 2, the amperage is 90, **Figure 1.18.** There are 100 small units on the fine adjusting knob. Dividing the amperage spread by the number of small units gives the amperage value for each unit ($200 \div 100 = 2$). Therefore, if the knob points to 6.1, the amperage is set at a value of $50 + 120 + 2 = 172$ A. This method provides a good starting place for the current setting, but if the welding is to be made in accordance with a welding procedure's specific amperage setting it will be necessary to use a calibrated meter to make the correct setting.

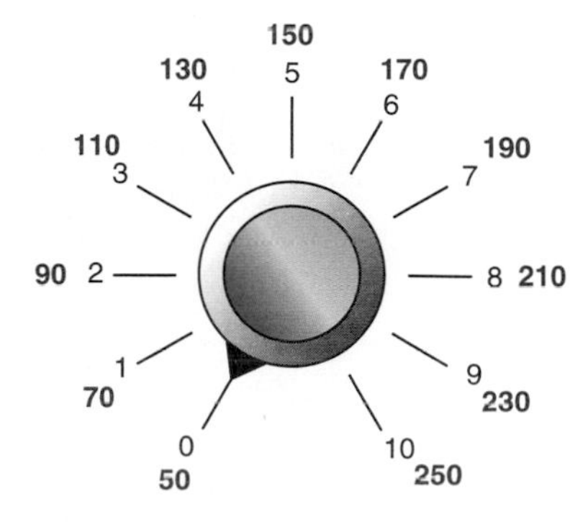

Figure 1.18
Fine adjusting knob

Table 1.1 Example of a Table Used to Calculate the Amperage Setting

Setting	Value in Amperes
0 = 50 + 0,	or 50 A
1 = 50 + 20,	or 70 A
2 = 50 + 40,	or 90 A
3 = 50 + 60,	or 110 A
4 = 50 + 80,	or 130 A
5 = 50 + 100,	or 150 A
6 = 50 + 120,	or 170 A
7 = 50 + 140,	or 190 A
8 = 50 + 160,	or 210 A
9 = 50 + 180,	or 230 A
10 = 50 + 200,	or 250 A

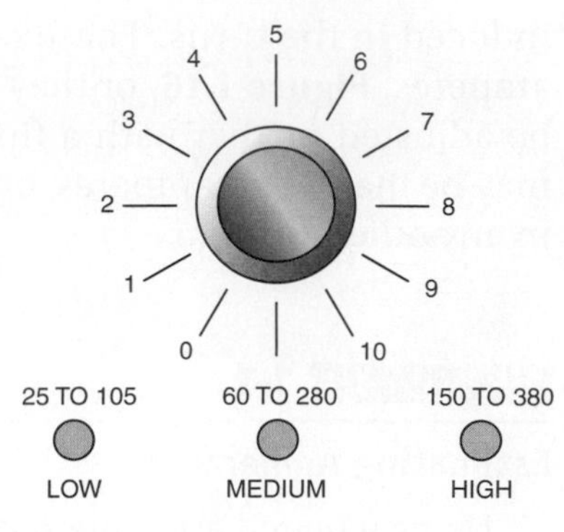

Figure 1.19 Practice 1.1

Complete a copy of the "Student Welding Report" listed in Appendix I or provided by your instructor.

PRACTICE 1-1

Estimating Amperages

Module 1 Key Indicator 1, 3, 4

Module 4 Key Indicator 3

Using a pencil and paper and the amperage ranges given in this practice (or from machines in the shop), you will estimate the amperage when the knob is at the 1/4, 1/2, and 3/4 settings, **Figure 1.19.**

Complete a copy of the "Student Welding Report" listed in Appendix I or provided by your instructor.

PRACTICE 1-2

Calculating Amperages

Module 1 Key Indicator 1, 3, 4

Module 4 Key Indicator 3

Using a pencil and paper or a calculator, and the amperage ranges given in this practice (or from machines in the shop), you will calculate the amperages for each of the following knob settings: 1, 4, 7, 9, 2.3, 5.7, and 8.5.

Complete a copy of the "Student Welding Report" listed in Appendix I or provided by your instructor.

Movable-coil or Movable-core Machines

Movable-coil or movable-core machines are adjusted by turning a handwheel that moves the internal parts closer together or farther apart. The adjustment may also be made by moving a lever, **Figure 1.20.** These machines may have a high and low range, but they do not have a fine adjusting knob. The closer the primary and secondary coils are, the greater is the induced current; the greater the distance between the coils, the smaller the induced current, **Figure 1.21.** Moving the core in concentrates more of the magnetic force on the secondary coil, thus increasing the current. Moving the core out allows the field to disperse, and the current is reduced, **Figure 1.22.**

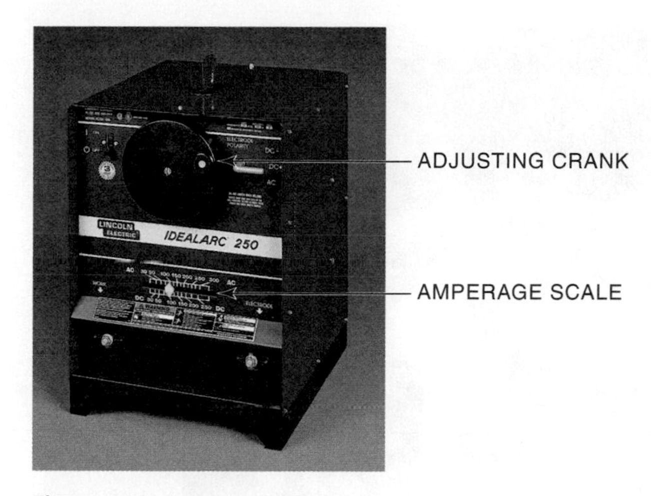

Figure 1.20
A movable core-type welding machine
Courtesy of Lincoln Electric Company

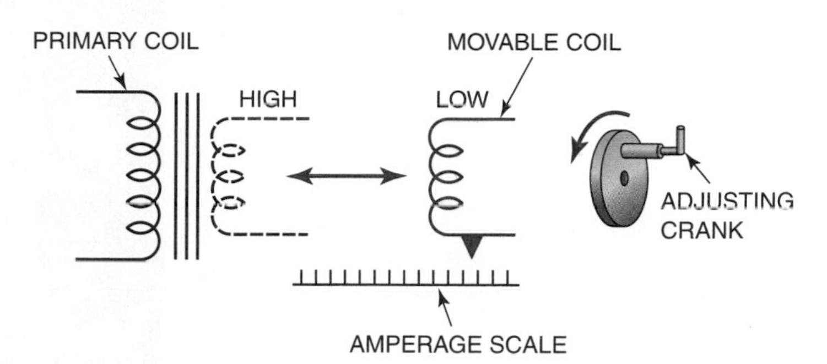

Figure 1.21
Movable coil

Inverter Machines

Inverter welding machines are much smaller than other types of machines of the same amperage range. This smaller size makes the welder much more portable and increases the energy efficiency. In a standard welding transformer, the iron core used to concentrate the magnetic field in the coils must be a specific size. The size of the iron core is determined by the length of time it takes for the magnetic field to build and collapse. By using solid-state electronic parts, the incoming power in an inverter welder is changed from 60 cycles a second to several thousand cycles a second. This higher frequency allows the use of a transformer that may be as light as 7 lb and still do the work of a standard transformer weighing 100 lb. Additional electronic parts remove the high frequency for the output welding power.

The use of electronics in the inverter-type welder allows it to produce any desired type of welding power. Before the invention of this machine, each type of welding required a separate machine. Now a single welding machine can produce the specific type of current needed for shielded metal arc welding, gas tungsten arc welding, gas metal arc welding, and plasma arc cutting. Because the machine can be light enough to be carried closer to work, shorter welding cables can be used. The welder does not have to walk as far to adjust the machine. Welding machine power wire is cheaper than welding cables. Some manufacturers produce machines that can be stacked so that when you need a larger machine all you have to do is add another unit to your existing welder, **Figure 1.23.**

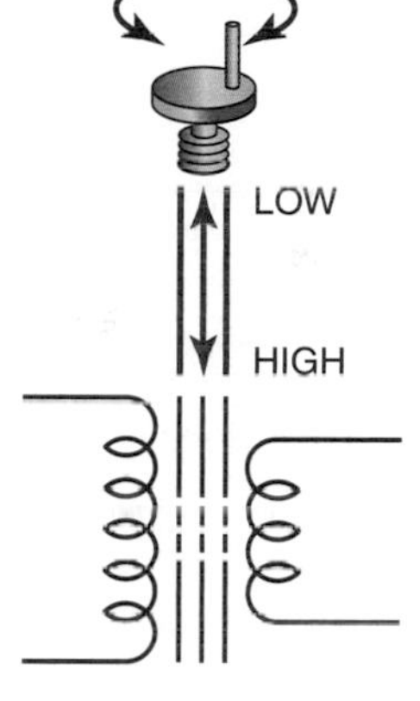

Figure 1.22
Movable core

GENERATORS AND ALTERNATORS

Generators and alternators both produce welding electricity from a mechanical power source. Both devices have an armature that rotates and a stator that is stationary. As a wire moves through a magnetic force field, electrons in the wire are made to move, producing electricity.

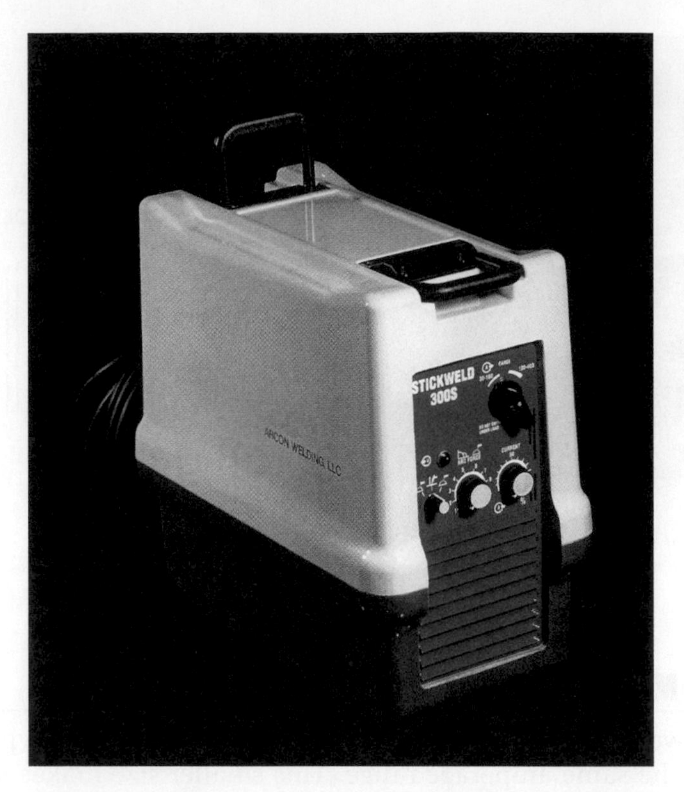

Figure 1.23
Typical 300-A inverter-type power supply weighing only 70 lb
Courtesy of Arcon Welding, L.L.C.

In an alternator, magnetic lines of force rotate inside a coil of wire, **Figure 1.24.** An alternator can produce AC only. In a generator, a coil of wire rotates inside a magnetic field. A generator produces DC. It is possible for alternators to use diodes to change the AC to DC for welding. In generators, the welding current is produced on the armature and is picked up with brushes, **Figure 1.25.** In alternators, the welding current is produced on the stator, and only the small current for the electromagnetic force field goes across the brushes. Therefore, the brushes in an alter-

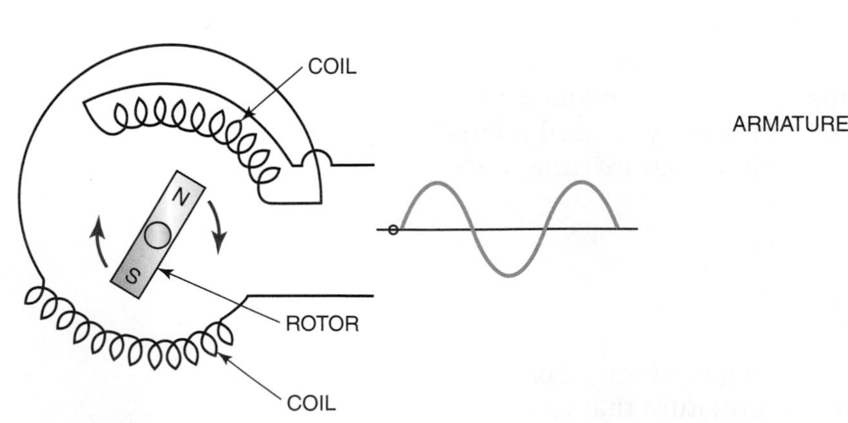

Figure 1.24
Schematic diagram of an alternator

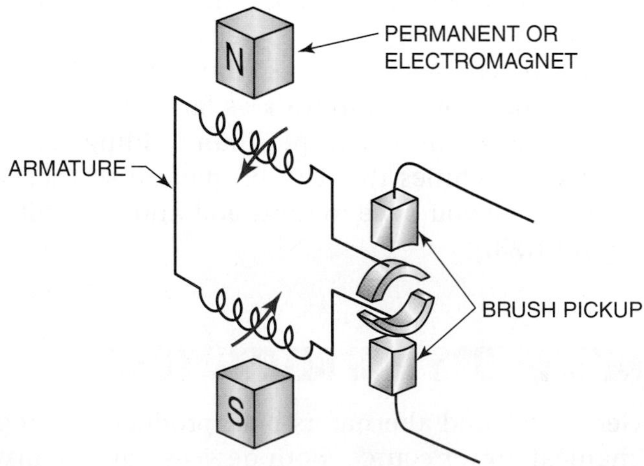

Figure 1.25
Diagram of a generator

(A)

(B)

Figure 1.26
Portable engine generator welder
Courtesy of Lincoln Electric Company

nator are smaller and last longer. Alternators can be smaller in size and lighter in weight than generators and still produce the same amount of power.

Engine-driven generators and alternators may run at the welding speed all the time, or they may have an option that reduces their speed to an idle when welding stops. This option saves fuel and reduces wear on the welding machine. To strike an arc when using this type of welder, stick the electrode to the work for a second. When you hear the welding machine (welder) pick up speed, remove the electrode from the work and strike an arc. In general, the voltage and amperage are too low to start a weld, so shorting the electrode to the work should not cause the electrode to stick. A timer can be set to control the length of time that the welder maintains speed after the arc is broken. The time should be set long enough to change electrodes without losing speed.

Portable welders often have 110-V or 220-V plug outlets, which can be used to run grinders, drills, lights, and other equipment. The power provided may be AC or DC. If DC is provided, only equipment with brush-type motors or tungsten light bulbs can be used. If the plug is not specifically labeled 110 V AC, check the owner's manual before using it for such devices as radios or other electronic equipment. A typical portable welder is shown in **Figure 1.26.**

It is recommended that a routine maintenance schedule for portable welders be set up and followed. By checking the oil, coolant, battery, filters, fuel, and other parts, the life of the equipment can be extended. A checklist can be posted on the welder, **Table 1.2.**

RECTIFIERS

Alternating-welding current can be converted to direct current by using a series of rectifiers. A **rectifier** allows current to flow in one direction only, **Figure 1.27.**

Table 1.2 Portable Welder Checklist. The owner's manual should be checked for any additional items that might need attention.

Check Each Day before Starting
Oil level
Water level
Fuel level
Check Each Monday
Battery level
Cables
Fuel line filter
Check at Beginning of Month
Air filter
Belts and hoses
Change oil and filter
Check Each Fall
Antifreeze
Test battery
Pack wheel bearings
Change gas filter

If one rectifier is added, the welding power appears as shown in **Figure 1.28.** It would be difficult to weld with pulsating power such as this. A series of rectifiers, known as a bridge rectifier, can modify the alternating current so that it appears as shown in **Figure 1.29.**

Rectifiers become hot as they change AC to DC. They must be attached to a heat sink and cooled by having air blown over them. The heat produced by a rectifier reduces the power efficiency of the welding machine. **Figure 1.30** shows the amperage dial of a typical machine. Notice that at the same dial settings for AC and DC, the DC is at a lower amperage. The difference in amperage (power) is due to heat lost in the rectifiers. The loss in power makes operation with AC more efficient and less expensive compared to DC.

A DC adapter for small AC machines is available from manufacturers. For some types of welding, AC does not work properly.

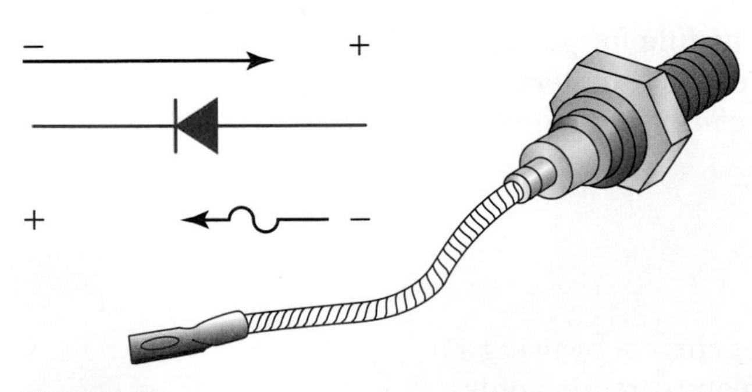

Figure 1.27
Rectifier

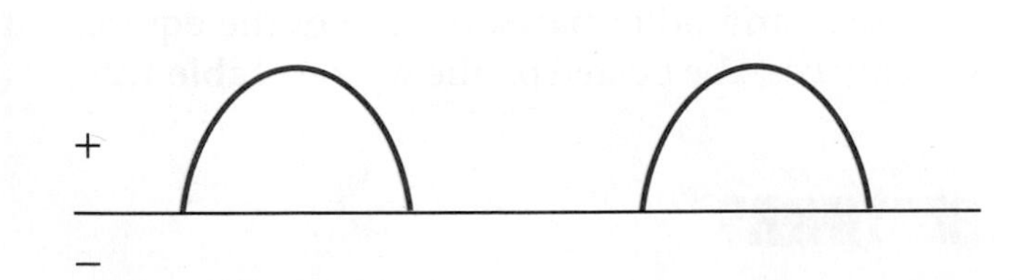

Figure 1.28
One rectifier in a welding power supply results in pulsating power

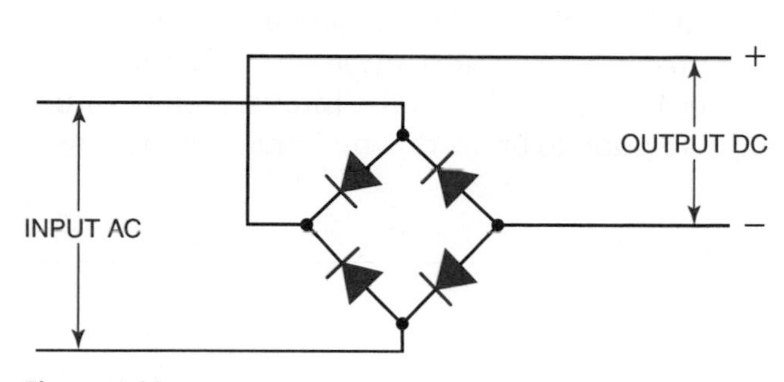

Figure 1.29
Bridge rectifier

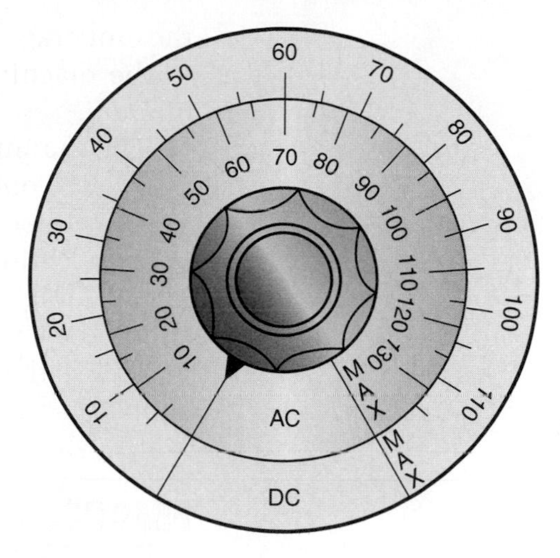

Figure 1.30
Typical dial on an AC-DC transformer rectifier welder

DUTY CYCLE

Welding machines produce internal heat at the same time that they produce the welding current. Except for automatic welding machines, welders are rarely used every minute for long periods of time. The welder must take time to change electrodes, change positions, or change parts. Shielded metal arc welding never continues for long periods of time.

The **duty cycle** is the percentage of time a welding machine can be used continuously. A 60% duty cycle means that out of any 10 minutes, the machine can be used for a total of six minutes at the maximum rated current. When providing power at this level, it must be cooled off for four minutes out of every 10 minutes. The duty cycle increases as the amperage is lowered and decreases for higher amperages, **Figure 1.31.** Most welding machines weld at a 60% rate or less. Therefore, most manufacturers list

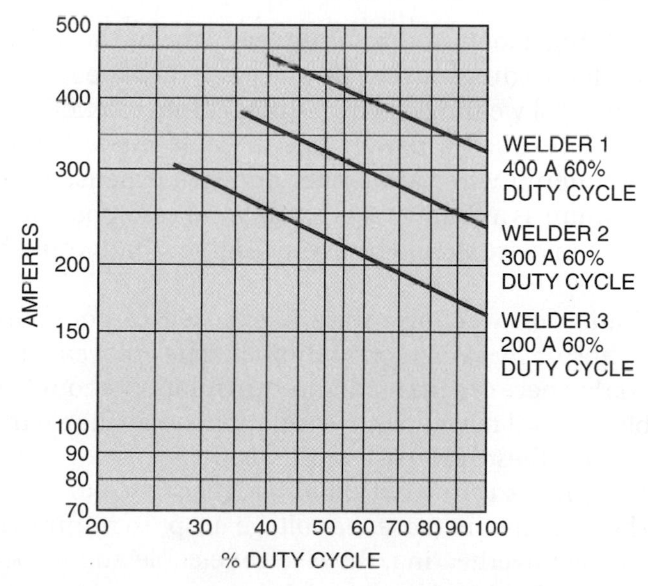

Figure 1.31
Duty cycle of a typical shielded metal arc welding machine

the amperage rating for a 60% duty cycle on the nameplate that is attached to the machine. Other duty cycles are given on a graph in the owner's manual.

The manufacturing cost of power supplies increases in proportion to their rated output and duty cycle. To reduce their price, it is necessary to reduce either their rating or their duty cycle. For this reason, some home-hobby welding machines may have duty cycles as low as 20% even at a low welding setting of 90 A to 100 A. The duty cycle on these machines should never be exceeded because a buildup of the internal temperature can cause the transformer insulation to break down, damaging the power supply.

PRACTICE 1-3

Reading Duty Cycle Chart

Module 1
Key Indicator 1, 3, 4

Module 4
Key Indicator 3

Using a pencil and paper and the duty cycle chart in Figure 1.31 (or one from machines in the shop), you will determine the following:

Welder 1: maximum welding amperage percent duty cycle at maximum amperage
Welder 2: maximum welding amperage percent duty cycle at maximum amperage
Welder 3: maximum welding amperage percent duty cycle at maximum amperage
Welder 1: maximum welding amperage at 100% duty cycle
Welder 2: maximum welding amperage at 100% duty cycle
Welder 3: maximum welding amperage at 100% duty cycle

Complete a copy of the "Student Welding Report" listed in Appendix I or provided by your instructor.

WELDING CABLES AND LEADS

The terms **welding cables** and **welding leads** mean the same thing. Cables to be used for welding must be flexible, well insulated, and the correct size for the job. Most welding cables are made from stranded copper wire. Some manufacturers sell a newer type of cable made from aluminum wires. The aluminum wires are lighter and less expensive than copper. Because aluminum as a conductor is not as good as copper for a given wire size, the aluminum wire should be one size larger than would be required for copper.

The insulation on welding cables is exposed to hot sparks, flames, grease, oils, sharp edges, impact, and other types of wear. To withstand such wear, only specially manufactured insulation should be used for welding cable. Several new types of insulation are available that give longer service against these adverse conditions.

As electricity flows through a cable, the resistance to the flow causes the cable to heat up and increase the voltage drop. To minimize the loss of power and prevent overheating, the electrode cable and work cable must be the correct size. **Table 1.3** lists the minimum size cable that is required for each amperage and length. Large welding lead sizes make electrode

Table 1.3 Copper and Aluminum Welding Lead Sizes

Amperes		Copper Welding Lead Sizes								
ft	m	100	150	200	250	300	350	400	450	500
50	15	2	2	2	2	1	1/0	1/0	2/0	2/0
75	23	2	2	1	1/0	2/0	2/0	3/0	3/0	4/0
100	30	2	1	1/0	2/0	3/0	4/0	4/0		
125	38	2	1/0	2/0	3/0	4/0				
150	46	1	2/0	3/0	4/0					
175	53	1/0	3/0	4/0						
200	61	1/0	3/0	4/0						
250	76	2/0	4/0							
300	91	3/0								
350	107	3/0								
400	122	4/0								

Length of Cable

Amperes		Aluminum Welding Lead Sizes								
ft	m	100	150	200	250	300	350	400	450	500
50	15	2	2	1/0	2/0	2/0	3/0	4/0		
75	23	2	1/0	2/0	3/0	4/0				
100	30	1/0	2/0	4/0						
125	38	2/0	3/0							
150	46	2/0	3/0							
175	53	3/0								
200	61	4/0								
225	69	4/0								

Length of Cable

manipulation difficult. Smaller cable can be spliced to the electrode end of a large cable to make it more flexible. This whip-end cable must not be more than 10 ft (3 m) long.

CAUTION

A splice in a cable should not be within 10 ft (3 m) of the electrode because of the possibility of electrical shock.

PRACTICE 1-4

Determining Welding Lead Sizes

Using a pencil and paper and Table 1.3, Copper and Aluminum Welding Lead Sizes, you will determine the following:

1. the minimum copper welding lead size for a 200-A welder with 100-ft (30-m) leads
2. the minimum copper welding lead size for a 125-A welder with 225-ft (69-m) leads
3. the maximum length aluminum welding lead that can carry 300 A

Module 1
Key Indicator 1, 3, 4

Module 4
Key Indicator 3

Splices and end lugs are available from suppliers. Be sure that a good electrical connection is made whenever splices or lugs are used. A poor electrical connection will result in heat buildup, voltage drop, and poor service from the cable. Splices and end lugs must be well insulated against possible electrical shorting, **Figure 1.32.**

Complete a copy of the "Student Welding Report" listed in Appendix I or provided by your instructor.

Figure 1.32
Power lug protection is provided by insulators
Courtesy of ESAB Welding & Cutting Products

CAUTION

Never dip a hot electrode holder in water to cool it off. The problem causing the holder to overheat should be repaired.

ELECTRODE HOLDERS

The electrode holder should be of the proper amperage rating and in good repair for safe welding. Electrode holders are designed to be used at their maximum amperage rating or less. Higher amperage values will cause the holder to overheat and burn up. If the holder is too large for the amperage range being used, manipulation is hard, and operator fatigue increases. Make sure that the correct amperage holder is chosen, **Figure 1.33.**

A properly sized electrode holder can overheat if the jaws are dirty or loose, or if the cable is loose. If the holder heats up, welding power is being lost. In addition, a hot electrode holder is uncomfortable to work with.

Replacement springs, jaws, insulators, handles, screws, and other parts are available to keep the holder in good working order, **Figure 1.34.** To prevent excessive damage to the holder, welding electrodes should not be burned too short. A 2-in. (51-mm) electrode stub is short enough to minimize electrode waste and save the holder.

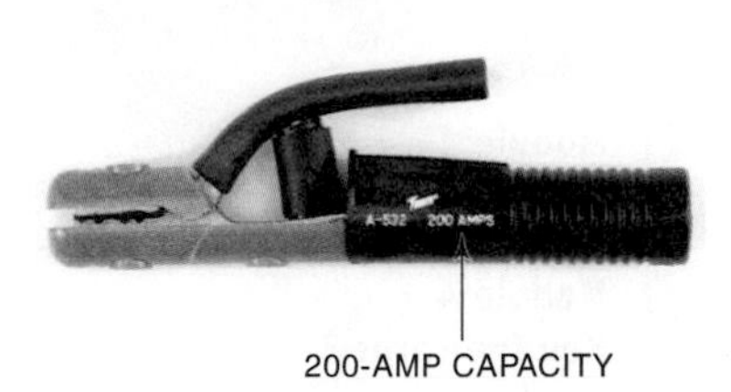

Figure 1.33
The amperage capacity of an electrode holder is often marked on its side
Courtesy of Tweco®, a Thermadyne® Company

PRACTICE 1-5

Repairing Electrode Holders

Module 1
Key Indicator 1, 3, 4

Module 4
Key Indicator 1, 2

Using the manufacturer's instructions for your type of electrode holder, required hand tools, and replacement parts, you will do the following:

1. Remove the electrode holder from the welding cable.
2. Remove the jaw insulating covers.
3. Replace the jaw insulating covers.

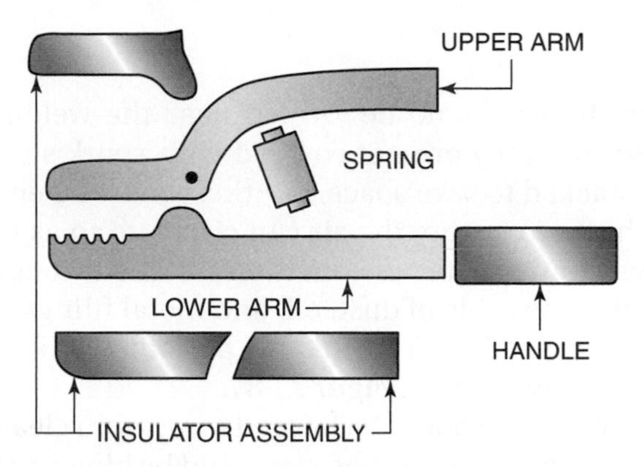

Figure 1.34
Replaceable parts of an electrode holder

4. Reconnect the electrode holder to the welding cable.
5. Turn on the welding power or reconnect the welding cable to the welder.
6. Make a weld to ensure that the repair was made correctly.

Complete a copy of the "Student Welding Report" listed in Appendix I or provided by your instructor.

CAUTION

Before starting any work, make sure that the power to the welder is off and locked off or the welding lead has been removed from the machine.

WORK CLAMPS

The work clamp must be the correct size for the current being used, and it must clamp tightly to the material. Heat can build up in the work clamp, reducing welding efficiency, just as was previously described for the electrode holder. Power losses in the work clamp are often overlooked. The clamp should be carefully touched occasionally to find out if it is getting hot.

In addition to power losses due to poor work lead clamping, a loose clamp may cause arcing that can damage a part. Improper work clamp placement may also contribute to arc blow. To avoid arc blow, keep the work clamp situated as close to the weld as practical, and weld away from it. If the part is to be moved during welding, a swivel-type work clamp may be needed, **Figure 1.35.** It may be necessary to weld a tab to thick parts so that the work lead can be clamped to the tab, **Figure 1.36.**

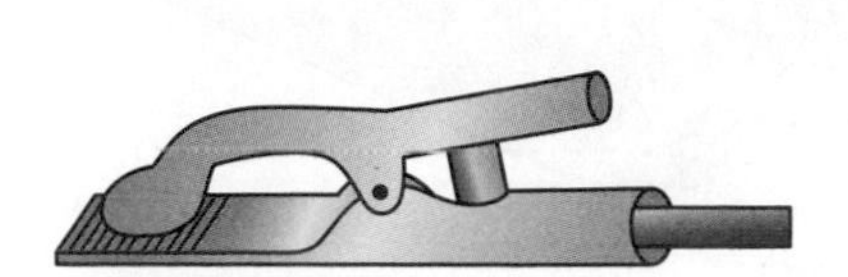

Figure 1.35
A work clamp may be attached to the workpiece

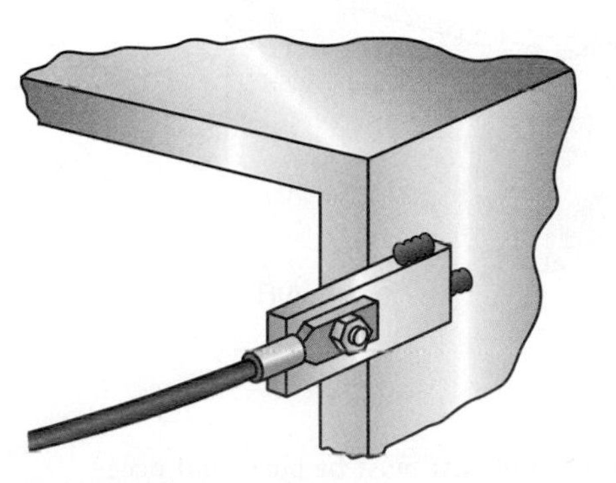

Figure 1.36
Tack welded ground to part

SETUP

Module 1
Key Indicator 1

Module 4
Key Indicator 3

Arc welding machines should be located near the welding site, but far enough away so that they are not covered with spark showers. The machines may be stacked to save space, but there must be enough room between the machines to ensure the air can circulate so as to keep the machines from overheating. The air that is circulated through the machine should be as free as possible of dust, oil, and metal filings. Even in a good location, the power should be turned off periodically and the machine blown out with compressed air, **Figure 1.37.**

The welding machine should be located away from cleaning tanks and any other sources of corrosive fumes that could be blown through it. Water leaks must be fixed and puddles cleaned up before a machine is used.

Power to the machine must be fused, and a power shut-off switch provided. The switch must be located so that it can be reached in an emergency without touching either the machine or the welding station. The machine case or frame must be grounded.

The welding cables should be sufficiently long to reach the work station but not so long that they must always be coiled. Cables should not be placed on the floor in aisles or walkways. If cables must cross a walkway, the cable must be installed overhead, or it must be protected by a ramp, **Figure 1.38.** The welding machine and its main power switch should be off while a person is installing or working on the cables.

The work station must be free of combustible materials. Screens or curtains should be provided to protect other workers from the arc light.

The welding cable should never be wrapped around arms, shoulders, waist, or any other part of the body. If the cable was caught by any moving equipment, such as a forklift, crane, or dolly, a welder could be pulled off balance or more seriously injured. If it is necessary to hold the weight off the cable so that the welding can more easily be done, a free hand can be used. The cable should be held so that if it is pulled it can be easily released.

Check the surroundings before starting to weld. If heavy materials are being moved in the area around you, there should be a safety watch. A safety watch can warn a person of danger while that person is welding.

CAUTION

The cable should never be tied to scaffolding or ladders. If the cable is caught by moving equipment, the scaffolding or ladder may be upset, causing serious personal injury.

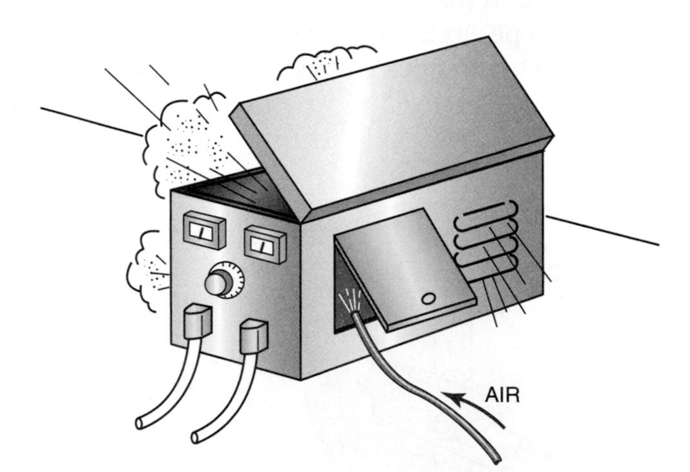

Figure 1.37
Slag, chips from grinding, and dust must be blown out occasionally so that they will not start a fire or cause a short-out or other types of machine failure

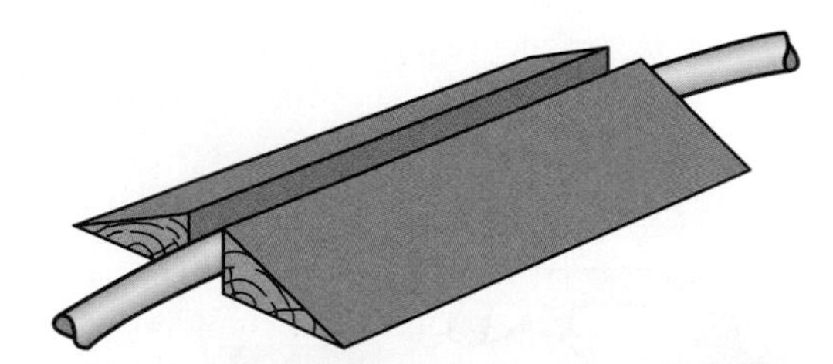

Figure 1.38
To prevent people from tripping, when cables must be placed in walkways, lay two blocks of wood beside the cables

SUMMARY

Understanding the scientific theory of electricity and magnetism will aid you in understanding how the welding currents are produced and their reactions to changes in their physical surroundings. Understanding electromagnetic phenomena will aid you in controlling arc blow. Failure to control arc blow can result in weld failures. In addition, understanding electricity will help you interpret information given on manufacturers' tables, charts, and equipment specifications.

Before starting any new job or welding operation, be sure to check with the equipment manufacturer's safety guidelines for proper operation and maintenance. Follow all recommended guidelines.

Keeping your work area clean and orderly will help prevent accidents.

REVIEW

1. Describe the welding current.
2. What produces the heat during a shielded metal arc weld?
3. Voltage can be described as _________.
4. Amperage can be described as _________.
5. Wattage can be described as _________.
6. What determines the exact temperature of the shielded metal welding arc?
7. Does all of the heat produced by an SMA weld stay in the weld? Why or why not?
8. What do the following abbreviations mean: AC, DCEN, DCEP, DCSP, and DCRP?
9. Sketch a welding machine, an electrode lead, an electrode holder, an electrode, a work lead, and work connected for DCEN welding.
10. Sketch a welding machine, an electrode lead, an electrode holder, an electrode, a work lead, and work connected for DCEP welding.
11. Why is SMA welding current referred to as *constant current*?
12. What is the higher voltage at the electrode before the arc is struck called? What is its advantage to welding?
13. Referring to the graph in Figure 1.9, what would the voltage be for the CC power supply at 110 A? What would the watts be?
14. How does arc blow affect welding?
15. How can arc blow be controlled?
16. How does a welding transformer work?
17. What are *taps* on a welding transformer?
18. What would the approximate amperage setting be if a welder were set to the high range (150 A to 350 A) and the fine adjustment knob were pointing at 5 on a 10-point scale?
19. What are the advantages of the inverter-type welding power supply?
20. What is the difference between the welding current produced by alternators and the one produced by generators?
21. What are the advantages of alternators over generators?
22. What must be checked before using the 110-V power plug on a portable welder?
23. What is meant by a *welder's duty cycle*?

24. Why must a welding machine's duty cycle never be exceeded?
25. Why is copper better than aluminum for welding cables?
26. A splice in a welding cable should never be any closer than _________ to the electrode holder.
27. Why must the electrode holder be correctly sized?
28. What can cause a properly sized electrode holder to overheat?
29. What problem can occur if welding machines are stacked or placed too closely together?
30. Why must welding cables never be tied to scaffolding or ladders?

CHAPTER

2

Shielded Metal Arc Welding of Plate

OBJECTIVES

After completing this chapter, the student should be able to

- demonstrate safe SMAW work practices
- strike an arc at a specific point
- contrast the problems that can result if the welding current is set too low or too high
- select the correct diameter of welding electrode for a weld
- compare a weld bead's shape for width, reinforcement, and appearance with differing heat inputs
- define arc length, and describe the effects of using too short or too long an arc length
- compare a leading electrode angle to a trailing electrode angle
- list three characteristics of the weld bead that can be controlled by the movement or weaving of the welding electrode
- demonstrate ten weave patterns for weld beads
- match various SMAW electrodes to the four filler metal groups
- define stringer beads and tell how they are used
- make a vertical up stringer bead and a horizontal stringer bead
- make a welded square butt joint in the flat, vertical up, and the horizontal positions
- on an edge joint, make a flat weld, a vertical down weld, a vertical up weld, a horizontal weld, and an overhead weld
- on an outside corner joint, make a flat weld, a vertical down weld, a vertical up weld, a horizontal weld, and an overhead weld
- make fillet welds in lap joints in all positions
- make fillet welds in tee joints in all positions
- make groove welds in plate in all positions without backing

KEY TERMS

amperage range
arc length
cellulose-based fluxes
electrode angle
lap joint
mineral-based fluxes
rutile-based fluxes
square butt joint
stringer bead
tee joint
weave pattern

AWS SENSE EG2.0

Key Indicators Addressed in this Chapter

Module 1: Occupational Orientation

Key Indicator 1: Prepares time or job cards, reports or records
Key Indicator 2: Performs housekeeping duties
Key Indicator 3: Follows verbal instructions to complete work assignments
Key Indicator 4: Follows written instructions to complete work assignments

Module 2: Safety and Health of Welders

Key indicator 1: Demonstrates proper use and inspection of personal protective equipment (PPE)

Key Indicator 2: Demonstrates safe operation practices in the work area

Key Indicator 3: Demonstrates proper use and inspection of ventilation equipment

Key Indicator 4: Demonstrates proper hot zone operation

Key Indicator 5: Demonstrates proper work actions for work in confined spaces

Key Indicator 6: Understands proper use of precautionary labeling and MSDS information

Module 4: Shielded Metal Arc Welding (SMAW)

Key Indicator 1: Performs safety inspections of SMAW equipment and accessories

Key Indicator 3: Sets up SMAW operations on carbon steel

Key Indicator 4: Operates SMAW equipment on carbon steel

Key Indicator 5: Makes SMAW fillet welds, in all positions, on carbon steel

Key Indicator 6: Makes SMAW groove welds, in all positions, on carbon steel

Module 9: Welding Inspection and Testing Principles

Key Indicator 2: Examines tacks, root passes, intermediate layers, and completed welds

INTRODUCTION

Shielded metal arc welding (SMAW), or stick welding, is a common method used to join plate. This method provides a high temperature and concentration of heat, which allows a small molten weld pool to be built up quickly. The addition of filler metal from the electrode adds reinforcement and increases the strength of the weld. SMAW can be performed on almost any type of metal 1/8 in. (3 mm) thick or thicker. A minimum amount of equipment is required, and it can be portable.

High-quality welds can be consistently produced on almost any type of metal and in any position. The quality of the welds produced depends largely upon the skill of the welder. Developing the necessary skill level requires practice. However, practicing the welds repeatedly without changing techniques will not aid in developing the required skills. Each time a weld is completed it should be evaluated, and then a change should be made in the technique to improve the next weld.

PRACTICE 2-1

Shielded Metal Arc Welding Safety

Using a welding workstation, welding machine, welding electrodes, welding helmet, eye and ear protection, welding gloves, proper work clothing, and any special protective clothing that may be required, demonstrate, to your instructor and other students, the safe way to prepare yourself and the welding workstation for welding. Include in your demonstration appropriate references to burn protection, eye and ear protection, material safety data sheets, ventilation, electrical safety, general work clothing, special protective clothing, and area cleanup. More information on welding safety can be found in Chapter 2 in *Welding Skills, Processes and Practices for Entry-Level Welders: Book One* and ANSI Z49.1, *Safety in Welding, Cutting, and Allied Processes.*

Complete a copy of the "Student Welding Report" listed in Appendix I or provided by your instructor.

Module 1
Key Indicator 1, 2, 3, 4

Module 2
Key Indicator 1, 2, 3, 4, 5, 6

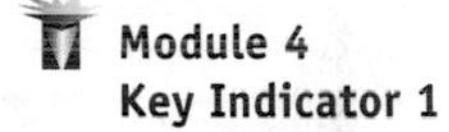

Module 4
Key Indicator 1

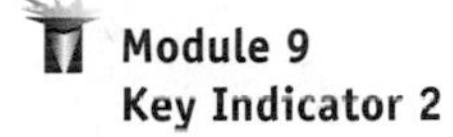

Module 9
Key Indicator 2

EXPERIMENT 2-1

Striking the Arc

Using a properly set up and adjusted arc welding machine, the proper safety protection as demonstrated in Practice 2-1, E6011 or E6103 welding electrodes having a 1/8-in. (3-mm) core wire diameter measured at the bare end of the electrode, and one piece of mild steel plate, 1/4 in. (6 mm) thick, you will practice striking an arc, **Figure 2.1.**

With the electrode held over the plate, lower your helmet. Scratch the electrode across the plate (like striking a large match), **Figure 2.2.** As the arc is established, slightly raise the electrode to the desired arc length. Hold the arc in one place until the molten weld pool builds to the desired size. Slowly lower the electrode as it burns off and move it forward to start the bead.

If the electrode sticks to the plate, quickly squeeze the electrode holder lever to release the electrode. Break the electrode free by bending it back and forth a few times. Do not touch the electrode without gloves because it will still be hot. If the flux breaks away very far from the end of the electrode, throw out the electrode because restarting the arc will be very difficult, **Figure 2.3.**

Module 1
Key Indicator 1, 2, 4

Module 4
Key Indicator 4

Module 9
Key Indicator 2

Once you are able to easily strike an arc and make a weld, try to strike the arc where it will be re-melted by the weld you are making. Arc strikes on the metal's surface that are not covered up by the weld are considered to be weld defects by most codes.

Break the arc by rapidly raising the electrode after completing a 1-in. (25-mm) weld bead. Restart the arc as you did before, and make another short weld. Repeat this process until you can easily start the arc each time.

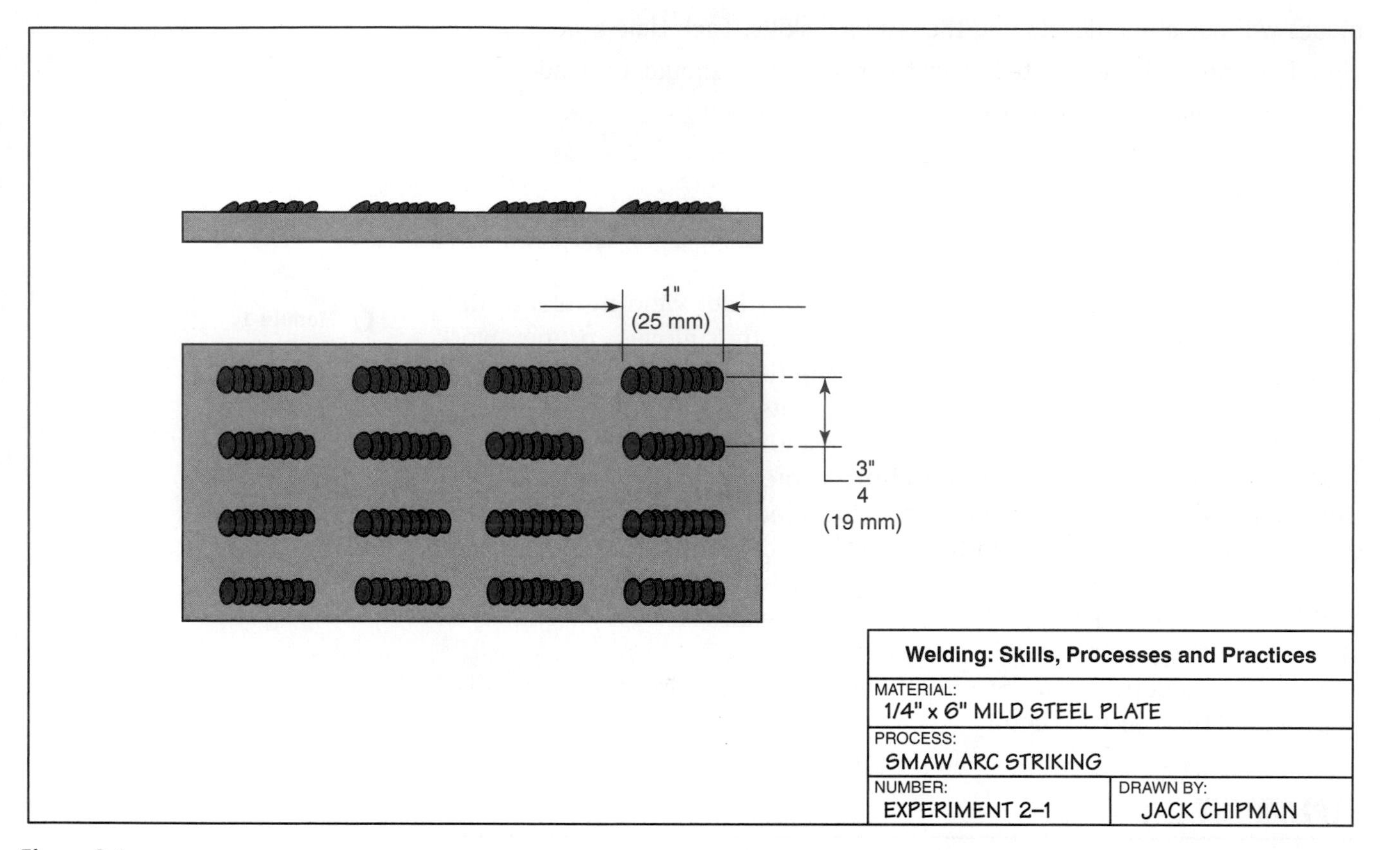

Figure 2.1
Striking an arc and running short beads

Figure 2.2
Striking the arc
Courtesy of Larry Jeffus

Figure 2.3
If the flux is broken off the end completely or on one side, the arc can be erratic or forced to the side
Courtesy of Larry Jeffus

Turn off the welding machine and clean up your work area when you are finished welding.

Complete a copy of the "Student Welding Report" listed in Appendix I or provided by your instructor.

EXPERIMENT 2-2

Striking the Arc Accurately

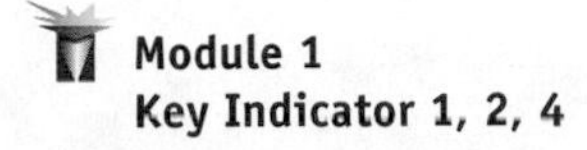

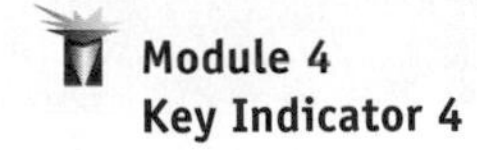

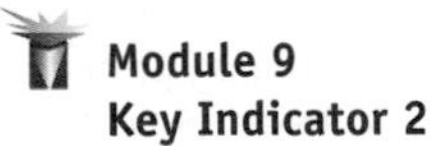

Using the same materials and setup as described in Experiment 2-1, you will start the arc at a specific spot in order to prevent damage to the surrounding plate.

Hold the electrode over the desired starting point. After lowering your helmet, swiftly bounce the electrode against the plate, **Figure 2.4.** A lot of practice is required to develop the speed and skill needed to prevent the electrode from sticking to the plate.

A more accurate method of starting the arc involves holding the electrode steady by resting it on your free hand like a pool cue. The electrode is rapidly pushed forward so that it strikes the metal exactly where it should. This is an excellent method of striking an arc. Striking an arc in an incorrect spot may cause damage to the base metal.

Practice starting the arc until you can start it within 1/4 in. (6 mm) of the desired location. Turn off the welding machine and clean up your work area when you are finished welding.

Complete a copy of the "Student Welding Report" listed in Appendix I or provided by your instructor.

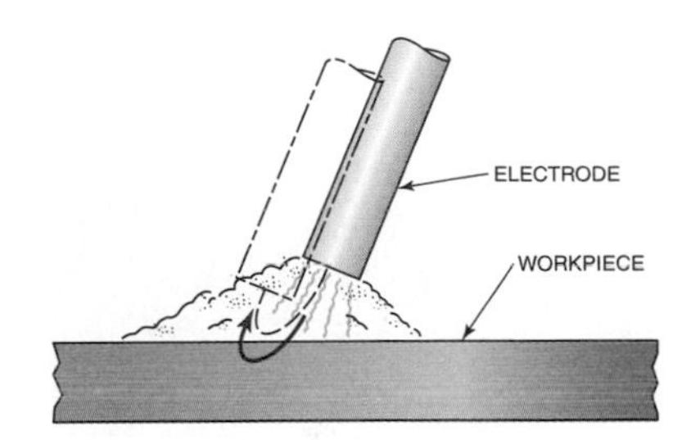

Figure 2.4
Striking the arc on a spot

EFFECT OF TOO-HIGH OR TOO-LOW CURRENT SETTINGS

Each welding electrode must be operated in a specified current (amperage) range, **Table 2.1.** Welding with the current set too low results in poor fusion and poor arc stability, **Figure 2.5.** The weld may have slag or gas inclusions because the molten weld pool was not fluid long enough for the flux to react. Little or no penetration of the weld into the base plate may also be evident. With the current set too low, the arc length is very short. A very short arc length results in frequent shorting and sticking of the electrode.

The core wire of the welding electrode is limited in the amount of current it can carry. As the current is increased, the wire heats up because of electrical resistance. This preheating of the wire causes some of the chemicals in the covering to be burned out too early, **Figure 2.6.** The loss of the proper balance of elements causes poor arc stability. This condition leads to spatter, porosity, and slag inclusions.

Table 2.1 Welding Amperage Range for Common Electrode Types and Sizes

Electrode	Classification					
Size	E6010	E6011	E6012	E6013	E7016	E7018
3/32 in. (2.4 mm)	40–80	50–70	40–90	40–85	75–105	70–110
1/8 in. (3.2 mm)	70–130	85–125	75–130	70–120	100–150	90–165
5/32 in. (4 mm)	110–165	130–160	120–200	130–160	140–190	125–220

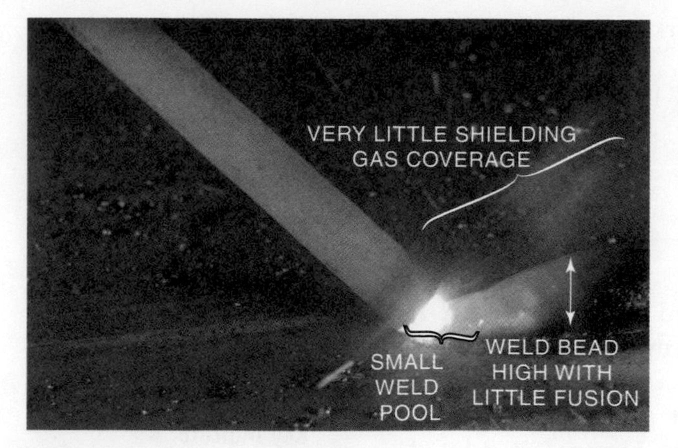

Figure 2.5
Welding with the amperage set too low
Courtesy of Larry Jeffus

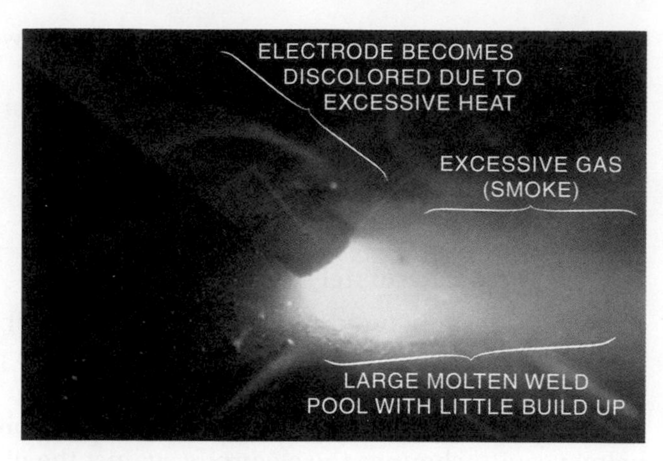

Figure 2.6
Welding with too high an amperage
Courtesy of Larry Jeffus

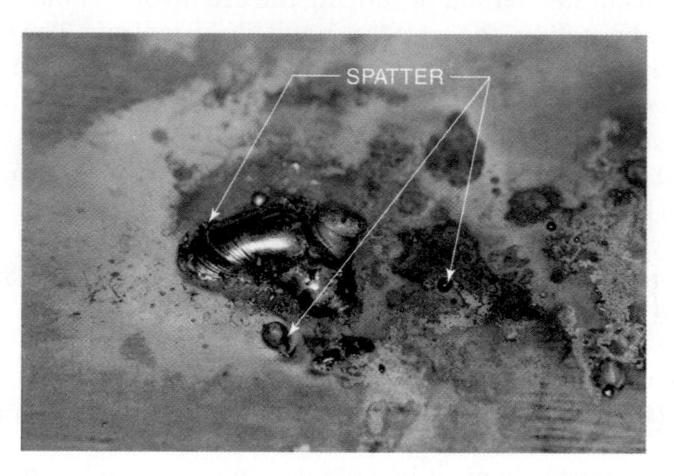

Figure 2.7
Hard weld spatter is fused to base metal and is difficult to remove
Courtesy of Larry Jeffus

An increase in the amount of spatter is also caused by longer arc lengths. The weld bead made at a high amperage setting is wide and flat with deep penetration. The spatter is excessive and is mostly hard. The spatter is called hard because it fuses to the base plate and is difficult to remove, **Figure 2.7.** The electrode covering is discolored more than 1/8 in. (3 mm) to 1/4 in. (6 mm) from the end of the electrode. Extremely high settings may also cause the electrode to discolor, crack, glow red, or burn.

EXPERIMENT 2-3

Effect of Amperage Changes on a Weld Bead

Module 1
Key Indicator 1, 2, 4

Module 4
Key Indicator 3, 4

Module 9
Key Indicator 2

For this experiment, you will need an arc welding machine, welding gloves, safety glasses, welding helmet, appropriate clothing, E6011 or E6013 welding electrodes having a 1/8-in. (3-mm) diameter, and one piece of mild steel plate, 1/4 in. (6 mm) to 1/2 in. (13 mm) thick. You will observe what happens to the weld bead when the amperage settings are raised and lowered.

Figure 2.8
Weld before cleaning and after cleaning
Courtesy of Larry Jeffus

Starting with the machine set at approximately 90 A AC or DCRP, strike an arc and make a weld 1 in. (25 mm) long. Break the arc. Raise the current setting by 10 A, strike an arc, and make another weld 1 in. (25 mm) long. Repeat this procedure until the machine amperage is set at the maximum value.

Replace the electrode and reset the machine to 90 A. Make a weld 1 in. (25 mm) long. Stop and lower the current setting by 10 A. Repeat this procedure until the machine amperage is set at a minimum value.

Cool and chip the plate, comparing the different welds for width, buildup, molten weld pool size, spatter, slag removal, and penetration, **Figure 2.8.** In addition, compare the electrode stubs. Turn off the welding machine and clean up your work area when you are finished welding.

Complete a copy of the "Student Welding Report" listed in Appendix I or provided by your instructor.

CAUTION

Do not change the current settings during welding. A change in the setting may cause arcing inside the machine, resulting in damage to the machine.

ELECTRODE SIZE AND HEAT

The selection of the correct size of welding electrode for a weld is determined by some or all of the following: the skill of the welder, the thickness of the metal to be welded, the size of the metal, and welding codes or standards. Using small diameter electrodes requires less skill than using large diameter electrodes. The deposition rate, or the rate at which the weld metal is added to the weld, is slower when small diameter electrodes are used. Small diameter electrodes will make acceptable welds on thick plate, but more time is required to make the weld.

Large diameter electrodes may overheat the metal if they are used with thin or small pieces of metal. To determine if a weld is too hot, watch the shape of the trailing edge of the molten weld pool, **Figure 2.9.** Rounded ripples indicate the weld is cooling uniformly and that the heat is not excessive. If the ripples are pointed, the weld is cooling too slowly because of excessive heat. Extreme overheating can cause a burn-through. Once a burn-through occurs, it is hard to repair.

To correct an overheating problem, a welder can turn down the amperage, use a shorter arc, travel at a faster rate, use a chill plate (a large piece of metal used to absorb excessive heat), or use a smaller electrode at a lower current setting.

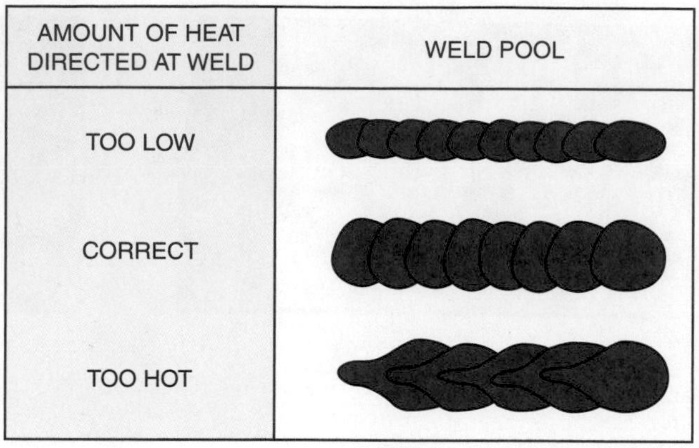

AMOUNT OF HEAT DIRECTED AT WELD	WELD POOL
TOO LOW	
CORRECT	
TOO HOT	

Figure 2.9
The effect on the shape of the molten weld pool caused by the heat input

EXPERIMENT 2-4

Excessive Heat

Module 1
Key Indicator 1, 2, 4

Module 4
Key Indicator 3, 4

Module 9
Key Indicator 2

Using a properly set up and adjusted arc welding machine, the proper safety protection, E6011 or E6013 welding electrodes having a 1/8-in. (3-mm) diameter, and three pieces of mild steel plate, 1/8 in. (3 mm), 3/16 in. (5 mm), and 1/4 in. (6 mm) thick, you will observe the effects of overheating on the weld. Make a stringer weld on each of the three plates using the same amperage setting, travel rate, and arc length for each weld. Cool and chip the welds. Then compare the weld beads for width, reinforcement, and appearance.

Using the same amperage settings, make additional welds on the 1/8-in. (3-mm) and 3/16-in. (5-mm) plates. Vary the arc lengths and travel speeds for these welds. Cool and chip each weld and compare the beads for width, reinforcement, and appearance. Make additional welds on the 1/8-in. (3-mm) and 3/16-in. (5-mm) plates, using the same arc length and travel speed as in the earlier part of this experiment, but at a lower amperage setting. Cool and chip the welds and compare the beads for width, reinforcement, and appearance.

Welders often use the terms *heat* and *amperage* as interchangeable when they are speaking about making changes to the welding current. For example, a welder may say, "Turn up the heat a little," or "Turn up the amperage a little." In both cases what's being asked is that the welding amperage be increased a little.

The plates should be cooled between each weld so that the heat from the previous weld does not affect the test results. Turn off the welding machine and clean up your work area when you are finished welding.

Complete a copy of the "Student Welding Report" listed in Appendix I or provided by your instructor.

ARC LENGTH

The **arc length** is the distance the arc must jump from the end of the electrode to the plate or weld pool surface. As the weld progresses, the electrode becomes shorter as it is consumed. To maintain a constant arc length, the electrode must be lowered continuously. Maintaining a con-

Figure 2.10
Welding with too short an arc length
Courtesy of Larry Jeffus

Figure 2.11
Welding with too long an arc length
Courtesy of Larry Jeffus

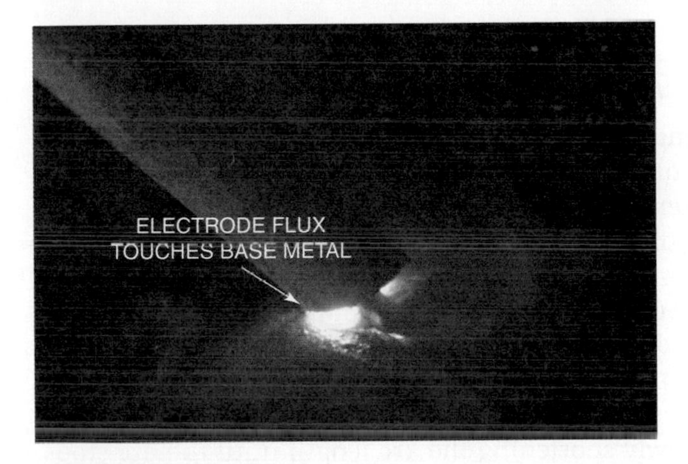

Figure 2.12
Welding with a drag technique
Courtesy of Larry Jeffus

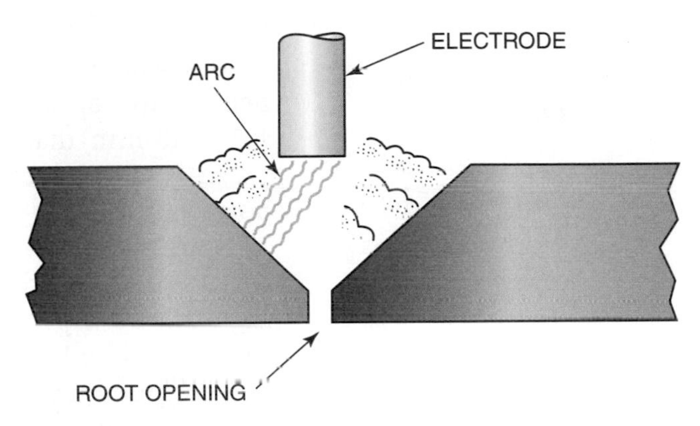

Figure 2.13
The arc may jump to the closest metal, reducing root penetration

stant arc length is important, as too great a change in the arc length will adversely affect the weld.

As the arc length is shortened, metal transferring across the gap may short out the electrode, causing it to stick to the plate. The weld that results from a short arc is narrow and has a high buildup, **Figure 2.10.**

Long arc lengths produce more spatter because the metal being transferred may drop outside of the molten weld pool. The weld is wider and has little buildup, **Figure 2.11.**

There is a narrow range for the arc length in which stability is maintained, metal transfer is smooth, spatter is minimized, and the bead shape is controlled. Factors affecting the length are the type of electrode, joint design, metal thickness, and current setting.

Some welding electrodes, such as E7024, have a thick flux covering. The rate at which the covering melts is slow enough to permit the electrode coating to be rested against the plate. The arc burns back inside the covering as the electrode is dragged along touching the joint, **Figure 2.12.** For that reason, electrodes like E7024 are sometimes referred to as drag electrodes, or drag rods. For this type of welding electrode, the arc length is maintained by the electrode covering.

An arc will jump to the closest metal conductor. On joints that are deep or narrow, the arc is pulled to one side and not to the root, **Figure 2.13.** As a result, the root fusion is reduced or may be nonexistent, thus causing a poor weld. If a very short arc is used, the arc is forced into the root for better fusion.

Because shorter arcs produce less heat and penetration, they are best suited for use on thin metal or thin-to-thick metal joints. Using this technique, metal as thin as 16 gauge can be arc welded easily. Higher amperage settings are required to maintain a short arc that gives good fusion with a minimum of slag inclusions. The higher settings, however, must be within the **amperage range** for the specific electrode.

Finding the correct arc length often requires some trial and adjustment. Most welding jobs require an arc length of 1/8 in. (3 mm) to 3/8 in. (10 mm) when using a 1/8-in. electrode, but this distance varies. It may be necessary to change the arc length when welding to adjust for varying welding conditions.

EXPERIMENT 2-5

Effect of Changing the Arc Length on a Weld

Module 1
Key Indicator 1, 2, 4

Module 4
Key Indicator 3, 4

Module 9
Key Indicator 2

Using an arc welding machine, welding gloves, safety glasses, welding helmet, appropriate clothing, E6011 or E6013 welding electrodes having a 1/8-in. (3-mm) diameter, and one piece of mild steel plate, 1/4 in. (6 mm) to 1/2 in. (13 mm) thick, you will observe the effect of changing the arc length on a weld.

Starting with the welding machine set at approximately 90 A AC or DCRP, strike an arc and make a weld 1 in. (25 mm) long. Continue welding while slowly increasing the arc length until the arc is broken. Restart the arc and make another weld 1 in. (25 mm) long. Welding should again be continued while slowly shortening the arc length until the arc stops. Quickly break the electrode free from the plate, or release the electrode by squeezing the lever on the electrode holder.

Cool and chip both welds. Compare both welding beads for width, reinforcement, uniformity, spatter, and appearance. Turn off the welding machine and clean up your work area when you are finished welding.

Complete a copy of the "Student Welding Report" listed in Appendix I or provided by your instructor.

ELECTRODE ANGLE

The **electrode angle** is measured from the electrode to the surface of the metal. The term used to identify the electrode angle is affected by the direction of travel, generally leading or trailing, **Figure 2.14.** The relative angle is important because there is a jetting force blowing the metal and flux from the end of the electrode to the plate.

Leading Angle

A leading electrode angle pushes molten metal and slag ahead of the weld, **Figure 2.15.** When welding in the flat position, caution must be taken to prevent overlap and slag inclusions. The solid metal ahead of the weld cools and solidifies the molten filler metal and slag before they can melt

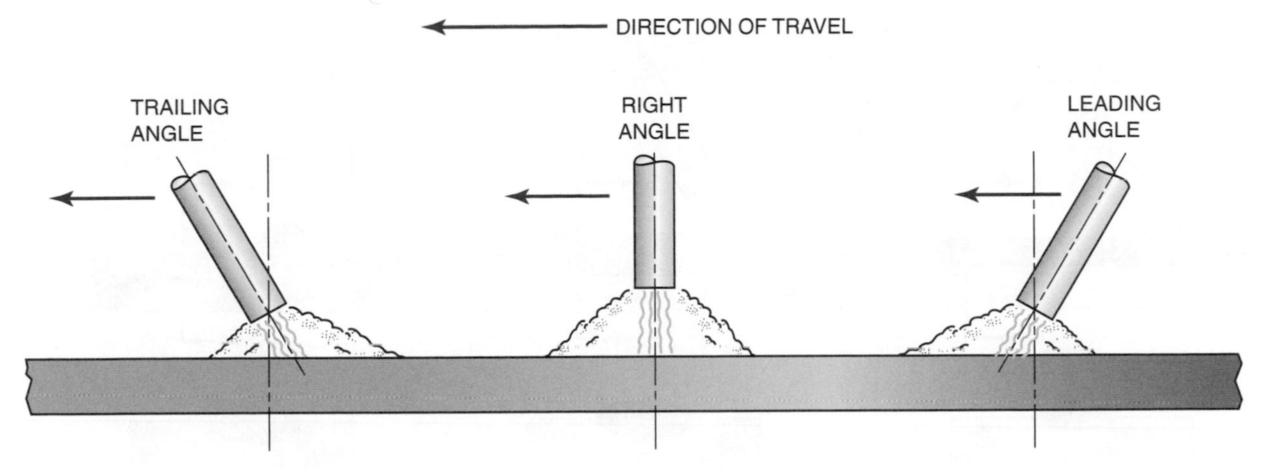

Figure 2.14
Direction of travel and electrode angle

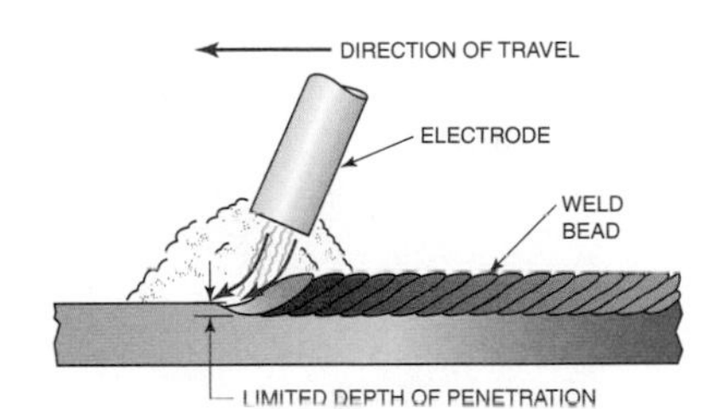

Figure 2.15
Leading, lag, or pushing electrode angle

the solid metal. This rapid cooling prevents the metals from fusing together, **Figure 2.16.** As the weld passes over this area, heat from the arc may not melt it. As a result, some overlap and slag inclusions are left.

The following are suggestions for preventing cold lap and slag inclusions:

- Use as little leading angle as possible.
- Ensure that the arc melts the base metal completely, **Figure 2.17.**
- Use a penetrating-type electrode that causes little buildup.
- Move the arc back and forth across the molten weld pool to fuse both edges.

A leading angle can be used to minimize penetration or to help hold molten metal in place for vertical welds, **Figure 2.18.**

Trailing Angle

A trailing electrode angle pushes the molten metal away from the leading edge of the molten weld pool toward the back where it solidifies, **Figure 2.19.** As the molten metal is forced away from the bottom of the

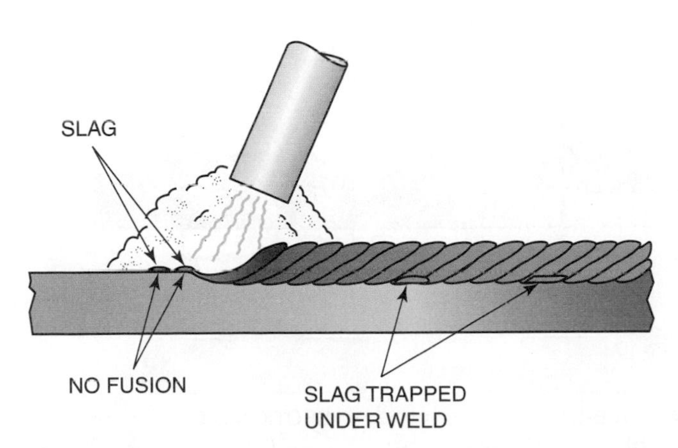

Figure 2.16
Some electrodes, such as E7018, may not remove the deposits ahead of the molten weld pool, resulting in discontinuities within the weld

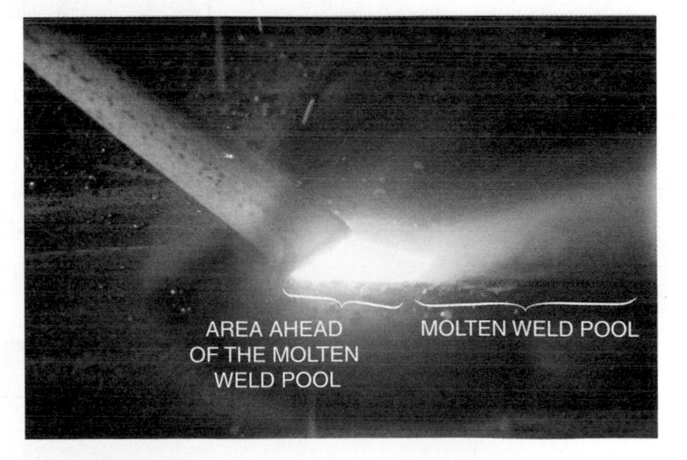

Figure 2.17
Metal being melted ahead of the molten weld pool helps to ensure good weld fusion
Courtesy of Larry Jeffus

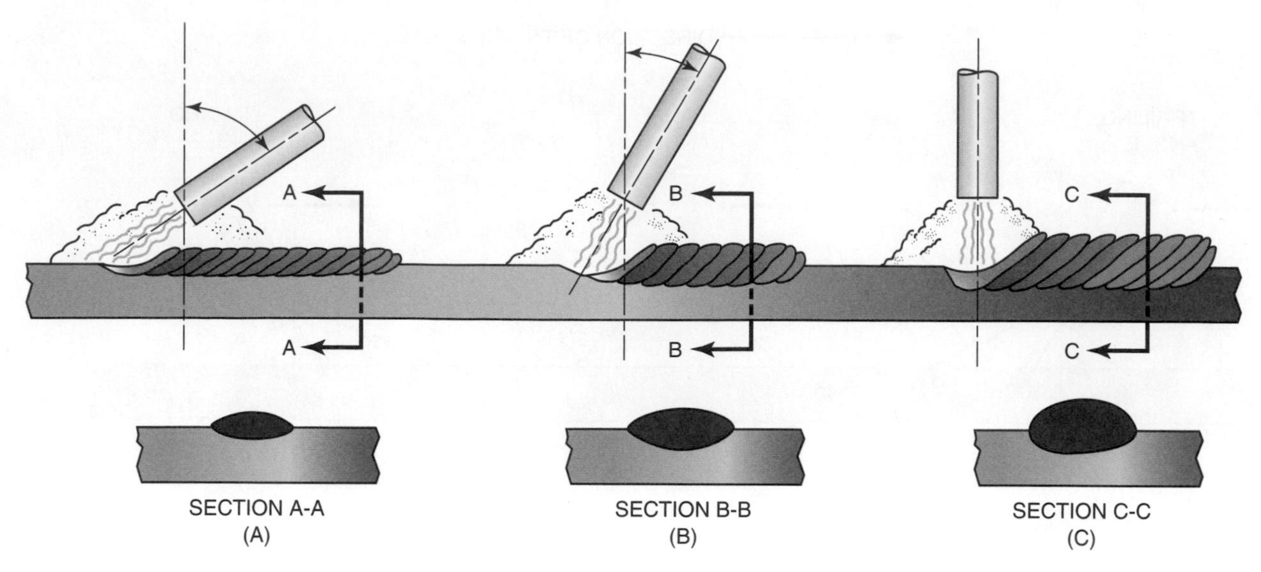

Figure 2.18
Effect of a leading angle on weld bead buildup, width, and penetration. As the angle increases toward the vertical position (C), penetration increases

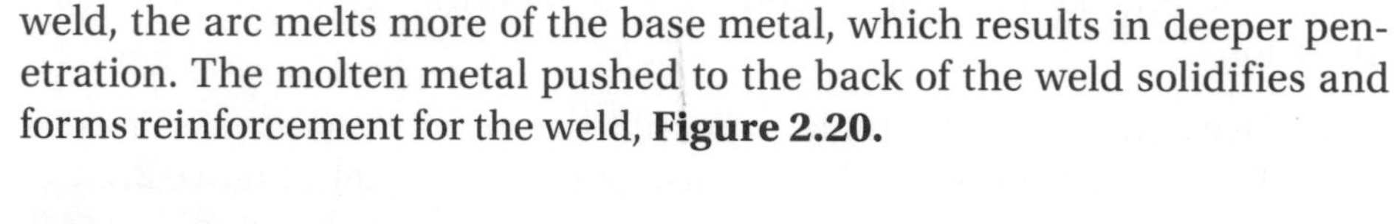
weld, the arc melts more of the base metal, which results in deeper penetration. The molten metal pushed to the back of the weld solidifies and forms reinforcement for the weld, **Figure 2.20.**

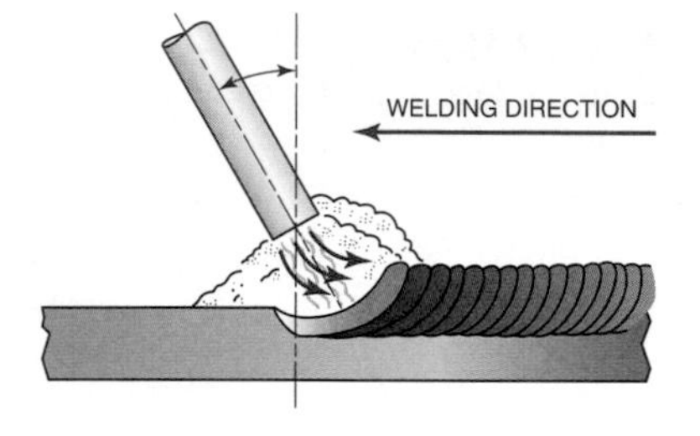

Figure 2.19
Trailing electrode angle

EXPERIMENT 2-6

Effect of Changing the Electrode Angle on a Weld

Using a properly set up and adjusted arc welding machine, the proper safety protection, E6011 welding electrodes having a 1/8-in. (3-mm) diameter, and one piece of mild steel plate, 1/4 in. (6 mm) to 1/2 in. (13 mm) thick, you will observe the effect of changes in the electrode angle on a weld.

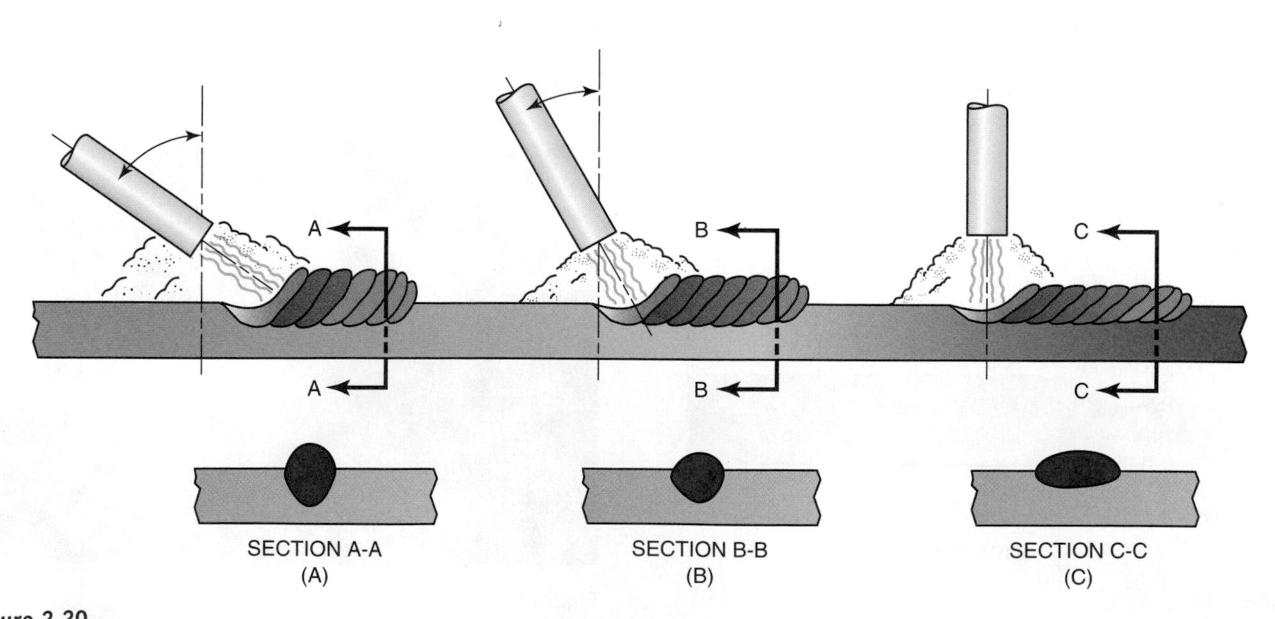

Figure 2.20
Effect of a trailing angle on weld bead buildup, width, and penetration. Section A-A shows more weld buildup due to a greater angle of the electrode

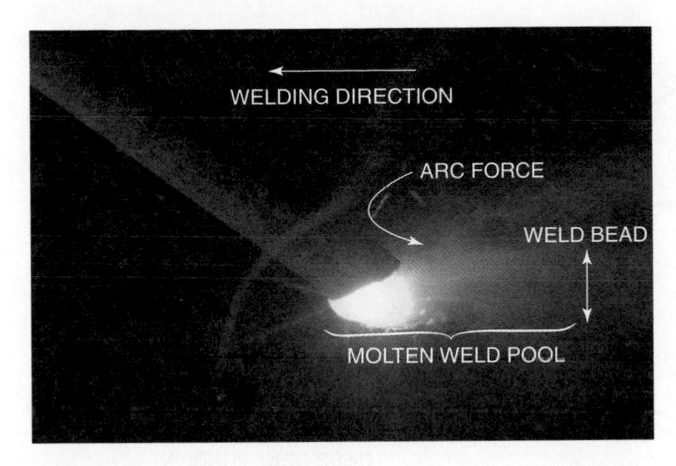

Figure 2.21
Welding with a trailing angle
Courtesy of Larry Jeffus

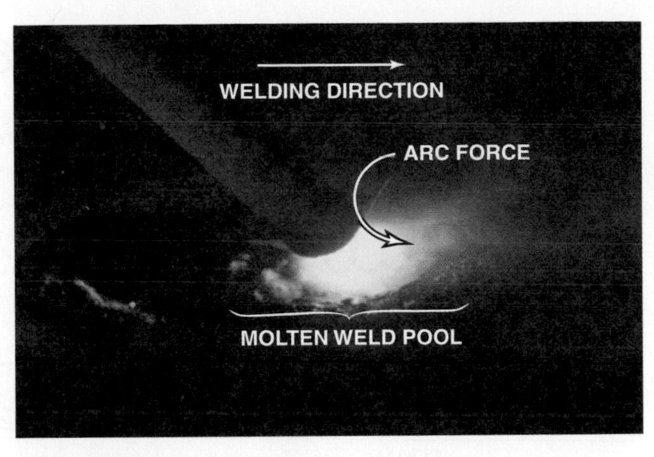

Figure 2.22
Welding with a leading angle
Courtesy of Larry Jeffus

Module 1
Key Indicator 1, 2, 4

Module 4
Key Indicator 3, 4

Module 9
Key Indicator 2

Start welding with a sharp trailing angle. Make a weld about 1 in. (25 mm) long. Closely observe the molten weld pool at the points shown in **Figure 2.21.** Slowly increase the electrode angle and continue to observe the weld.

When you reach a 90° electrode angle, make a weld about 1 in. (25 mm) long. Observe the parts of the molten weld pool as shown in Figure 2.21.

Continue welding and change the electrode angle to a sharp leading angle. Observe the molten weld pool at the points shown in **Figure 2.22.**

During this experiment, you must maintain a constant arc length, travel speed, and weave pattern if the observations and results are to be accurate.

Cool and chip the weld. Compare the weld bead for uniformity in width, reinforcement, and appearance. Turn off the welding machine and clean up your work area when you are finished welding.

Complete a copy of the "Student Welding Report" listed in Appendix I or provided by your instructor.

ELECTRODE MANIPULATION

The movement or weaving of the welding electrode can control the following characteristics of the weld bead: penetration, buildup, width, porosity, undercut, overlap, and slag inclusions. The exact **weave pattern** for each weld is often the personal choice of the welder. However, some patterns are especially helpful for specific welding situations. The pattern selected for a flat (1G) butt joint is not as critical as is the pattern selection for other joints and other positions.

Many weave patterns are available for the welder to use. **Figure 2.23** shows ten different patterns that can be used for most welding conditions.

The circular pattern is often used for flat position welds on butt, tee, outside corner joints, and for buildup or surfacing applications. The circle can be made wider or longer to change the bead width or penetration, **Figure 2.24.**

The "C" and square patterns are both good for most 1G (flat) welds, but can also be used for vertical (3G) positions. These patterns can also be

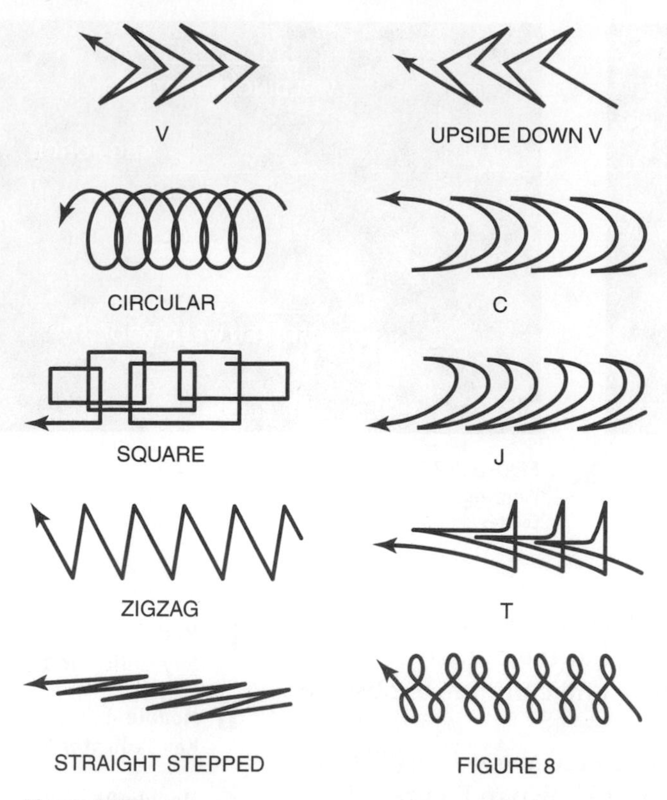

Figure 2.23
Weave patterns

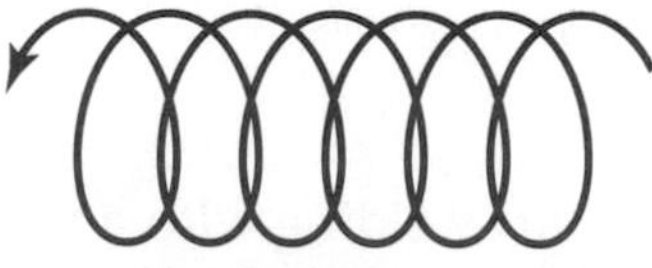

Figure 2.24
Changing the weave pattern width to change the weld bead characteristics

used if there is a large gap to be filled when both pieces of metal are nearly the same size and thickness.

The "J" pattern works well on flat (1F) lap joints, all vertical (3G) joints, and horizontal (2G) butt and lap (2F) welds. This pattern allows the heat to be concentrated on the thicker plate, **Figure 2.25.** It also allows the reinforcement to be built up on the metal deposited during the first part of the pattern. As a result, a uniform bead contour is maintained during out-of-position welds.

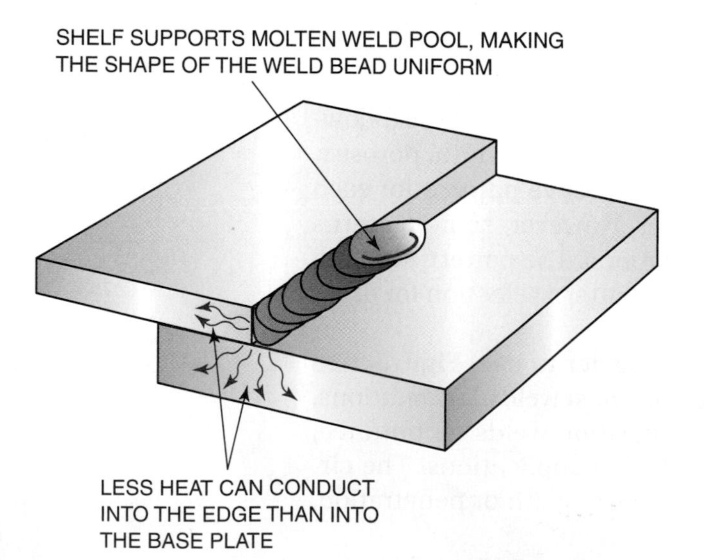

Figure 2.25
The "J" pattern allows the heat to be concentrated on the thicker plate

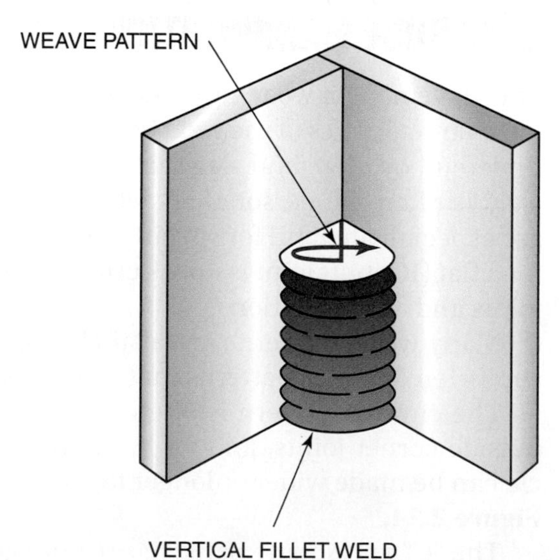

Figure 2.26
"Inverted T" pattern

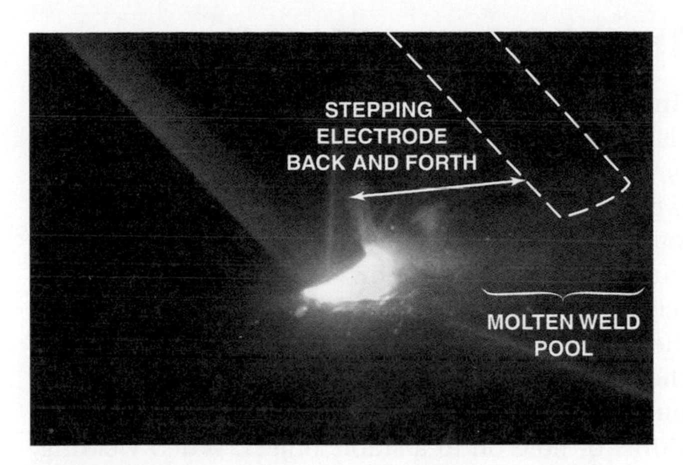

Figure 2.27
The electrode is moved slightly forward and then returned to the weld pool
Courtesy of Larry Jeffus

Figure 2.28
The electrode does not deposit metal or melt the base metal
Courtesy of Larry Jeffus

The inverted "T" pattern works well with fillet welds in the vertical (3F) and overhead (4F) positions, **Figure 2.26.** It also can be used for deep groove welds for the hot pass. The top of the "T" can be used to fill in the toe of the weld to prevent undercutting.

The straight step pattern can be used for stringer beads, root pass welds, and multiple pass welds in all positions. For this pattern, the smallest quantity of metal is molten at one time as compared to other patterns. Therefore, the weld is more easily controlled. At the same time that the electrode is stepped forward, the arc length is increased so that no metal is deposited ahead of the molten weld pool, **Figure 2.27** and **Figure 2.28.** This action allows the molten weld pool to cool to a controllable size. In addition, the arc burns off any paint, oil, or dirt from the metal before it can contaminate the weld.

The figure-eight pattern and the zigzag pattern are used as cover passes in the flat and vertical positions. Do not weave more than 2-1/2 times the width of the electrode. These patterns deposit a large quantity of metal at one time. A shelf can be used to support the molten weld pool when making vertical welds using either of these patterns, **Figure 2.29.**

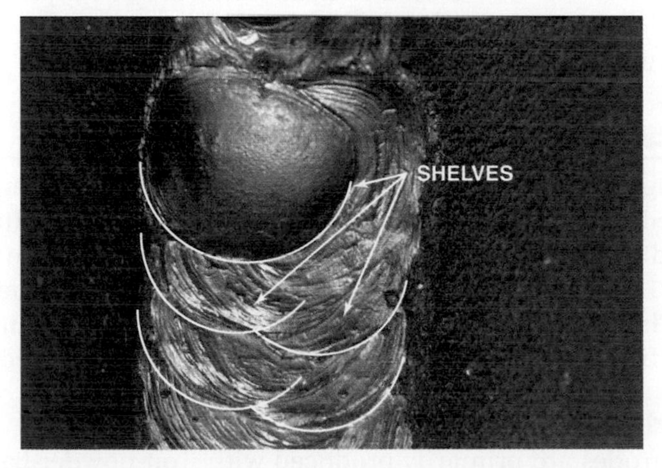

Figure 2.29
Using the shelf to support the molten pool for vertical welds
Courtesy of Larry Jeffus

POSITIONING OF THE WELDER AND THE PLATE

The welder should be in a relaxed, comfortable position before starting to weld. A good position is important for both the comfort of the welder and the quality of the welds. Welding in an awkward position can cause welder fatigue, which leads to poor welder coordination and poor-quality welds. Welders must have enough freedom of movement so that they do not need to change position during a weld. Body position changes should be made only during electrode changes.

When the welding helmet is down, the welder is blind to the surroundings. Due to the arc, the field of vision of the welder is also very limited. These factors often cause the welder to sway. To stop this swaying, the welder should lean against or hold on to a stable object. When welding, even if a welder is seated, touching a stable object will make that welder more stable and will make welding more relaxing.

Welding is easier if the welder can find the most comfortable angle. The welder should be in either a seated or a standing position in front of the welding table. The welding machine should be turned off. With an electrode in place in the electrode holder, the welder can draw a straight line along the plate to be welded. By turning the plate to several different angles, the welder should be able to determine which angle is most comfortable for welding, **Figure 2.30.**

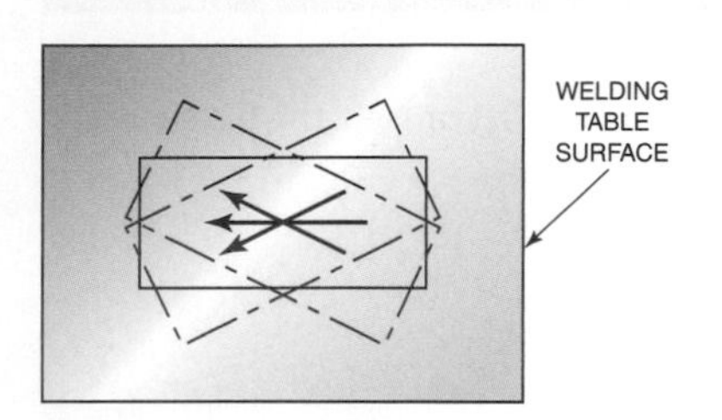

Figure 2.30
Change the plate angle to find the most comfortable welding position

PRACTICE WELDS

Practice welds are grouped according to the type of joint and the type of welding electrode. The welder or instructor should select the order in which the welds are made. The stringer beads should be practiced first in each position before the welder tries the different joints in each position. Some time can be saved by starting with the stringer beads. If this is done, it is not necessary to cut or tack the plate together, and a number of beads can be made on the same plate.

Students will find it easier to start with butt joints. The lap, tee, and outside corner joints are all about the same level of difficulty. Starting with the flat position allows the welder to build skills slowly so that out-of-position welds become easier to do. The horizontal tee and lap welds are almost as easy to make as the flat welds. Overhead welds are as simple to make as vertical welds, but they are harder to position. Horizontal butt welds are more difficult to perform than most other welds.

Electrodes

Arc welding electrodes used for practice welds in the following chapters are grouped into three filler metal (F-number) classes according to their major welding characteristics. E6012 and E6013 fall into the F2 group, E6010 and E6011 fall into the F3 group, and E7016 and E7018 fall into the F4 group. The F1 electrode group (not used in this chapter) is limited flat and horizontal fillet welds, E7024 and E7028 are examples of F1 electrodes.

F1 E7024 and E7028 Electrodes

These electrodes are primarily produced with iron-powder-based fluxes that are substantially thicker than the other groups. As a result, they have very high deposition rates and produce welds with medium to low penetration, but are restricted to groove welds in the flat position and fillet welds in the flat or horizontal positions.

F2 E6012 and E6013 Electrodes

These electrodes have rutile-based fluxes, giving a smooth, easy arc with a thick slag left on the weld bead. They may be used in all positions and do not require special rod ovens. These electrodes are popular because slag is easily removed, and they may be used with all polarities to produce welds with medium to low penetration.

F3 E6010 and E6011 Electrodes

Both of these electrodes have cellulose-based fluxes. As a result, these electrodes have a forceful deep penetrating arc with little slag left on the weld bead. They do not require a rod oven, and they are often the electrode of choice when surface conditions on the base metal are less than optimal. F3 electrodes are also commonly used for open root welds on plate and pipe.

F4 E7016 and E7018 Electrodes

Both of these electrodes have a mineral-based flux. The resulting medium penetration arc is smooth and easy, with a very heavy slag left on the weld bead. F4 electrodes are also referred to as low hydrogen or low-hi electrodes. They require special handling and storage in a rod oven after being removed from their factory packaging. Refer to manufacturer's requirements or the applicable welding code for specific handling directions.

The cellulose- and rutile-based groups of electrodes have characteristics that make them the best electrodes for starting specific welds. The electrodes with the **cellulose-based fluxes** do not have heavy slags that may interfere with the welder's view of the weld. This feature is an advantage for flat tee and lap joints. Electrodes with the **rutile-based fluxes** (giving an easy arc with low spatter) are easier to control and are used for flat stringer beads and butt joints.

Unless a specific electrode has been required by a welding procedure specification (WPS), welders can select what they consider to be the best electrode for a specific weld. Without a WPS, the welder has the final choice. An accomplished welder can make defect-free welds on all types of joints using all types of electrodes in any weld position.

Electrodes with **mineral-based fluxes** should be the last choice. Welds with a good appearance are more easily made with these electrodes, but strong welds are hard to obtain. Without special care being taken during the start of the weld, porosity will be formed in the weld. **Figure 2.31** shows a starting tab used to prevent this porosity from becoming part of the finished weld.

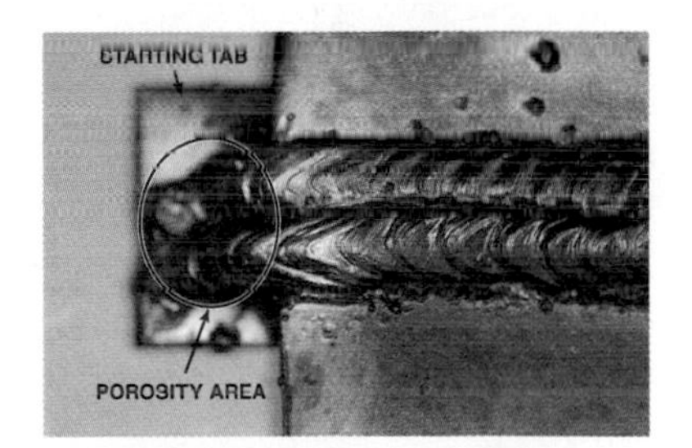

Figure 2.31
Porosity is found on the starting tab where it will not affect the weld
Courtesy of Larry Jeffus

STRINGER BEADS

A straight weld bead on the surface of a plate, with little or no side-to-side electrode movement, is known as a **stringer bead.** Stringer beads are used by students to practice maintaining arc length and electrode angle so that their welds will be straight, uniform, and free from defects. Stringer beads,

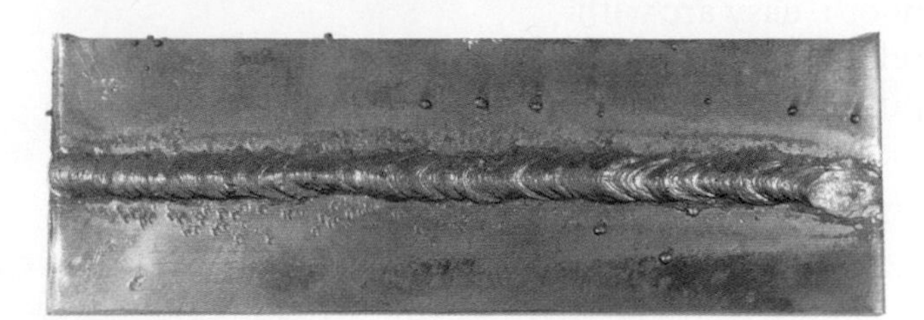

Figure 2.32
Stringer bead
Courtesy of Larry Jeffus

Figure 2.33
New welders frequently see only the arc and sparks from the electrode
Courtesy of Larry Jeffus

Figure 2.32, are also used to set the machine amperage and for buildup or surfacing applications. Stringer beads are the most commonly used type of bead for vertical, horizontal, and overhead welds.

The stringer bead should be straight. A beginning welder needs time to develop the skill of viewing the entire welding area. At first, the welder sees only the arc, **Figure 2.33.** With practice, the welder begins to see parts of the molten weld pool. After much practice, the welder will see the molten weld pool (front, back, and both sides), slag, buildup, and the surrounding plate, **Figure 2.34.** Often, at this skill level, the welder may not even notice the arc.

A straight weld is easily made once the welder develops the ability to view the entire welding zone. The welder will occasionally glance around to ensure that the weld is straight. In addition, it can be noted if the weld is uniform and free from defects. The ability of the welder to view the entire weld area is demonstrated by making consistently straight and uniform stringer beads.

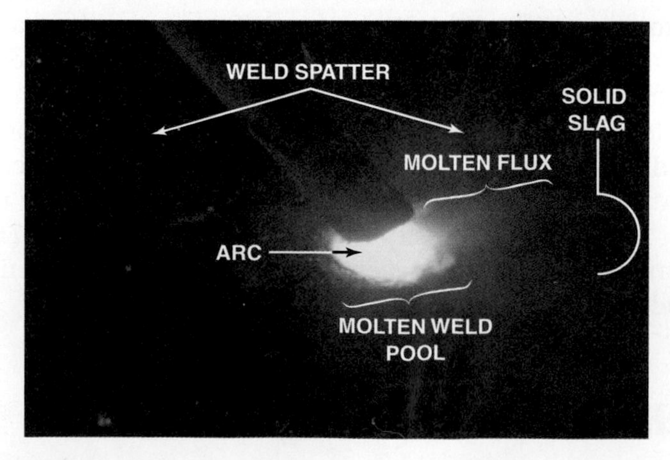

Figure 2.34
More experienced welders can see the molten pool, metal being transferred across the arc, and penetration into the base metal
Courtesy of Larry Jeffus

After making practice stringer beads, a variety of weave bead patterns should be practiced to gain the ability to control the molten weld pool when welding out of position.

PRACTICE 2-2

Straight Stringer Beads in the Flat Position Using E6010 or E6011 Electrodes, E6012 or E6013 Electrodes, and E7016 or E7018 Electrodes

Using a properly set up and adjusted arc welding machine, proper safety protection as demonstrated in Practice 2-1, arc welding electrodes with a 1/8-in. (3-mm) diameter, and one piece of mild steel plate, 6 in. (152 mm) long × 1/4 in. (6 mm) thick, you will make straight stringer beads.

Module 1
Key Indicator 1, 2, 3, 4

Module 4
Key Indicator 1, 3, 4

Module 9
Key Indicator 2

- Starting at one end of the plate, make a straight weld the full length of the plate.
- Watch the molten weld pool at this point, not the end of the electrode. As you become more skillful, it is easier to watch the molten weld pool.
- Repeat the beads with all three (F) groups of electrodes until you have consistently good beads.
- Cool, chip, and inspect the bead for defects after completing it. Turn off the welding machine and clean up your work area when you are finished welding.

Complete a copy of the "Student Welding Report" listed in Appendix I or provided by your instructor.

PRACTICE 2-3

Stringer Beads in the Vertical Up Position Using E6010 or E6011 Electrodes, E6012 or E6013 Electrodes, and E7016 or E7018 Electrodes

Using the same setup, materials, and electrodes as listed in Practice 2-2, you will make vertical up stringer beads. Start with the plate at a 45° angle.

Module 1
Key Indicator 1, 2, 3, 4

Module 4
Key Indicator 1, 3, 4

Module 9
Key Indicator 2

This technique is the same as that used to make a vertical weld. However, a lower level of skill is required at 45°, and it is easier to develop your skill. After the welder masters the 45° angle, the angle is increased successively until a vertical position is reached, **Figure 2.35.**

Before the molten metal drips down the bead, the back of the molten weld pool will start to bulge, **Figure 2.36.** When this happens, increase the speed of travel and the weave pattern.

Cool, chip, and inspect each completed weld for defects. Repeat the beads as necessary with all three (F) groups of electrodes until consistently good beads are obtained in this position. Turn off the welding machine and clean up your work area when you are finished welding.

Complete a copy of the "Student Welding Report" listed in Appendix I or provided by your instructor.

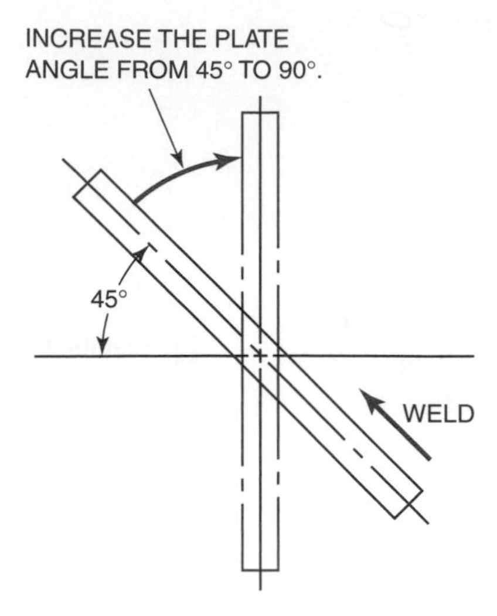

Figure 2.35
Once the 45° angle is mastered, the plate angle is increased successively until a vertical position (90°) is reached

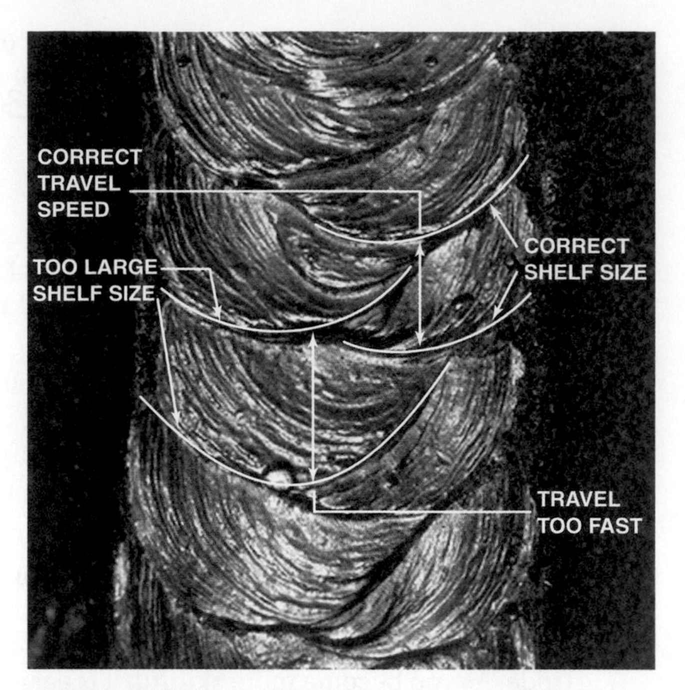

Figure 2.36
E7018 vertical up weld
Courtesy of Larry Jeffus

PRACTICE 2-4

Horizontal Stringer Beads Using E6010 or E6011 Electrodes, E6012 or E6013 Electrodes, and E7016 or E7018 Electrodes

Module 1
Key Indicator 1, 2, 3, 4

Module 4
Key Indicator 1, 3, 4

Module 9
Key Indicator 2

Using the same setup, materials, and electrodes as listed in Practice 2-2, you will make horizontal stringer beads on a plate.

When the welder begins to practice the horizontal stringer bead, the plate may be reclined slightly, **Figure 2.37.** This placement allows the welder to build the required skill by practicing the correct techniques successfully. The "J" weave pattern is suggested for this practice. As the electrode is drawn along the straight back of the "J," metal is deposited. This metal supports the molten weld pool, resulting in a bead with a uniform contour, **Figure 2.38.**

Angling the electrode up and back toward the weld causes more metal to be deposited along the top edge of the weld. Keeping the bead small allows the surface tension to hold the molten weld pool in place.

Gradually increase the angle of the plate until it is vertical and the stringer bead is horizontal. Repeat the beads as needed with all three (F) groups of electrodes until consistently good beads are obtained in this position. Turn off the welding machine and clean up your work area when you are finished welding.

Complete a copy of the "Student Welding Report" listed in Appendix I or provided by your instructor.

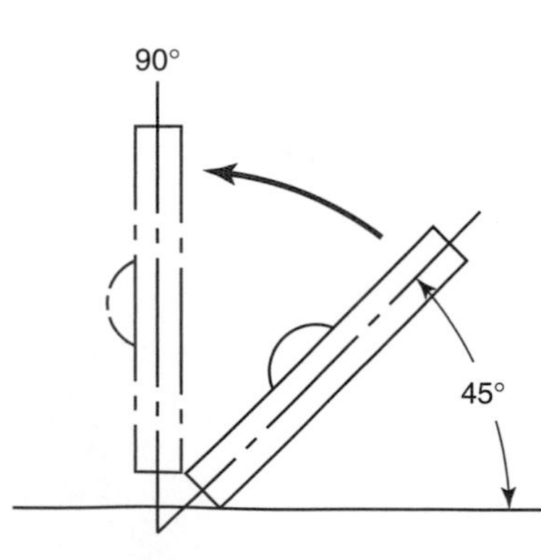

Figure 2.37
Change the plate angle as welding skill improves

SQUARE BUTT JOINT

The **square butt joint** is made by tack welding two flat pieces of plate together, **Figure 2.39.** The space between the plates is called the root opening or root gap. Changes in the root opening will affect penetration. As the

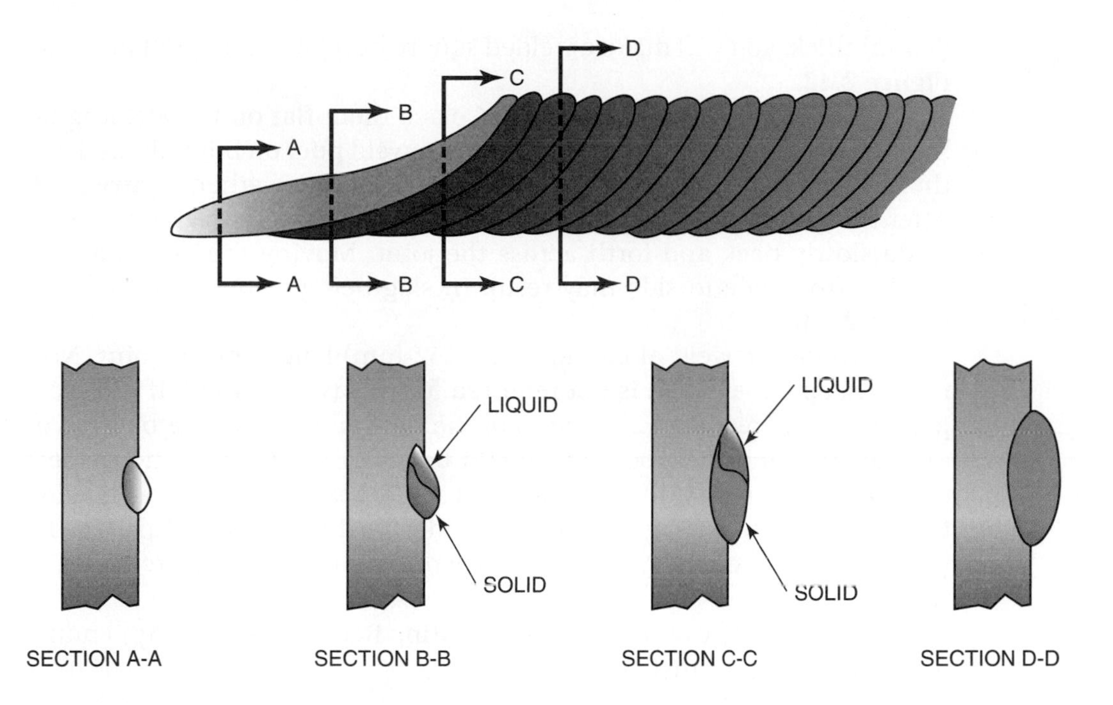

Figure 2.38
The progression of a horizontal bead

space increases, the weld penetration also increases. The root opening for most butt welds will vary from 0 in. (0 mm) to 1/8 in. (3 mm). Excessively large openings can cause burn-through or a cold lap at the weld root, **Figure 2.40.**

After a butt weld is completed, the plate can be cut apart so it can be used for rewelding. The strips for butt welding should be no smaller than 1 in. (25 mm) wide. If they are too narrow, there will be a problem with heat buildup.

If the plate strips are no longer flat after the weld has been cut out, they can be tack welded together and flattened with a hammer, **Figure 2.41.**

Figure 2.39
The tack weld should be small and uniform to minimize its effect on the final weld
Courtesy of Larry Jeffus

PRACTICE 2-5

Welded Square Butt Joint in the Flat Position (1G) Using E6010 or E6011 Electrodes, E6012 or E6013 Electrodes, and E7016 or E7018 Electrodes

Using a properly set up and adjusted arc welding machine, proper safety protection, arc welding electrodes having a 1/8-in. (3-mm) diameter, and two or more pieces of mild steel plate, 6 in. (152 mm) long × 1/4 in.

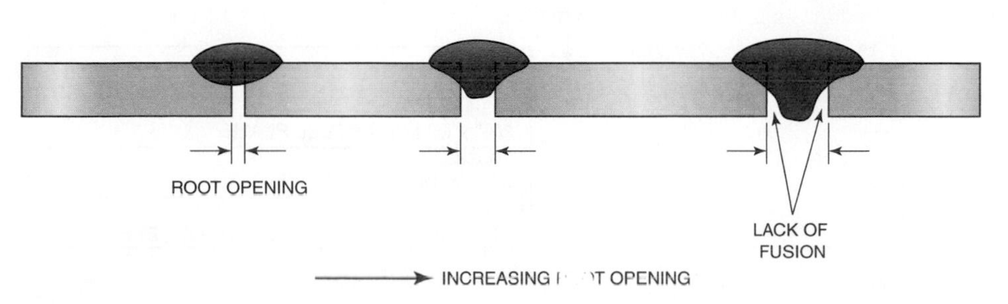

Figure 2.40
Effect of root opening on weld penetration

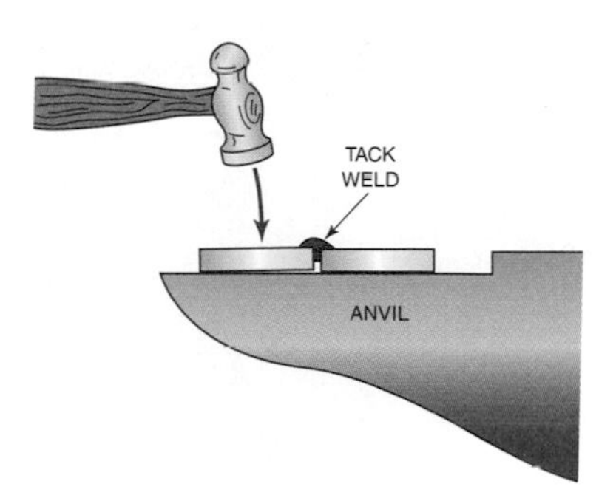

Figure 2.41
After the plates are tack welded together, they can be forced into alignment by striking them with a hammer

Module 1
Key Indicator 1, 2, 3, 4

Module 4
Key Indicator 1, 6
This practice addresses the "Flat" position portion of the all-position requirement of 6.

Module 9
Key Indicator 2

(6 mm) thick, you will make a welded square butt joint in the flat position, **Figure 2.42**.

Tack weld the plates together and place them flat on the welding table. Starting at one end, establish a molten weld pool on both plates. Hold the electrode in the molten weld pool until it flows together, **Figure 2.43.** After the gap is bridged by the molten weld pool, start weaving the electrode slowly back and forth across the joint. Moving the electrode too quickly from side to side may result in slag being trapped in the joint, **Figure 2.44.**

Continue the weld along the 6-in. (152-mm) length of the joint. Normally, deep penetration is not required for this type of weld. If full plate penetration is required, the edges of the butt joint should be beveled or a larger than normal root gap should be used. Cool, chip, and inspect the weld for uniformity and soundness. Repeat the welds as needed to master all three (F) groups of electrodes in this position. Turn off the welding machine and clean up your work area when you are finished welding.

Complete a copy of the "Student Welding Report" listed in Appendix I or provided by your instructor.

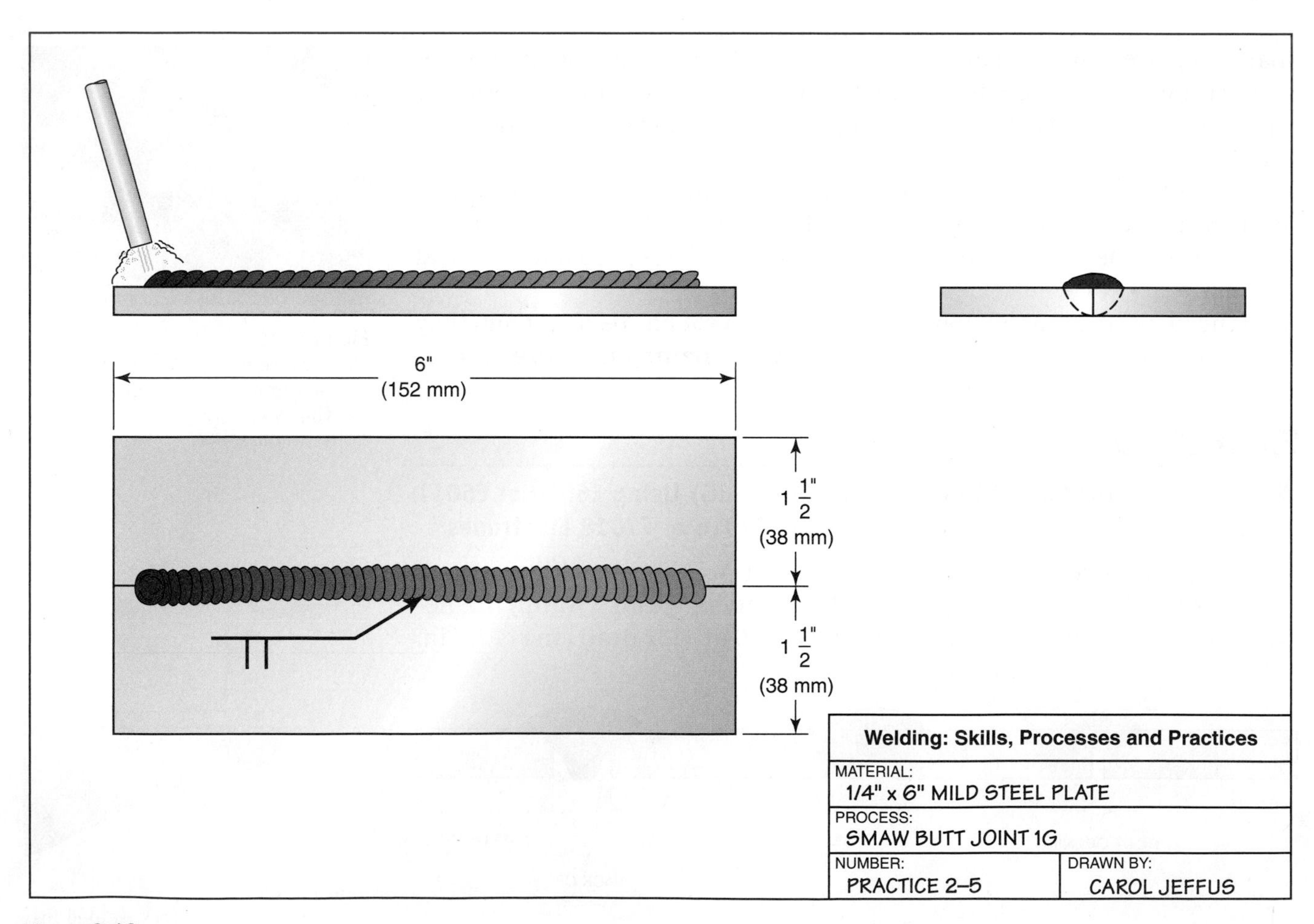

Figure 2.42
Square butt joint in the flat position

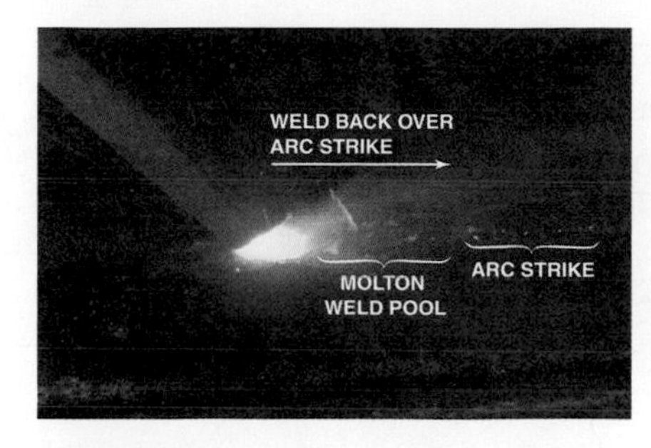

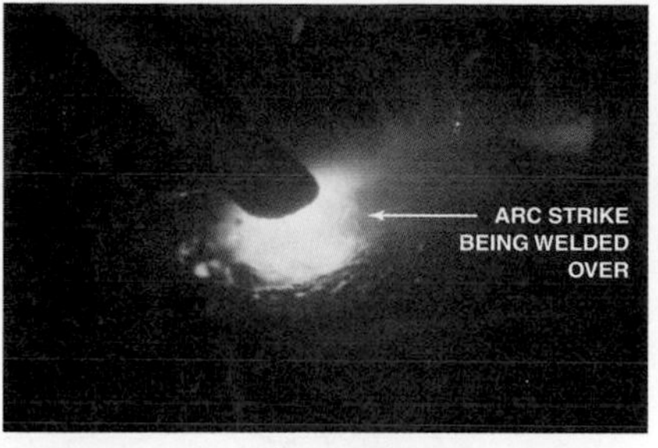

Figure 2.43
(A) After the arc is established, hold it in one area long enough to establish the size of molten weld pool desired. (B) Weld back over the arc strike to melt into the weld.
Courtesy of Larry Jeffus

Figure 2.44
Moving the electrode from side to side too quickly can result in slag being trapped between the plates
Courtesy of Larry Jeffus

PRACTICE 2-6

Welded Vertical Up (3G) Square Butt Joint Using E6010 or E6011 Electrodes, E6012 or E6013 Electrodes, and E7016 or E7018 Electrodes

Using the same setup, materials, and electrodes as listed in Practice 2-5, you will make welded vertical up square butt joints.

With the plates at a 45° angle, start at the bottom and make the molten weld pool bridge the gap between the plates, **Figure 2.45.** Build the bead size slowly so that the molten weld pool has a shelf for support. The "C," "J," or square weave pattern works well for this joint.

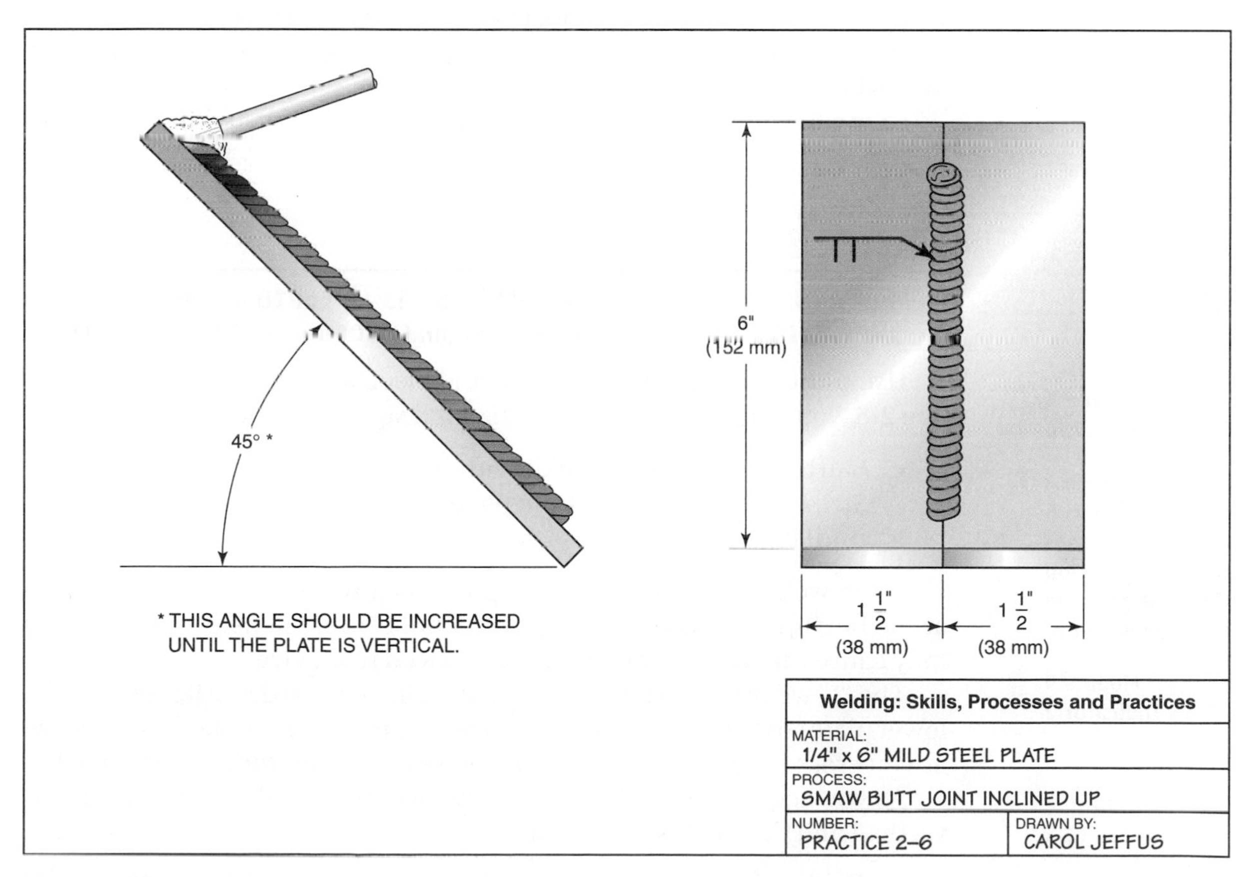

Figure 2.45
Square butt joint in the vertical up position

Module 1
Key Indicator 1, 2, 3, 4

Module 4
Key Indicator 1, 6
This practice addresses the "Vertical" position portion of the all-position requirement of 6.

Module 9
Key Indicator 2

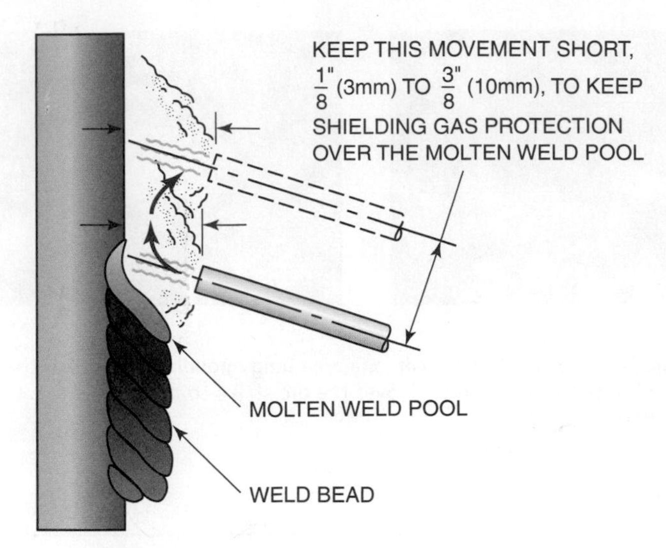

Figure 2.46
Electrode movement for vertical up welds when using F3 electrodes

As the electrode is moved up the weld, the arc is lengthened slightly so that little or no metal is deposited ahead of the molten weld pool. When the electrode is brought back into the molten weld pool, it should be lowered to deposit metal, **Figure 2.46.**

As skill is developed, increase the plate angle until it is vertical. Cool, chip, and inspect the weld for uniformity and defects. Repeat the welds with all three (F) groups of electrodes until you can consistently make welds free of defects. Turn off the welding machine and clean up your work area when you are finished welding.

Complete a copy of the "Student Welding Report" listed in Appendix I or provided by your instructor.

PRACTICE 2-7

Welded Horizontal (2G) Square Butt Joint Using E6010 or E6011 Electrodes, E6012 or E6013 Electrodes, and E7016 or E7018 Electrodes

Module 1
Key Indicator 1, 2, 3, 4

Module 4
Key Indicator 1, 6
This practice addresses the "Horizontal" position portion of the all-position requirement of 6.

Module 9
Key Indicator 2

Using the same setup, materials, and electrodes as described in Practice 2-5, you will make a welded horizontal square butt joint.

- Start practicing these welds with the plate at a slight angle.
- Strike the arc on the bottom plate and build the molten weld pool until it bridges the gap.

If the weld is started on the top plate, slag will be trapped in the root at the beginning of the weld because of poor initial penetration. The slag may cause the weld to crack when it is placed in service.

The "J" weave pattern is recommended in order to deposit metal on the lower plate so that it can support the bead. By pushing the electrode inward as you cross the gap between the plates, deeper penetration is achieved.

As you acquire more skill, gradually increase the plate angle until it is vertical and the weld is horizontal.

- Cool, chip, and inspect the weld for uniformity and defects.
- Repeat the welds with all three (F) groups of electrodes until you can consistently make welds free of defects. Turn off the welding

machine and clean up your work area when you are finished welding.

Complete a copy of the "Student Welding Report" listed in Appendix I or provided by your instructor.

EDGE WELD

An edge weld joint is made by placing the edges of the plate evenly, **Figure 2.47.** When assembling the edge joint the plates should be clamped tightly together; there should not be any gap between the plates. Both edges of the plate assembly can be welded. Make the tack welds to hold the plates together along the ends of the joint, **Figure 2.48.**

The size of the weld should equal the thickness of the plate being joined. A good indication the weld is being made large enough is when the weld bead width is equal to the width of the joint, **Figure 2.49.** The weld bead should also have a slight buildup.

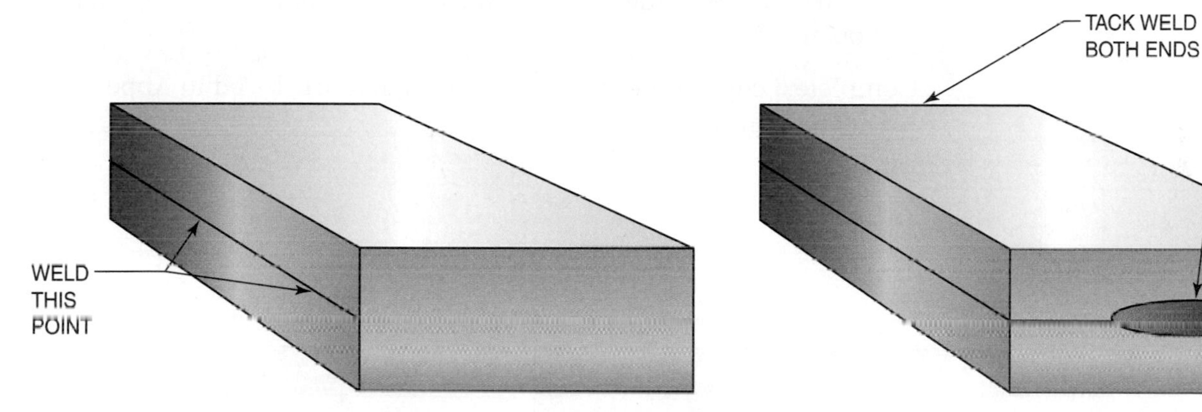

Figure 2.47
Edge joint

Figure 2.48
Make tack welds at the ends of the joint

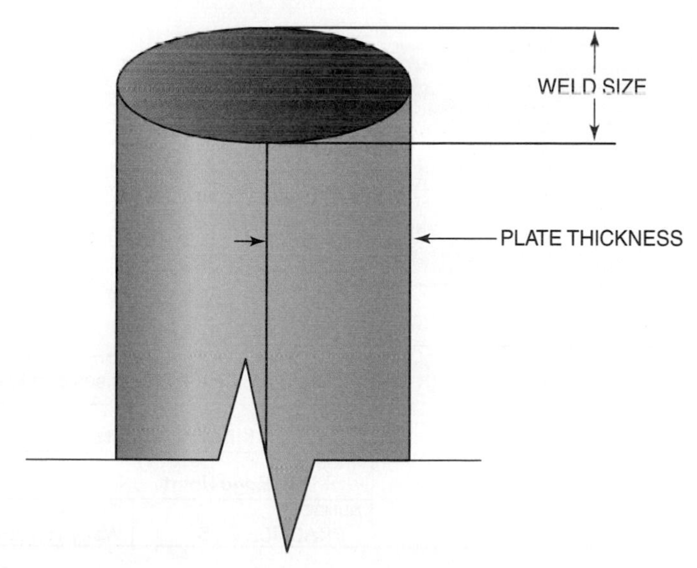

Figure 2.49
Edge weld size

PRACTICE 2-8

Edge Weld in the Flat Position Using E6010 or E6011 Electrodes, E6012 or E6013 Electrodes, and E7016 or E7018 Electrodes

Module 1
Key Indicator 1, 2, 3, 4

Module 4
Key Indicator 1, 3, 4

Module 9
Key Indicator 2

Using a properly set up and adjusted arc welding machine, proper safety protection as demonstrated in Practice 2-1, arc welding electrodes with a 1/8-in. (3-mm) diameter, and two pieces of mild steel plate, 6 in. (152 mm) long × 1/4 in. (6 mm) thick, you will make a weld on an edge joint, **Figure 2.50.**

- Clamp the plates flat together and make a tack weld along each end of the plates.
- Starting at one end of the plate, make a straight weld the full length of the plate. Make the weld bead as wide as the width of the edge joint.
- Watch the molten weld pool, not the end of the electrode.
- Cool, chip, and inspect the weld for uniformity and defects.
- Repeat the welds as needed with all three (F) groups of electrodes until you can consistently make welds free of defects.
- Turn off the welding machine and clean up your work area when you are finished welding.

Complete a copy of the "Student Welding Report" listed in Appendix I or provided by your instructor.

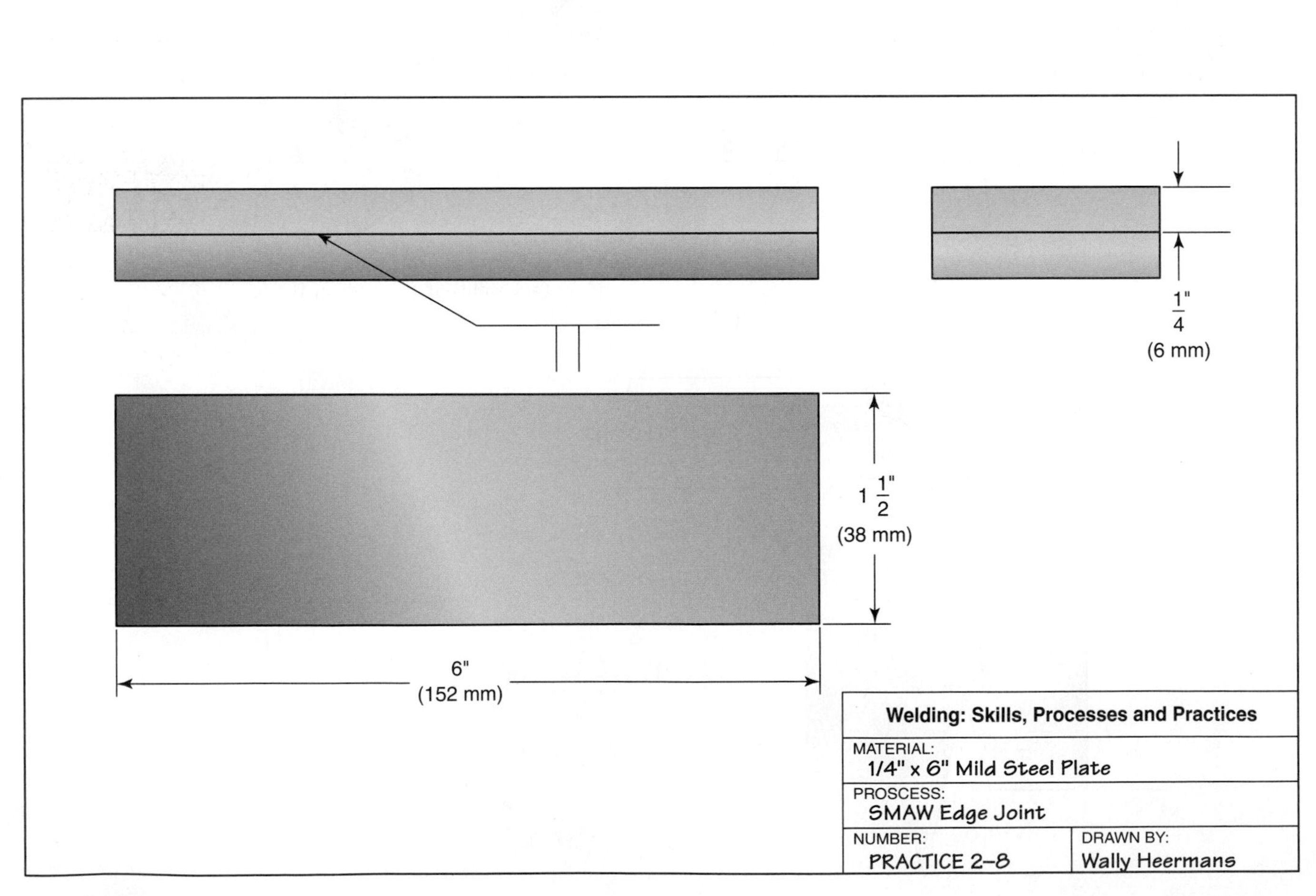

Figure 2.50
Practice 2.8 edge joint

PRACTICE 2-9

Edge Joint in the Vertical Down Position Using E6010 or E6011 Electrodes and E6012 or E6013 Electrodes

Using the same setup, materials, and electrodes as listed in Practice 2-8, you will make a vertical down weld on an edge joint. Start with the plates at a 45° angle.

This technique is the same as that used to make vertical down welds. However, a lower level of skill is required at 45°, and it is easier to develop your skill. After you master the 45° angle, the angle is increased successively until a vertical position is reached, **Figure 2.51.**

- Make the weld bead as wide as the joint. Controlling a weld bead this size is more difficult, but you must develop the skill required to control this larger molten weld pool.
- Cool, chip, and inspect the weld for uniformity and defects.
- Repeat the welds as needed with all three (F) groups of electrodes until you can consistently make welds free of defects. Turn off the welding machine and clean up your work area when you are finished welding.

Complete a copy of the "Student Welding Report" listed in Appendix I or provided by your instructor.

Module 1
Key Indicator 1, 2, 3, 4

Module 4
Key Indicator 1, 3, 4

Module 9
Key Indicator 2

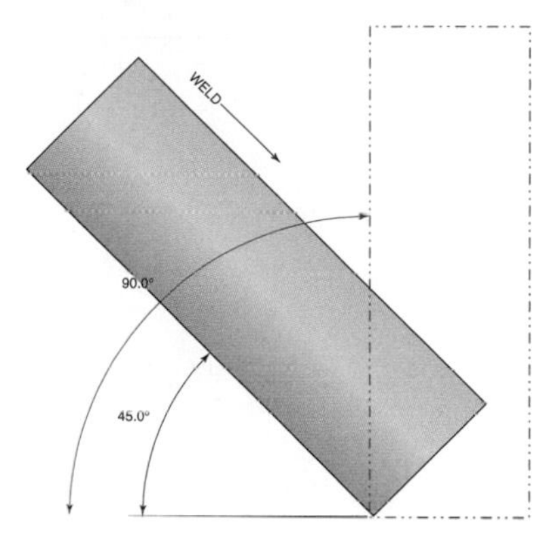

Figure 2.51
Vertical down

PRACTICE 2-10

Edge Joint in the Vertical Up Position Using E6010 or E6011 Electrodes, E6012 or E6013 Electrodes, and E7016 or E7018 Electrodes

Using the same setup, materials, and electrodes as listed in Practice 2-8, you will make a vertical up weld on an edge joint. Start with the plates at a 45° angle.

This technique is the same as that used to make vertical up welds. However, a lower level of skill is required at 45°, and it is easier to develop your skill. After you master the 45° angle, the angle is increased successively until a vertical position is reached, **Figure 2.52.**

Before the molten metal drips down the bead, the back of the molten weld pool will start to bulge, **Figure 2.53.** When this happens, increase the speed of travel and the weave pattern.

- Cool, chip, and inspect the weld for uniformity and defects.
- Repeat the welds as needed with all three (F) groups of electrodes until you can consistently make welds free of defects. Turn off the welding machine and clean up your work area when you are finished welding.

Complete a copy of the "Student Welding Report" listed in Appendix I or provided by your instructor.

Module 1
Key Indicator 1, 2, 3, 4

Module 4
Key Indicator 1, 3, 4

Module 9
Key Indicator 2

PRACTICE 2-11

Edge Joint in the Horizontal Position Using E6010 or E6011 Electrodes, E6012 or E6013 Electrodes, and E7016 or E7018 Electrodes

Using the same setup, materials, and electrodes as listed in Practice 2-8, you will make a horizontal weld on an edge joint. When you begin

Module 1
Key Indicator 1, 2, 3, 4

Module 4
Key Indicator 1, 3, 4

Module 9
Key Indicator 2

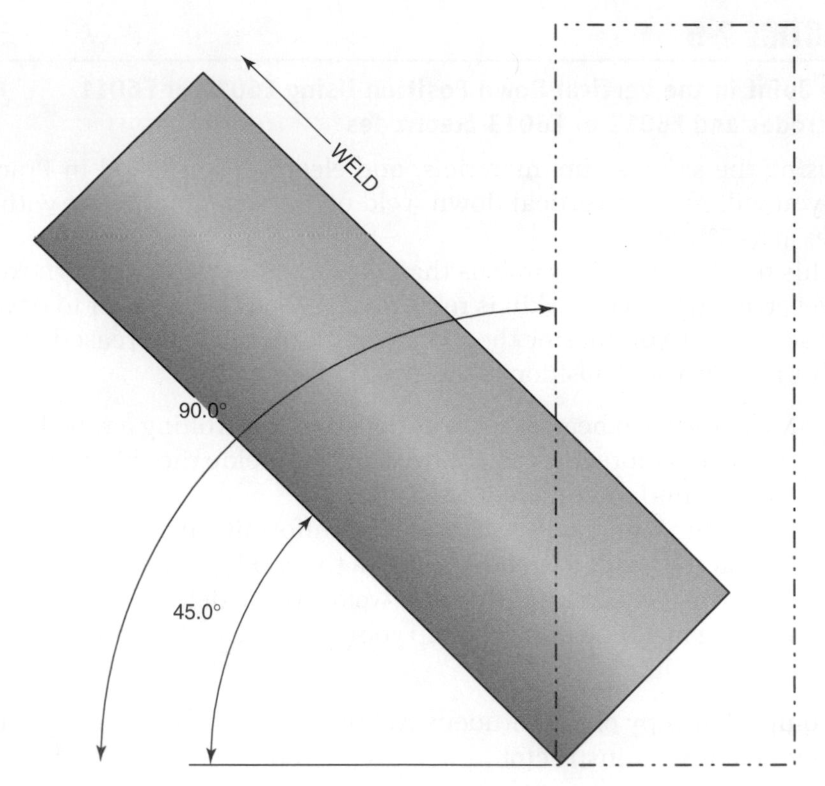

Figure 2.52
Vertical up

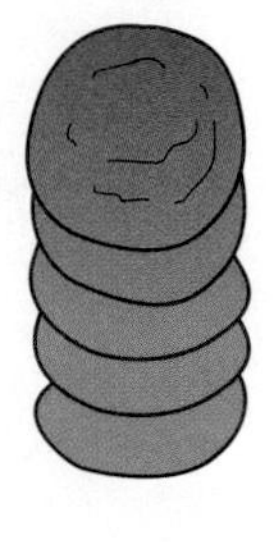

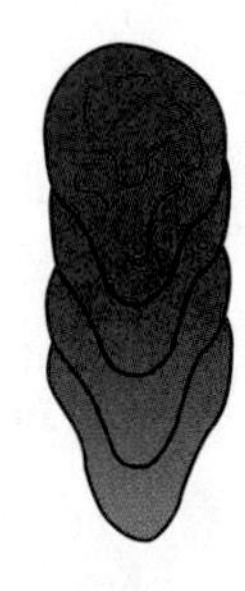

Figure 2.53
Watch the trailing edge of the weld pool to judge the correct travel speed

to practice the horizontal weld, the plate may be reclined slightly, **Figure 2.54.** This placement allows the welder to build the required skill by practicing the correct techniques successfully. The "J" weave or stepped pattern is suggested for this practice. As the electrode is drawn back to the back edge of the weld pool, metal is deposited. Use the metal that's being deposited to support the molten weld pool.

Angling the electrode up and back toward the weld causes more metal to be deposited along the top edge of the weld. Keeping the bead small allows the surface tension to hold the molten weld pool in place.

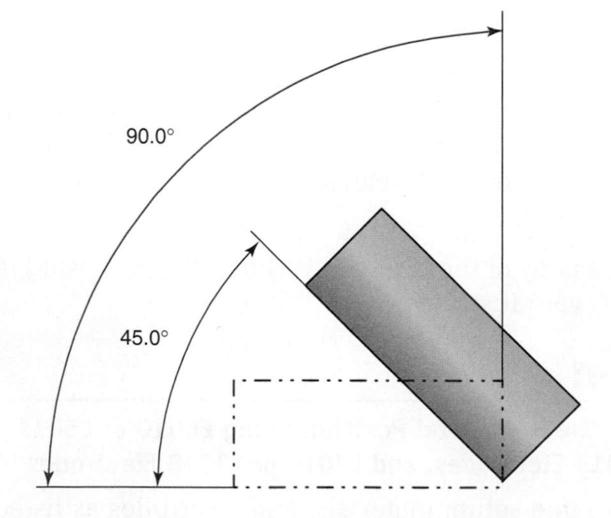

Figure 2.54
Incline angle

Gradually increase the angle of the plate until it and the weld bead are horizontal.

- Cool, chip, and inspect the weld for uniformity and defects.
- Repeat the welds as needed with all three (F) groups of electrodes until you can consistently make welds free of defects. Turn off the welding machine and clean up your work area when you are finished welding.

Complete a copy of the "Student Welding Report" listed in Appendix I or provided by your instructor.

PRACTICE 2-12

Edge Joint in the Overhead Position Using E6010 or E6011 Electrodes, E6012 or E6013 Electrodes, and E7016 or E7018 Electrodes

Using the same setup, materials, and electrodes as listed in Practice 2-8, you will make an overhead weld on an edge joint.

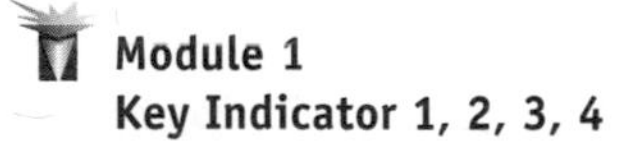

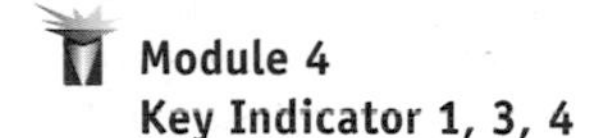

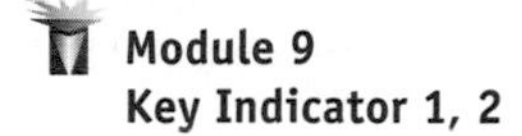

- With the electrode pointed in a slightly trailing angle, **Figure 2.55,** strike the arc in the joint.
- Keep a very short arc length.
- Use the stepped pattern and move the electrode forward slightly when the molten weld pool grows to the correct size, **Figure 2.56.**

As the molten weld pool gets larger, it has a tendency to quickly become convex. If you keep the arc in the molten weld pool once the joint is filled and the weld face is flat, it will quickly overfill and become convex. This can result in the weld face forming drips of metal that hang from the weld like icicles, **Figure 2.57.**

- When the molten weld pool cools and begins to shrink, move the arc back near the center of the weld.

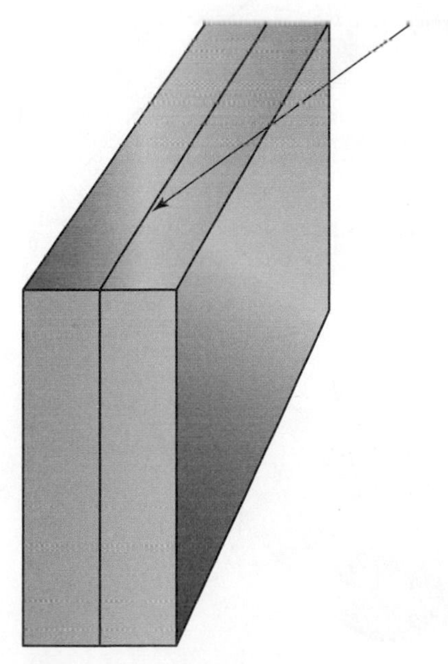

Figure 2.55
Strike the arc in the joint

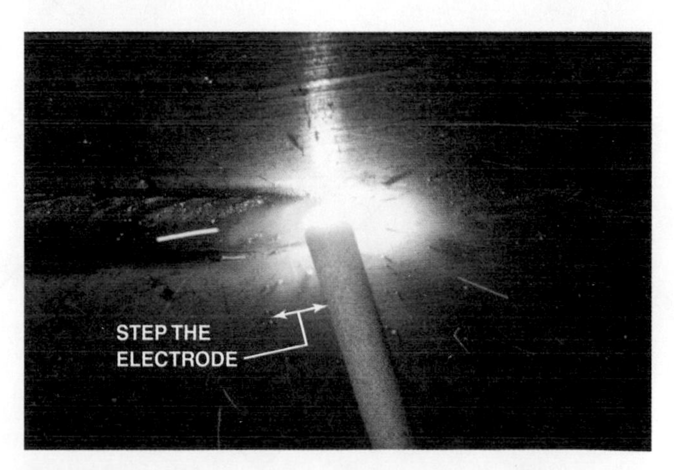

Figure 2.56
Step the electrode
Courtesy of Larry Jeffus

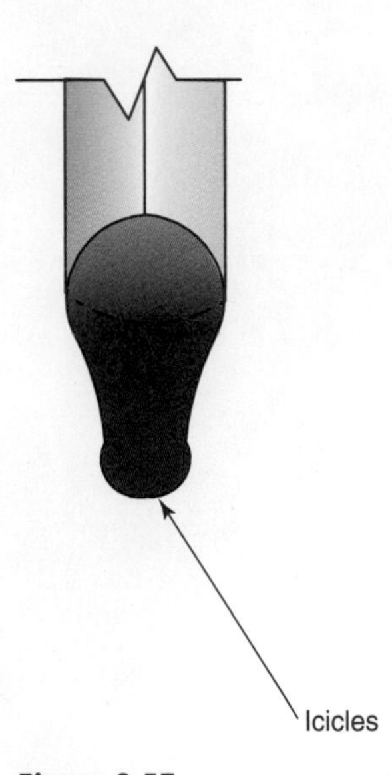

Figure 2.57
Welding too slow or with too high of an amperage setting will result in the weld metal dripping down like icicles

- Hold the arc in this new location until the molten weld pool again grows to the correct size.
- Step the electrode forward again and keep repeating this pattern until the weld progresses along the entire weld joint length.
- Cool, chip, and inspect the weld for uniformity and defects.
- Repeat the welds as needed with all three (F) groups of electrodes until you can consistently make welds free of defects. Turn off the welding machine and clean up your work area when you are finished welding.

Complete a copy of the "Student Welding Report" listed in Appendix I or provided by your instructor.

OUTSIDE CORNER JOINT

An outside corner joint is made by placing the plates at a 90° angle to each other, with the edges forming a V-groove, **Figure 2.58.** There may or may not be a slight root opening left between the plate edges. Small tack welds should be made approximately 1/2 in. (13 mm) from both ends of the joint.

The weld bead should completely fill the V-groove formed by the plates and may have a slightly convex surface buildup. The back side of an outside corner joint can be used to practice fillet welds, or four plates can be made into a box tube shape, **Figure 2.59.**

Figure 2.58
V formed by an outside corner joint

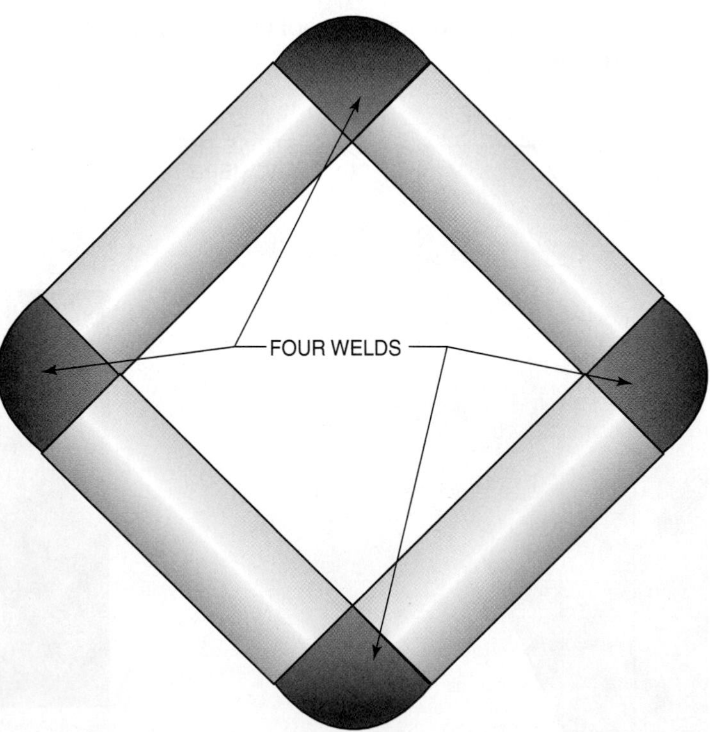

Figure 2.59
Box tube made from four outside corner joint welds

PRACTICE 2-13

Outside Corner Joint in the Flat Position Using E6010 or E6011 Electrodes, E6012 or E6013 Electrodes, and E7016 or E7018 Electrodes

Using a properly set up and adjusted arc welding machine, proper safety protection as demonstrated in Practice 2-1, arc welding electrodes with a 1/8-in. (3-mm) diameter, and two pieces of mild steel plate, 6 in. (152 mm) long × 1/4 in. (6 mm) thick, you will make a weld on an outside corner joint.

Module 1
Key Indicator 1, 2, 3, 4

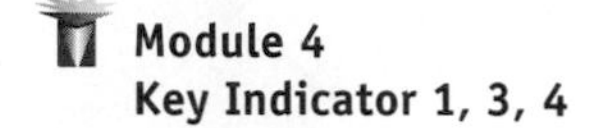

Module 4
Key Indicator 1, 3, 4

Module 9
Key Indicator 1, 2

- Starting at one end of the plate, make a straight weld the full length of the plate.
- Watch the molten weld pool at this point, not the end of the electrode. As you become more skillful, it is easier to watch the molten weld pool.
- Cool, chip, and inspect the weld for uniformity and defects.
- Repeat the welds as needed with all three (F) groups of electrodes until you can consistently make welds free of defects. Turn off the welding machine and clean up your work area when you are finished welding.

Complete a copy of the "Student Welding Report" listed in Appendix I or provided by your instructor.

PRACTICE 2-14

Outside Corner Joint in the Vertical Down Position Using E6010 or E6011 Electrodes, E6012 or E6013 Electrodes.

Using the same setup, materials, and electrodes as listed in Practice 2-13, you will make a vertical down weld on an outside corner joint. Start with the plate at a 45° angle.

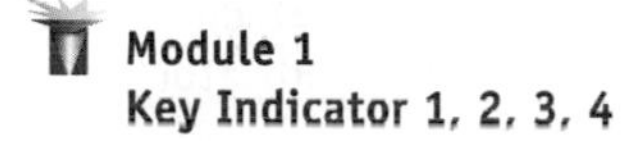

Module 1
Key Indicator 1, 2, 3, 4

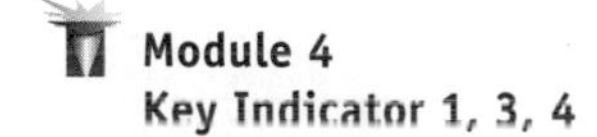

Module 4
Key Indicator 1, 3, 4

Module 9
Key Indicator 1, 2

This technique is the same as that used to make vertical down welds. However, a lower level of skill is required at 45°, and it is easier to develop your skill. After you master the 45° angle, the angle is increased successively until a vertical position is reached, **Figure 2.60.**

- Cool, chip, and inspect the weld for uniformity and defects.
- Repeat the welds as needed with all three (F) groups of electrodes until you can consistently make welds free of defects. Turn off the welding machine and clean up your work area when you are finished welding.

Complete a copy of the "Student Welding Report" listed in Appendix I or provided by your instructor.

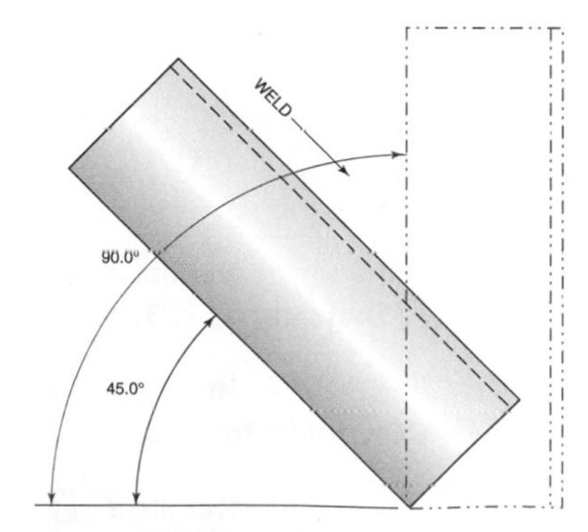

Figure 2.60
Start a 45° angle with and increase it to 90°

PRACTICE 2-15

Outside Corner Joint in the Vertical Up Position Using E6010 or E6011 Electrodes, E6012 or E6013 Electrodes, and E7016 or E7018 Electrodes

Using the same setup, materials, and electrodes as listed in Practice 2-13, you will make a vertical up weld on an outside corner joint. Start with the plate at a 45° angle.

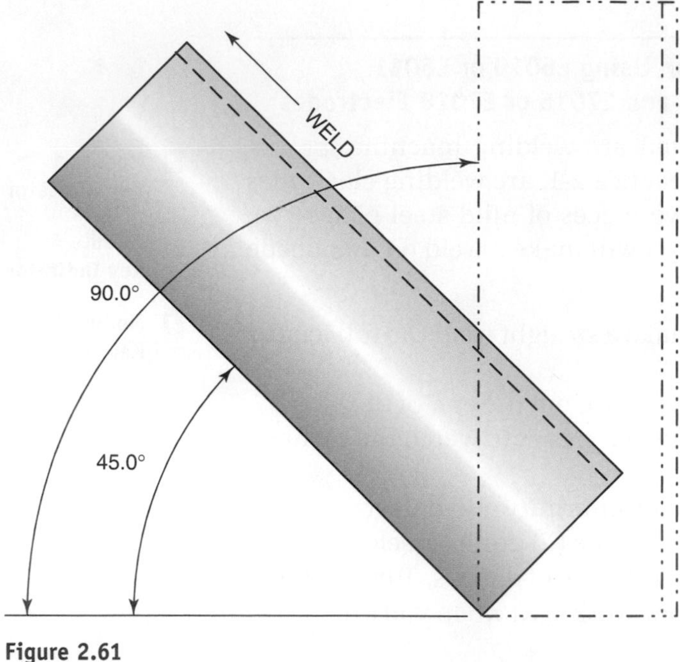

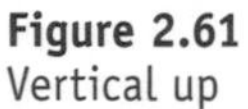

Figure 2.61
Vertical up

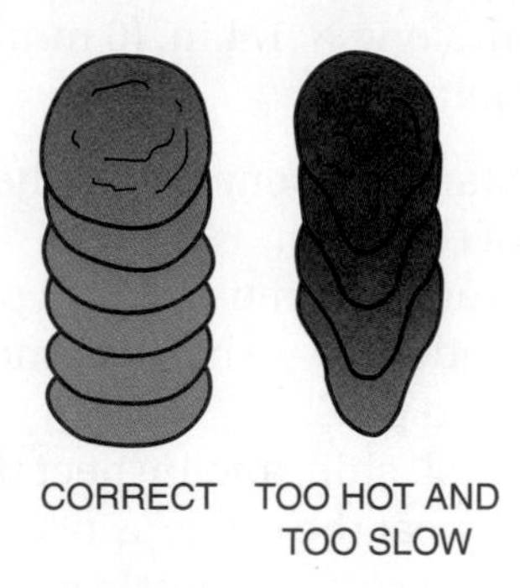

Figure 2.62
Watch the trailing edge of the weld pool to judge the correct travel speed

Module 1
Key Indicator 1, 2, 3, 4

Module 4
Key Indicator 1, 3, 4

Module 9
Key Indicator 2

This technique is the same as that used to make vertical up welds. However, a lower level of skill is required at 45°, and it is easier to develop your skill. After the welder masters the 45° angle, the angle is increased successively until a vertical position is reached, **Figure 2.61.**

Before the molten metal drips down the bead, the back of the molten weld pool will start to bulge, **Figure 2.62.** When this happens, increase the speed of travel and the weave pattern.

- Cool, chip, and inspect the weld for uniformity and defects.
- Repeat the welds as needed with all three (F) groups of electrodes until you can consistently make welds free of defects. Turn off the welding machine and clean up your work area when you are finished welding.

Complete a copy of the "Student Welding Report" listed in Appendix I or provided by your instructor.

PRACTICE 2-16

Outside Corner Joint in the Horizontal Position Using E6010 or E6011 Electrodes, E6012 or E6013 Electrodes, and E7016 or E7018 Electrodes

Module 1
Key Indicator 1, 2, 3, 4

Module 4
Key Indicator 1, 3, 4

Module 9
Key Indicator 2

Using the same setup, materials, and electrodes as listed in Practice 2-13, you will make a horizontal weld on an outside corner joint. When the welder begins to practice the horizontal weld, the joint may be reclined slightly, **Figure 2.63.** This placement allows the welder to build the required skill by practicing the correct techniques successfully. The "J" weave or stepped pattern is suggested for this practice. As the electrode is drawn back into the weld pool, metal is deposited. This metal supports the molten weld pool, resulting in a bead with a uniform contour, **Figure 2.64.**

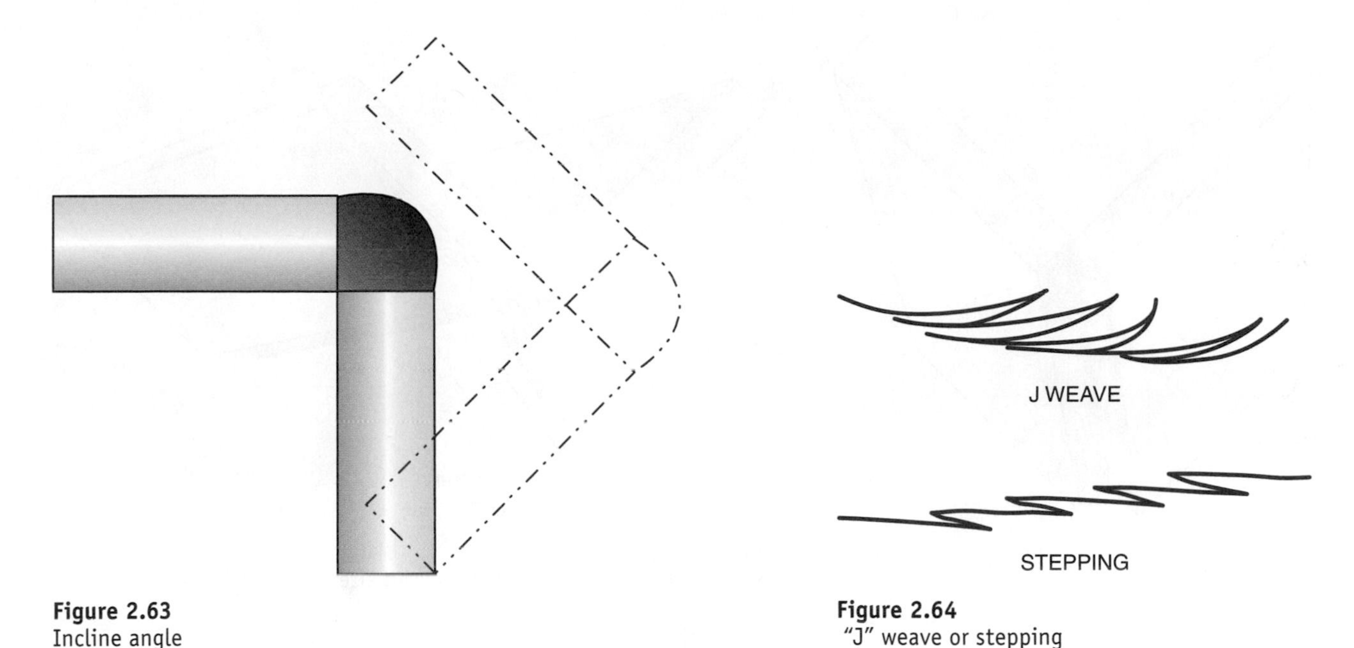

Figure 2.63
Incline angle

Figure 2.64
"J" weave or stepping

Angling the electrode up and back toward the weld causes more metal to be deposited along the top edge of the weld. Keeping the bead small allows the surface tension to hold the molten weld pool in place.

Gradually increase the angle of the plate until it is vertical and the weld bead is horizontal.

- Cool, chip, and inspect the weld for uniformity and defects.
- Repeat the welds as needed with all three (F) groups of electrodes until you can consistently make welds free of defects. Turn off the welding machine and clean up your work area when you are finished welding.

Complete a copy of the "Student Welding Report" listed in Appendix I or provided by your instructor.

PRACTICE 2-17

Outside Corner Joint in the Overhead Position Using E6010 or E6011 Electrodes, E6012 or E6013 Electrodes, and E7016 or E7018 Electrodes

Using the same setup, materials, and electrodes as listed in Practice 2-13, you will make an overhead weld on an outside corner joint.

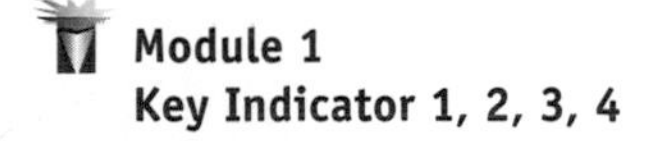
Module 1
Key Indicator 1, 2, 3, 4

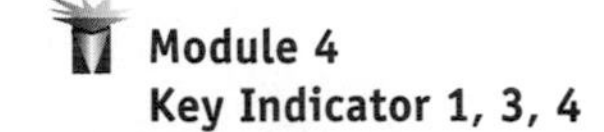
Module 4
Key Indicator 1, 3, 4

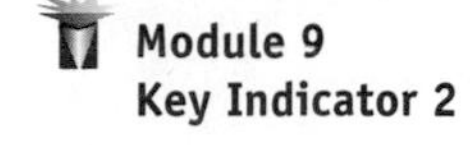
Module 9
Key Indicator 2

- With the electrode pointed slightly into the joint, **Figure 2.65,** strike the arc in the joint.
- Keep a very short arc length.
- Use the stepped pattern and move the electrode forward slightly when the molten weld pool grows to the correct size, **Figure 2.66.**

As the molten weld pool gets larger, it has a tendency to quickly become convex. If you keep the arc in the molten weld pool once the joint is filled and the weld face is flat, it will quickly overfill and become convex. This can result in the weld face forming drips of metal that hang from the weld like icicles, **Figure 2.67.**

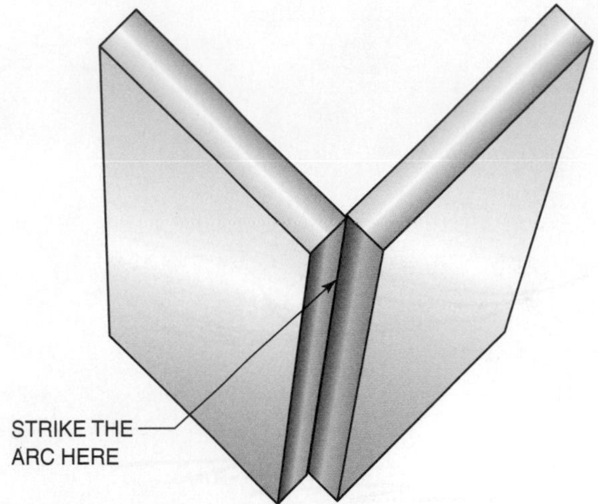

Figure 2.65
Strike arc in the joint

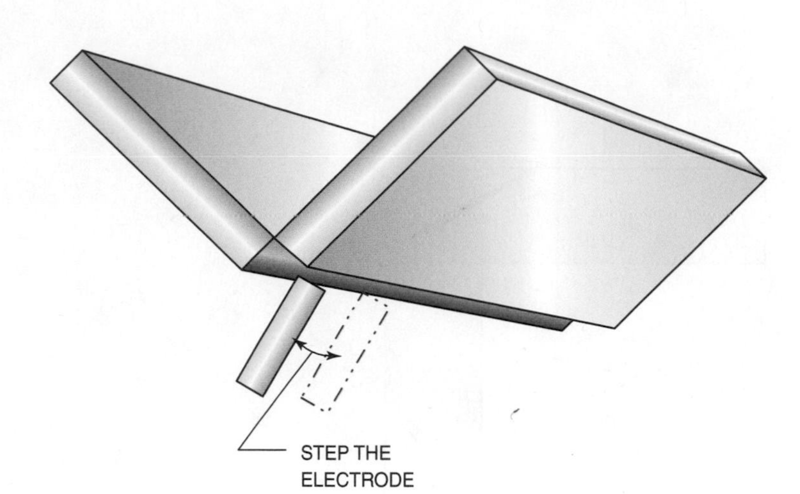

Figure 2.66
Stepping the electrode to control weld size

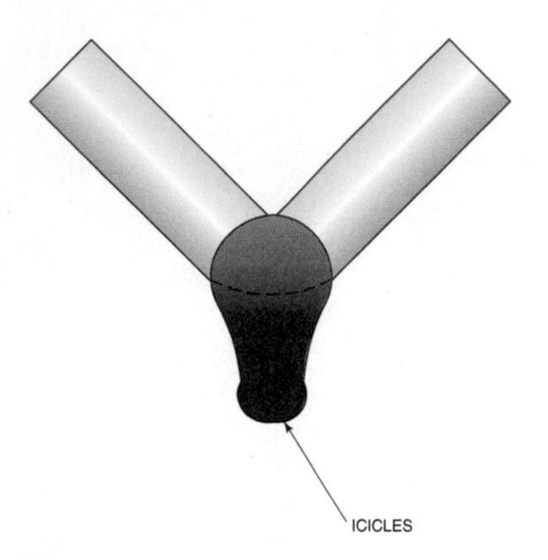

Figure 2.67
Welding too slowly or with too high of an amperage setting will result in the weld metal dripping down like icicles

- When the molten weld pool cools and begins to shrink, move the arc back near the center of the weld.
- Hold the arc in this new location until the molten weld pool again grows to the correct size.
- Step the electrode forward again and keep repeating this pattern until the weld progresses along the entire weld joint length.
- Cool, chip, and inspect the weld for uniformity and defects.
- Repeat the welds as needed with all three (F) groups of electrodes until you can consistently make welds free of defects. Turn off the welding machine and clean up your work area when you are finished welding.

Complete a copy of the "Student Welding Report" listed in Appendix I or provided by your instructor.

LAP JOINT

A **lap joint** is made by overlapping the edges of the two plates, **Figure 2.68.** The joint can be welded on one side or both sides with a fillet weld. In Practice 2-7, both sides should be welded unless otherwise noted.

As the fillet weld is made on the lap joint, the buildup should equal the thickness of the plate, **Figure 2.69.** A good weld will have a smooth transition from the plate surface to the weld. If this transition is abrupt, it can cause stresses that will weaken the joint.

Penetration for lap joints does not improve their strength; complete fusion is required. The root of fillet welds must be melted to ensure a completely fused joint. If the molten weld pool shows a notch during the weld, **Figure 2.70,** this is an indication that the root is not being fused together. The weave pattern will help prevent this problem, **Figure 2.71.**

Figure 2.68
Lap joint

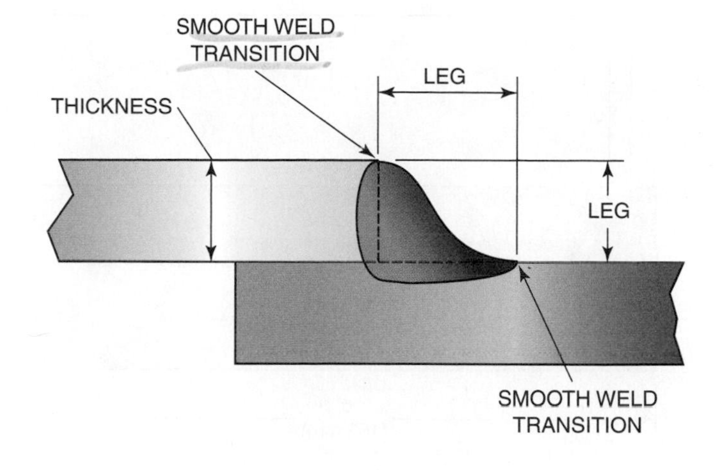

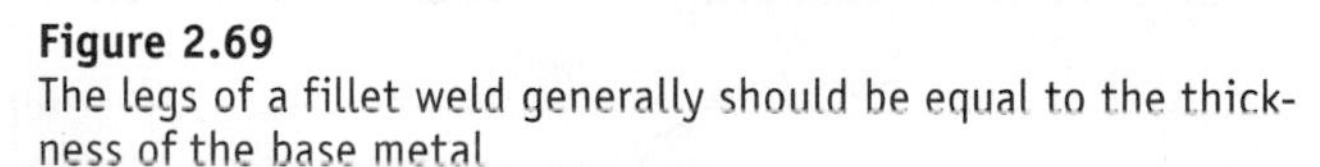
Figure 2.69
The legs of a fillet weld generally should be equal to the thickness of the base metal

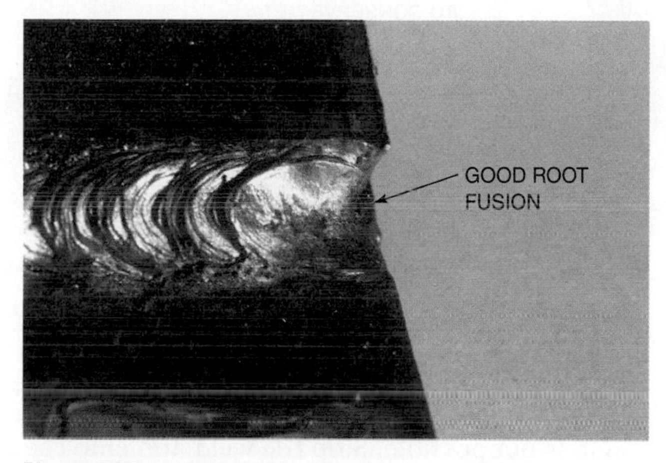

Figure 2.70
Watch the root of the weld bead to be sure there is a complete fusion
Courtesy of Larry Jeffus

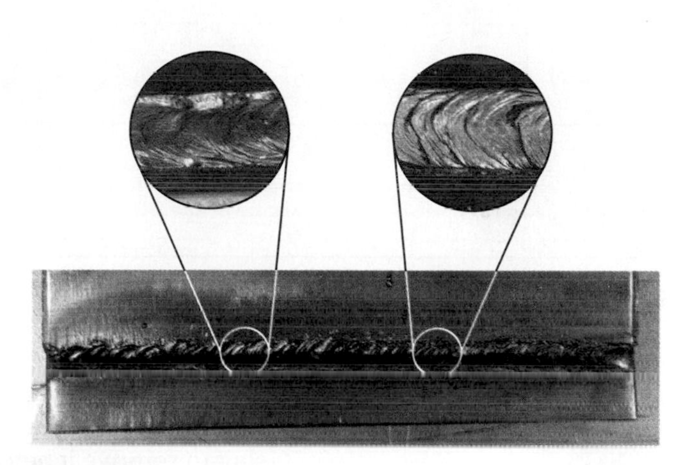

Figure 2.71
Lap joint
Courtesy of Larry Jeffus

PRACTICE 2-18

Welded Lap Joint in the Flat Position (1F) Using E6010 or E6011 Electrodes, E6012 or E6013 Electrodes, and E7016 or E7018 Electrodes

Using a properly set up and adjusted arc welding machine, proper safety protection, arc welding electrodes having a 1/8-in. (3-mm) diameter, and two or more pieces of mild steel plate, 6 in. (152 mm) long × 1/4 in. (6 mm) thick, you will make a welded lap joint in the flat position, **Figure 2.72.**

Hold the plates together tightly with an overlap of no more than 1/4 in. (6 mm). Tack weld the plates together. A small tack weld may be added in the center to prevent distortion during welding, **Figure 2.73.** Chip the tacks before you start to weld.

The "J," "C," or zigzag weave pattern works well on this joint. Strike the arc and establish a molten pool directly in the joint. Move the electrode out on the bottom plate and then onto the weld to the top edge of the top plate, **Figure 2.74.** Follow the surface of the plates with the arc. Do not follow the trailing edge of the weld bead. Following the molten weld pool will not allow for good root fusion and will cause slag to collect in the root. If

Module 1
Key Indicator 1, 2, 3, 4

Module 4
Key Indicator 1, 5

Module 9
Key Indicator 1, 2
This practice addresses the "Flat" position portion of the all-position requirement of 5.

Module 9
Key Indicator 2

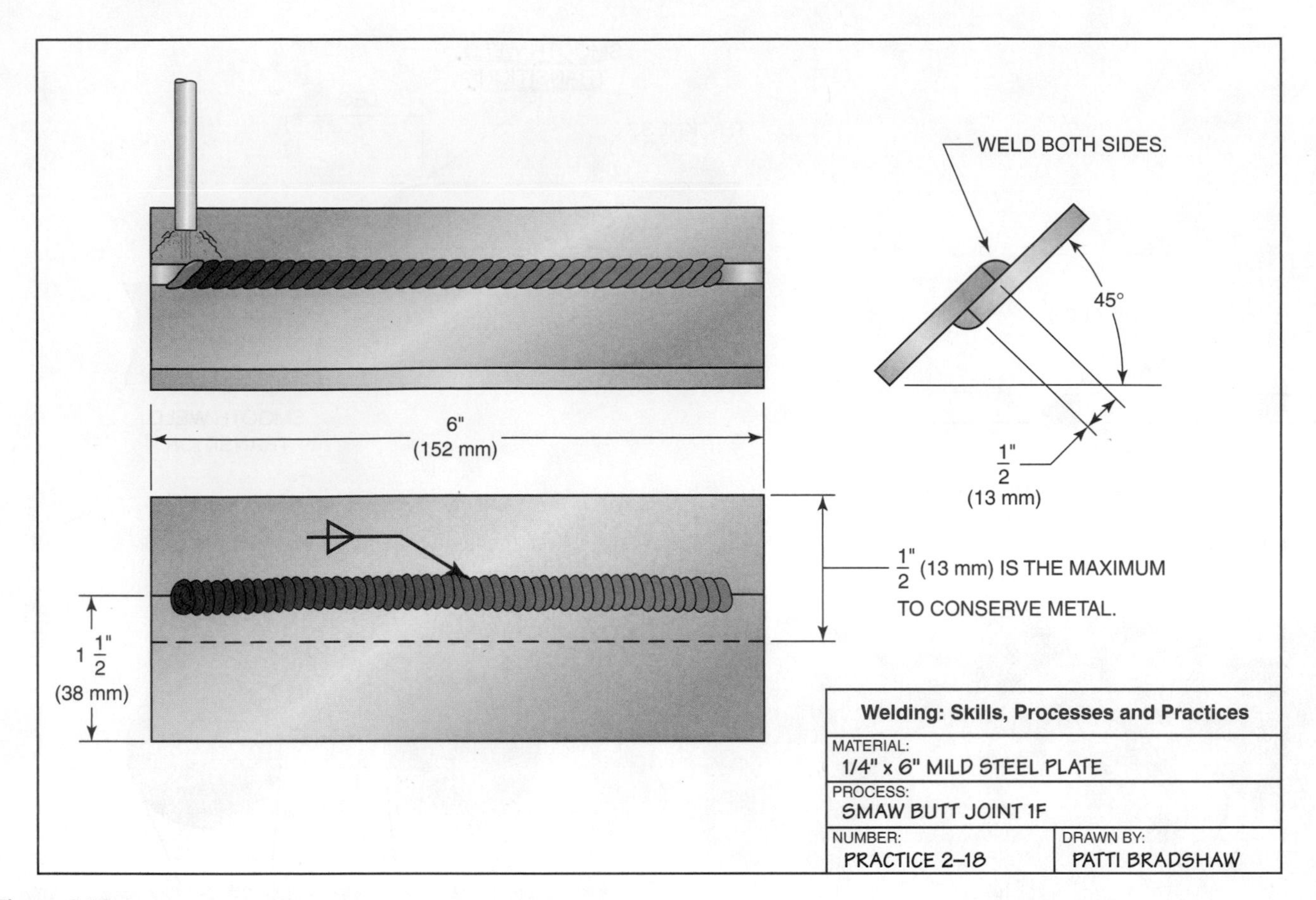

Figure 2.72
Lap joint in the flat position

slag does collect, a good weld is not possible. Stop the weld and chip the slag to remove it before the weld is completed. Cool, chip, and inspect the weld for uniformity and defects. Repeat the welds with all three (F) groups of electrodes until you can consistently make welds free of defects. Turn off the welding machine and clean up your work area when you are finished welding.

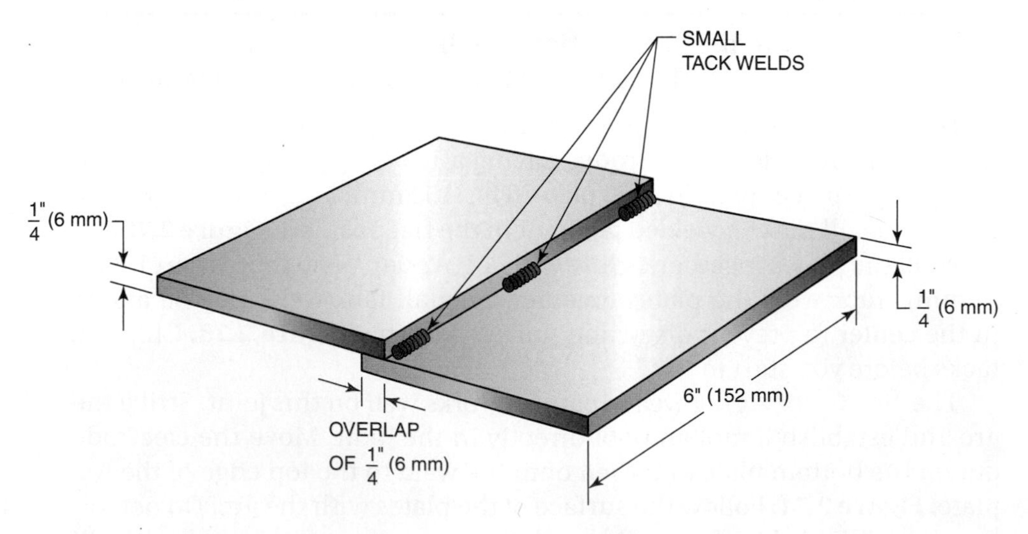

Figure 2.73
Tack welding the plates together

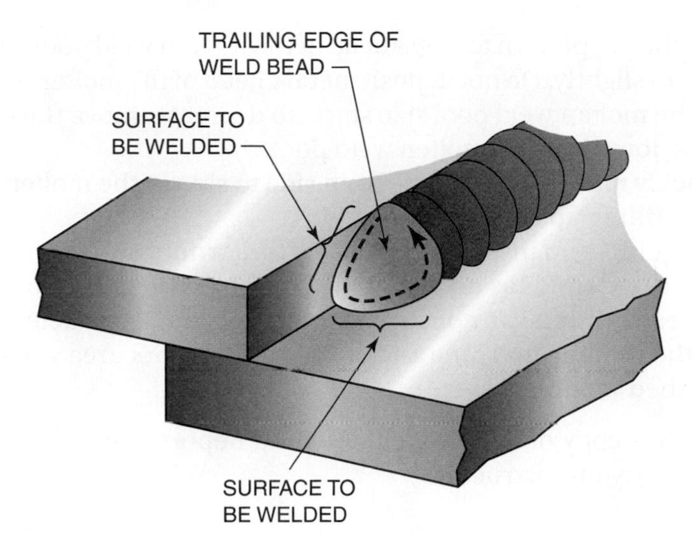

Figure 2.74
Follow the surface of the plate to ensure good fusion

Complete a copy of the "Student Welding Report" listed in Appendix I or provided by your instructor.

Module 1
Key Indicator 1, 2, 3, 4

Module 4
Key Indicator 1, 5

Module 9
Key Indicator 1, 2
This practice addresses the "Horizontal" position portion of the all-position requirement of 5.

Module 9
Key Indicator 2

PRACTICE 2-19

Welded Lap Joint in the Horizontal Position (2F) Using E6010 or E6011 Electrodes, E6012 or E6013 Electrodes, and E7016 or E7018 Electrodes

Using the same setup, materials, and electrodes as listed in Practice 2-18, you will make a welded horizontal lap joint.

The horizontal lap joint and the flat lap joint require nearly the same technique and skill to achieve a proper weld, **Figure 2.75.** Use the "J," "C," or zigzag weave pattern to make the weld. Do not allow slag to collect in the root. The fillet must be equally divided between both plates for good strength. After completing the weld, cool, chip, and inspect the weld for uniformity and defects. Repeat the welds using all three (F) groups of electrodes until you can consistently make welds free of defects. Turn off the welding machine and clean up your work area when you are finished welding.

Complete a copy of the "Student Welding Report" listed in Appendix I or provided by your instructor.

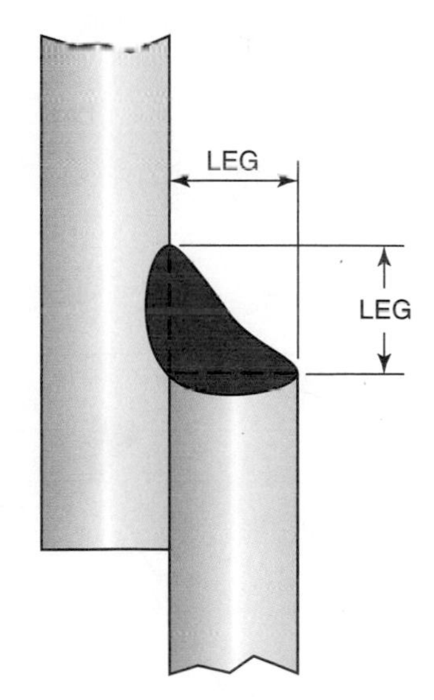

Figure 2.75
The horizontal lap joint should have a fillet weld that is equal on both plates

PROACTICE 2-20

Lap Joint in the Vertical Up Position (3F) Using E6010 or E6011 Electrodes, E6012 or E6013 Electrodes, and E7016 or E7018 Electrodes

Using the same setup, materials, and electrodes as listed in Practice 2-18, you will make a vertical up weld in a lap joint.

- Start practicing this weld with the plate at a 45° angle.
- Gradually increase the angle of the plate to vertical as skill is gained in welding this joint. The "J" or "T" weave pattern works well on this joint.
- Establish a molten weld pool in the root of the joint.

Module 1
Key Indicator 1, 2, 3, 4

Module 4
Key Indicator 1, 5
This practice addresses the "Vertical" position portion of the all-position requirement of 5.

Module 9
Key Indicator 2

- Use the "T" pattern to step ahead of the molten weld pool, allowing it to cool slightly. Do not deposit metal ahead of the molten weld pool.
- As the molten weld pool size starts to decrease, move the electrode back down into the molten weld pool.
- Quickly move the electrode from side to side in the molten weld pool, filling up the joint.
- Cool, chip, and inspect the weld for uniformity and defects.
- Repeat the welds as necessary with all three (F) groups of electrodes until you can consistently make welds free of defects. Turn off the welding machine and clean up your work area when you are finished welding.

Complete a copy of the "Student Welding Report" listed in Appendix I or provided by your instructor.

PRACTICE 2-21

Lap Joint in the Overhead Position (4F) Using E6010 or E6011 Electrodes, E6012 or E6013 Electrodes, and E7016 or E7018 Electrodes

Module 1
Key Indicator 1, 2, 3, 4

Module 4
Key Indicator 1, 5
This practice addresses the "Overhead" position portion of the all-position requirement of 5.

Module 9
Key Indicator 2

Using the same setup, materials, and electrodes as listed in Practice 2-18, you will make an overhead weld in a lap joint.

- With the electrode pointed slightly into the joint, **Figure 2.76,** strike the arc in the inside corner of the lap joint.
- Keep a very short arc length.
- Use the stepped pattern and move the electrode forward slightly when the molten weld pool grows to the correct size, **Figure 2.77.**

As the molten weld pool gets larger, it has a tendency to quickly become convex. If you keep the arc in the molten weld pool once the joint is filled and the weld face is flat, it will quickly overfill and become convex. This can result in the weld face forming drips of metal that hang from the weld like icicles, **Figure 2.78.**

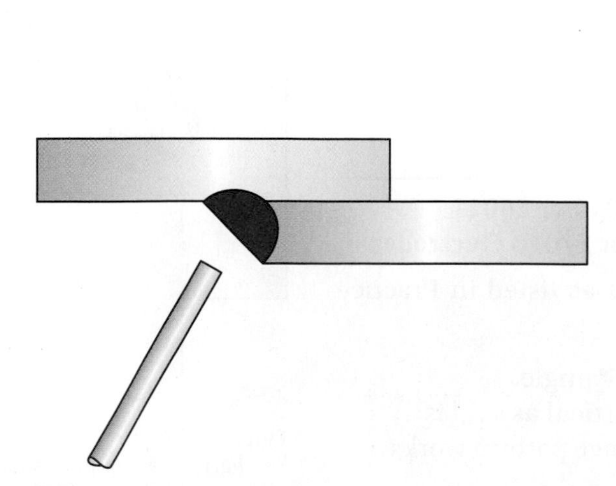

Figure 2.76
Point the electrode slightly toward the root of the joint

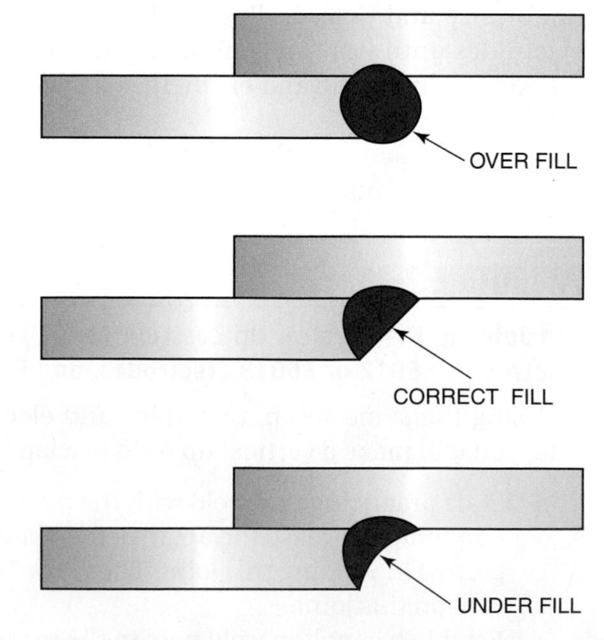

Figure 2.77
Correct fillet weld size for overhead welds

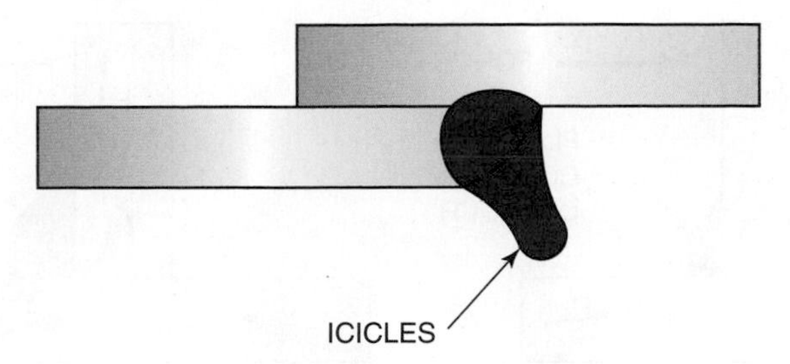

Figure 2.78
Overfilling the molten weld pool will result in drips of metal called icicles

- When the molten weld pool cools and begins to shrink, move the arc back near the center of the weld.
- Hold the arc in this new location until the molten weld pool again grows to the correct size.
- Step the electrode forward again and keep repeating this pattern until the weld progresses along the entire weld joint length.
- Cool, chip, and inspect the weld for uniformity and defects.
- Repeat the welds as needed with all three (F) groups of electrodes until you can consistently make welds free of defects. Turn off the welding machine and clean up your work area when you are finished welding.

Complete a copy of the "Student Welding Report" listed in Appendix I or provided by your instructor.

TEE JOINT

The **tee joint** is made by tack welding one piece of metal on another piece of metal at a right angle, **Figure 2.79.** After the joint is tack welded together, the slag is chipped from the tack welds. If the slag is not removed, it will cause a slag inclusion in the final weld.

The heat is not distributed uniformly between both plates during a tee weld. Because the plate that forms the stem of the tee can conduct heat away from the arc in only one direction, it will heat up faster than the base plate. Heat escapes into the base plate in two directions. When using a weave pattern, most of the heat should be directed to the base plate to keep the weld size more uniform and to help prevent undercut.

A welded tee joint can be strong if it is welded on both sides, even without having deep penetration, **Figure 2.80.** The weld will be as strong as the

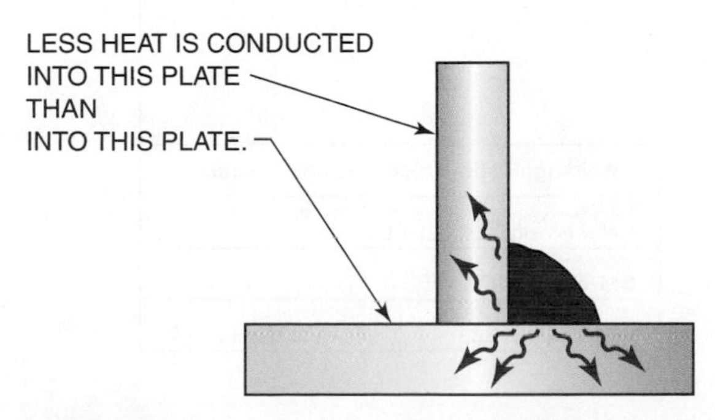

Figure 2.79
Tee joint

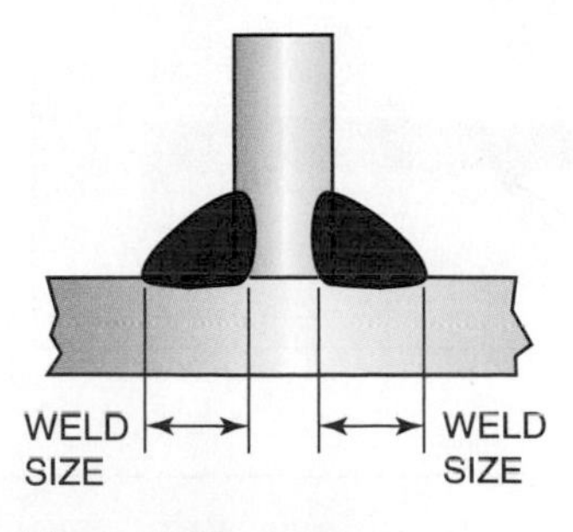

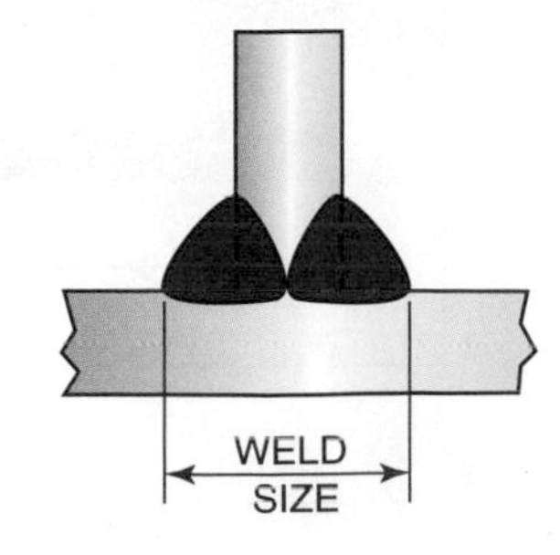

Figure 2.80
If the total weld sizes are equal, then both tee joints would have equal strength

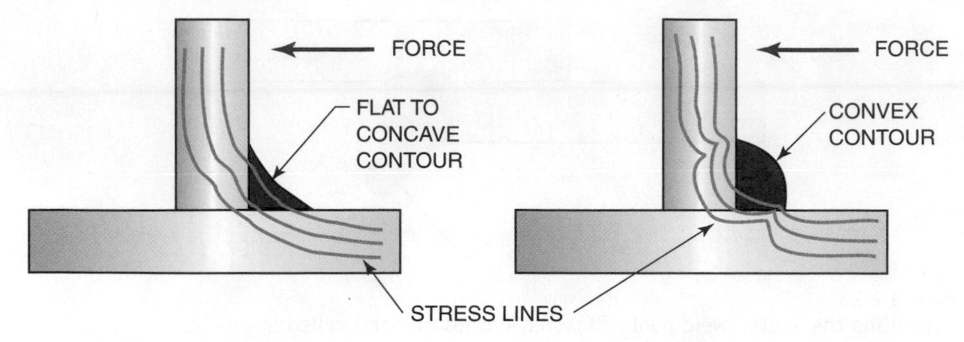

Figure 2.81
The stresses are distributed more uniformly through a flat or concave fillet weld

base plate if the size of the two welds equals the total thickness of the base plate. The weld bead should have a flat or slightly concave appearance to ensure the greatest strength and efficiency, **Figure 2.81.**

Module 1
Key Indicator 1, 2, 3, 4

Module 4
Key Indicator 1, 5
This practice addresses the "Flat" position portion of the all-position requirement of 5.

Module 9
Key Indicator 2

PRACTICE 2-22

Tee Joint in the Flat Position (1F) Using E6010 or E6011 Electrodes, E6012 or E6013 Electrodes, and E7016 or E7018 Electrodes

Using a properly set up and adjusted arc welding machine, proper safety protection, arc welding electrodes having a 1/8-in. (3-mm) diameter, and two or more pieces of mild steel plate, 6 in. (152 mm) long × 1/4 in. (6 mm) thick, you will make a welded tee joint in the flat position, **Figure 2.82.**

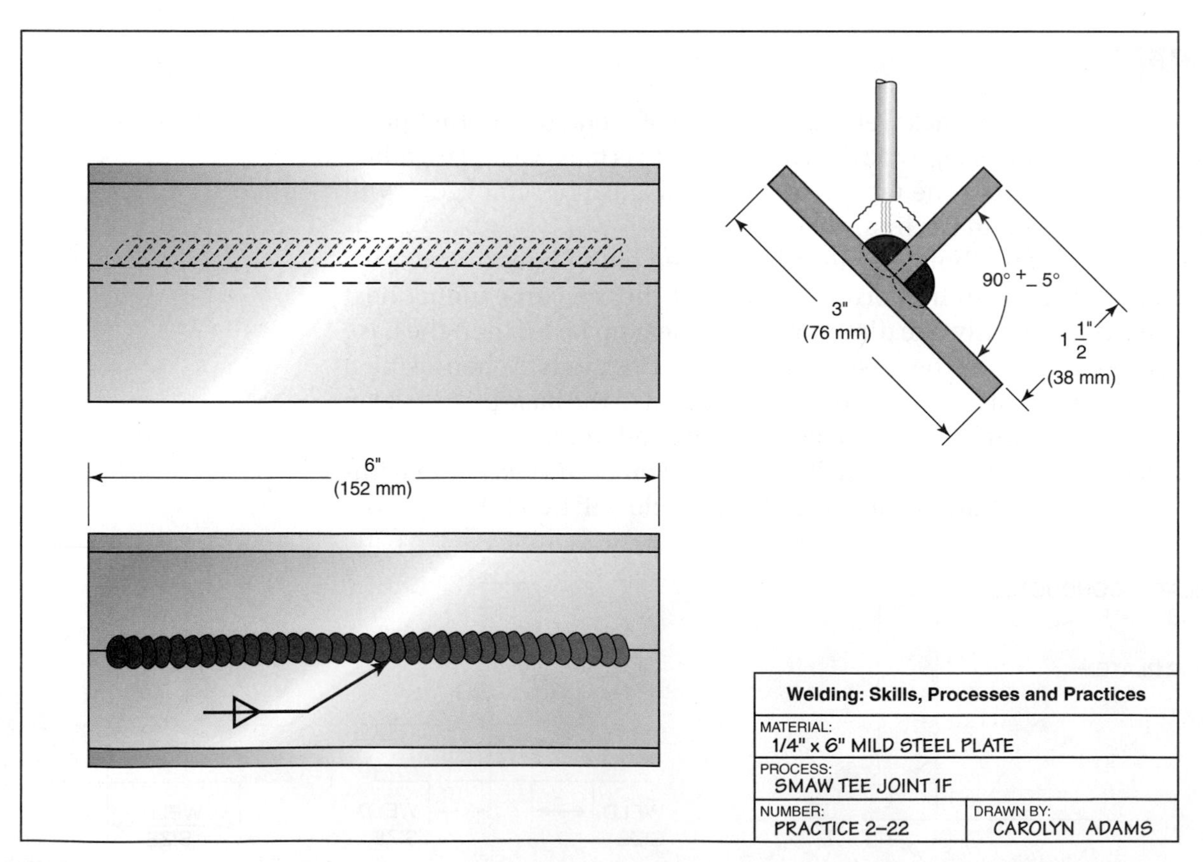

Figure 2.82
Tee joint in the flat position

After the plates are tack welded together, place them on the welding table so the weld will be flat. Start at one end and establish a molten weld pool on both plates. Allow the molten weld pool to flow together before starting the bead. Any of the weave patterns will work well on this joint. To prevent slag inclusions, use a slightly higher than normal amperage setting.

When the 6-in. (152-mm)-long weld is completed, cool, chip, and inspect it for uniformity and soundness. Repeat the welds as needed for all these groups of electrodes until you can consistently make welds free of defects. Turn off the welding machine and clean up your work area when you are finished welding.

Complete a copy of the "Student Welding Report" listed in Appendix I or provided by your instructor.

PRACTICE 2-23

Tee Joint in the Horizontal Position (2F) Using E6010 or E6011 Electrodes, E6012 or E6013 Electrodes, and E7016 or E7018 Electrodes

Using the same setup, materials, and electrodes as listed in Practice 2-22, you will make a welded tee joint in the horizontal position.

Place the tack welded tee plates flat on the welding table so that the weld is horizontal and the plates are flat and vertical, **Figure 2.83.** Start the arc on the flat plate and establish a molten weld pool in the root on both plates. Using the "J" or "C" weave pattern, push the arc into the root and slightly up the vertical plate. You must keep the root of the joint fusing together with the weld metal. If the metal does not fuse, a notch will appear on the leading edge of the weld bead. Poor or incomplete root fusion will cause the weld to be weak and easily cracked under a load.

When the weld is completed, cool, chip, and inspect it for uniformity and defects. Undercut on the vertical plate is the most common defect. Repeat the welds with all three (F) groups of electrodes until you can consistently make welds free of defects. Turn off the welding machine and clean up your work area when you are finished welding.

Complete a copy of the "Student Welding Report" listed in Appendix I or provided by your instructor.

Module 1
Key Indicator 1, 2, 3, 4

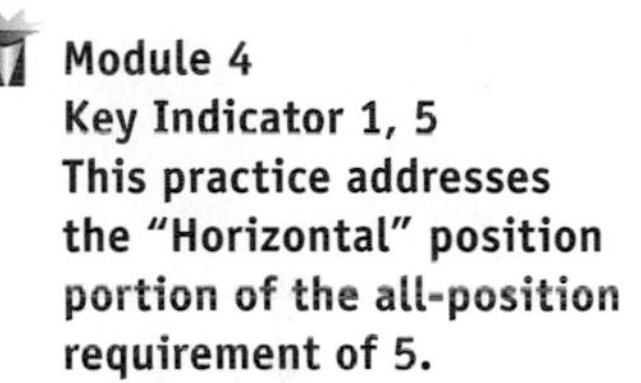

Module 4
Key Indicator 1, 5
This practice addresses the "Horizontal" position portion of the all-position requirement of 5.

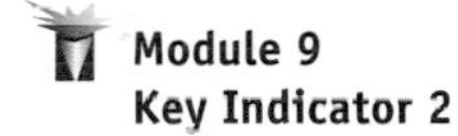

Module 9
Key Indicator 2

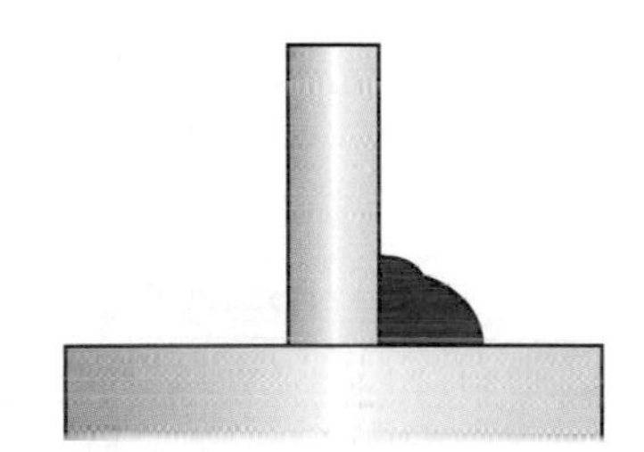

Figure 2.83
Horizontal tee

PRACTICE 2-24

Tee Joint in the Vertical Up Position (3F) Using E6010 or E6011 Electrodes, E6012 or E6013 Electrodes, and E7016 or E7018 Electrodes

Using the same setup, materials, and electrodes as listed in Practice 2-22, you will make a welded tee joint in the vertical up position.

Practice this weld with the plate at a 45° angle. This position will allow you to develop your skill for the vertical up position. Start the arc and molten weld pool deep in the root of the joint. Build a shelf large enough to support the bead as it progresses up the joint. The square, "J," or "C" pattern can be used, but the "T" or stepped pattern will allow deeper root penetration.

For this weld, undercut is a problem on both sides of the weld. It can be controlled by holding the arc on the side long enough for filler metal

Module 1
Key Indicator 1, 2, 3, 4

Module 4
Key Indicator 1, 5
This practice addresses the "Vertical" position portion of the all-position requirement of 5.

Module 9
Key Indicator 2

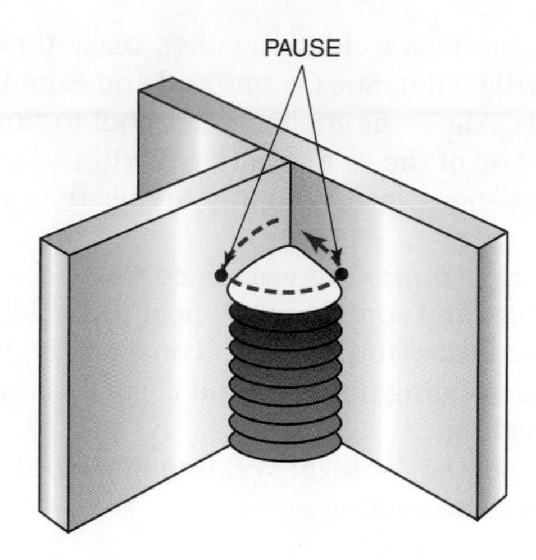

Figure 2.84
Pausing just above the undercut will fill it. This action also causes undercut, but that will be filled on the next cycle.

to flow down and fill it, **Figure 2.84.** Cool, chip, and inspect the weld for uniformity and defects. Repeat the welds as necessary with all three (F) groups of electrodes until you can consistently make welds free of defects. Turn off the welding machine and clean up your work area when you are finished welding.

Complete a copy of the "Student Welding Report" listed in Appendix I or provided by your instructor.

PRACTICE 2-25

Tee Joint in the Overhead Position (4F) Using E6010 or E6011 Electrodes, E6012 or E6013 Electrodes, and E7016 or E7018 Electrodes

Module 1
Key Indicator 1, 2, 3, 4

Module 4
Key Indicator 1, 5
This practice addresses the "Overhead" position portion of the all-position requirement of 5.

Module 9
Key Indicator 2

Using the same setup, materials, and electrodes as listed in Practice 2-12, you will make a welded tee joint in the overhead position.

Start the arc and molten weld pool deep in the root of the joint. Keep a very short arc length. The stepped pattern will allow deeper root penetration.

For this weld, undercut is a problem on both sides of the weld with a high buildup in the center. It can be controlled by holding the arc on the side long enough for filler metal to flow in and fill it. Cool, chip, and inspect the weld for uniformity and defects. Repeat the welds as necessary with all three (F) groups of electrodes until you can consistently make welds free of defects. Turn off the welding machine and clean up your work area when you are finished welding.

Complete a copy of the "Student Welding Report" listed in Appendix I or provided by your instructor.

SUMMARY

The shielded metal arc welding process is most often referred to in welding shops as stick welding. Some people say that it gets this name for one of two reasons. The first is most obviously a result of the stick shape of the electrode. The second reason is experienced by all new welders; it is the tendency for the electrode to stick to the workpiece. All new welders experience this, and your ability to control the sticking of the electrode can be improved as you develop the proper arc-striking techniques.

For a new welder, it is often difficult to concentrate on anything other than the bright sparks and glow at the end of the electrode. But, with time, as you develop your skills, your visual field will increase, allowing you to see a much larger welding zone. This skill comes with time and practice. Developing this skill is essential for you to become a highly proficient welder. Nothing enhances your welding skills more than time under the hood, actually welding, cleaning off the weld, inspecting it, determining the necessary corrections to be made, and immediately trying to produce the next weld with a higher level of quality.

REVIEW

1. Describe two methods of striking an arc with an electrode.
2. Why is it important to strike the arc only in the weld joint?
3. What problems may result by using an electrode at too low a current setting?
4. What problems may result by using an electrode at too high a current setting?
5. According to **Table 2.1,** what would the amperage range be for the following electrodes?
   a. 1/8 in. (3 mm), E6010
   b. 5/32 in. (4 mm), E7018
   c. 3/32 in. (2.4 mm), E7016
   d. 1/8 in. (3 mm), E6011
6. What makes some spatter "hard"?
7. Why should you never change the current setting during a weld?
8. What factors should be considered when selecting an electrode size?
9. What can a welder do to control overheating of the metal pieces being welded?
10. What effect does changing the arc length have on the weld?
11. What arc problems can occur in deep or narrow weld joints?
12. Describe the difference between using a leading and a trailing electrode angle.
13. Can all electrodes be used with a leading angle? Why or why not?
14. What characteristics of the weld bead does the weaving of the electrode cause?
15. What are some of the applications for the circular pattern in the flat position?
16. Using a pencil and paper, draw two complete lines of the weave patterns you are most comfortable making.
17. Why is it important to find a good welding position?
18. Which electrodes would be grouped in the following F-numbers: F2, F3, F4?
19. Give one advantage of using electrodes with cellulose-based fluxes.
20. What are stringer beads?
21. Describe an ideal tack weld.

22. What effect does the root opening or root gap have on a butt joint?
23. What can happen if the fillet weld on a lap joint does not have a smooth transition?
24. Which plate heats up faster on a tee joint? Why?
25. Can a tee weld be strong if the welds on both sides do not have deep penetration? Why or why not?

CHAPTER 3

Advanced Shielded Metal Arc Welding

OBJECTIVES

After completing this chapter, the student should be able to

- prepare metal before welding
- make the root pass, filler weld, and cover pass in all positions and techniques
- explain the purpose of a hot pass
- make a visual inspection, and describe the appearance of an acceptable weld
- make a root pass on plate in all positions
- make a root pass on plate with an open root in all positions
- make an open root weld on plate using the step technique in all positions
- make a multiple pass filler weld on a V-joint in all positions using E7018 electrodes
- make groove welds on a limited thickness plate with a backing strip using E6010 root and hot passes and E7018 fill passes in all positions
- make an SMAW workmanship sample and welder qualification test plate for limited thickness horizontal 2G and 3G positions with E7018 electrodes
- make a cover bead in all positions
- make limited thickness welder performance qualification test plate without backing in all positions
- make a single V-groove open root butt joint with an increasing root opening
- make a single V-groove open root butt joint with a decreasing root opening
- make SMAW welds of plate to plate

KEY TERMS

back gouging
burn-through
cover pass
filler pass
guided bend specimen
hot pass
interpass temperature
key hole
molten weld pool
multiple pass weld
postheating
preheating
root pass
wagon tracks
weld groove
weld specimen

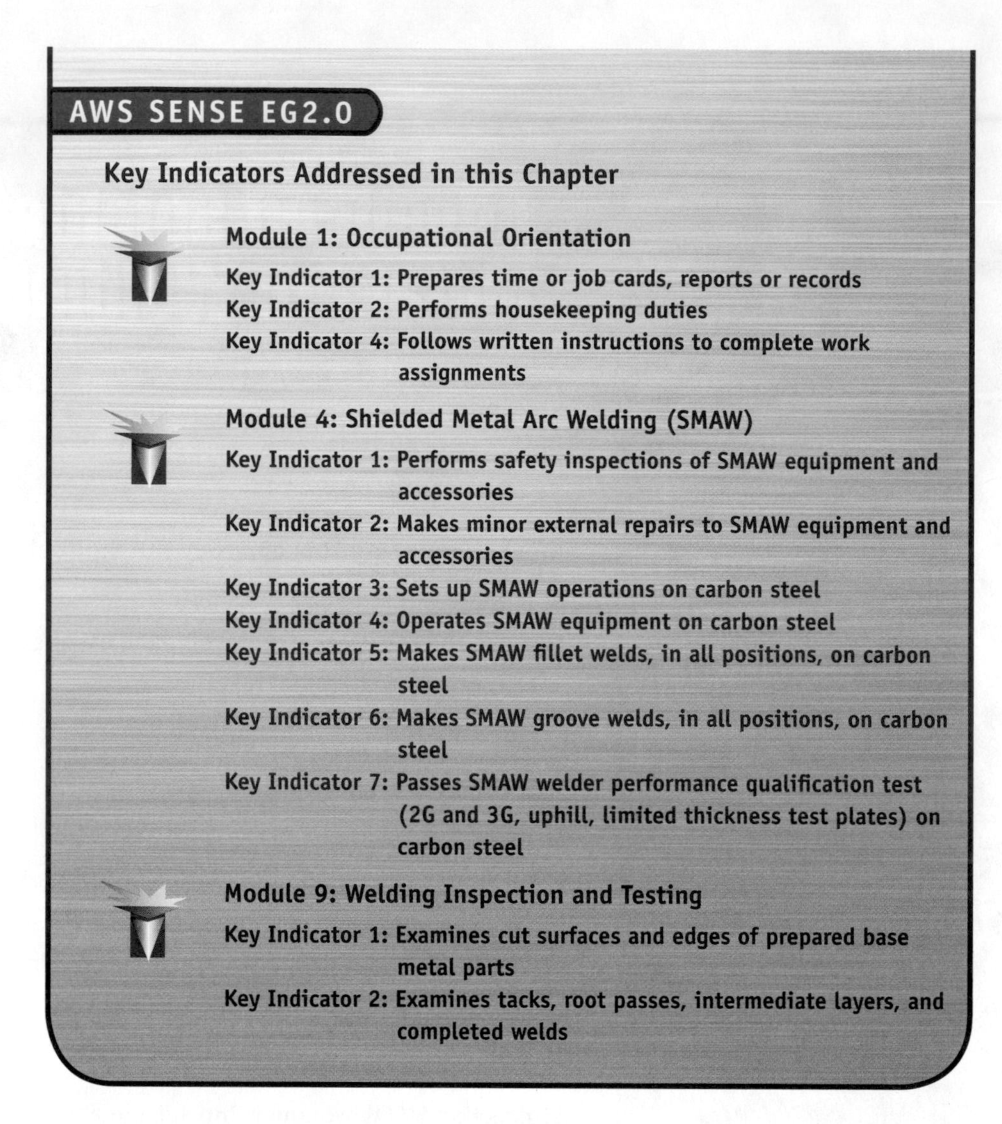

AWS SENSE EG2.0

Key Indicators Addressed in this Chapter

Module 1: Occupational Orientation

Key Indicator 1: Prepares time or job cards, reports or records
Key Indicator 2: Performs housekeeping duties
Key Indicator 4: Follows written instructions to complete work assignments

Module 4: Shielded Metal Arc Welding (SMAW)

Key Indicator 1: Performs safety inspections of SMAW equipment and accessories
Key Indicator 2: Makes minor external repairs to SMAW equipment and accessories
Key Indicator 3: Sets up SMAW operations on carbon steel
Key Indicator 4: Operates SMAW equipment on carbon steel
Key Indicator 5: Makes SMAW fillet welds, in all positions, on carbon steel
Key Indicator 6: Makes SMAW groove welds, in all positions, on carbon steel
Key Indicator 7: Passes SMAW welder performance qualification test (2G and 3G, uphill, limited thickness test plates) on carbon steel

Module 9: Welding Inspection and Testing

Key Indicator 1: Examines cut surfaces and edges of prepared base metal parts
Key Indicator 2: Examines tacks, root passes, intermediate layers, and completed welds

INTRODUCTION

The SMAW process can be used to consistently produce high-quality welds. Sometimes it is necessary to make welds in less than ideal conditions. Knowing how to produce a weld of high strength in an out-of-position, difficult situation or on an unusual metal takes both practice and knowledge. A welder is frequently required to make these types of welds to a code or standard. This chapter covers the high-quality welding of plate. The practices are designed to give you the experience of taking code-type tests in a variety of materials and positions as well as to develop good workmanship.

Any time a code-quality weld requiring 100% joint penetration is to be made on metal thicker than 1/4 in. (6 mm), the metal edges must be prepared before welding. A joint is prepared for welding by cutting a groove in the metal along the edge. The preparation is done to allow deeper penetration into the joint of the weld for improved strength. Prepared joints often require more than one weld pass to complete them. By preparing the joint, metal

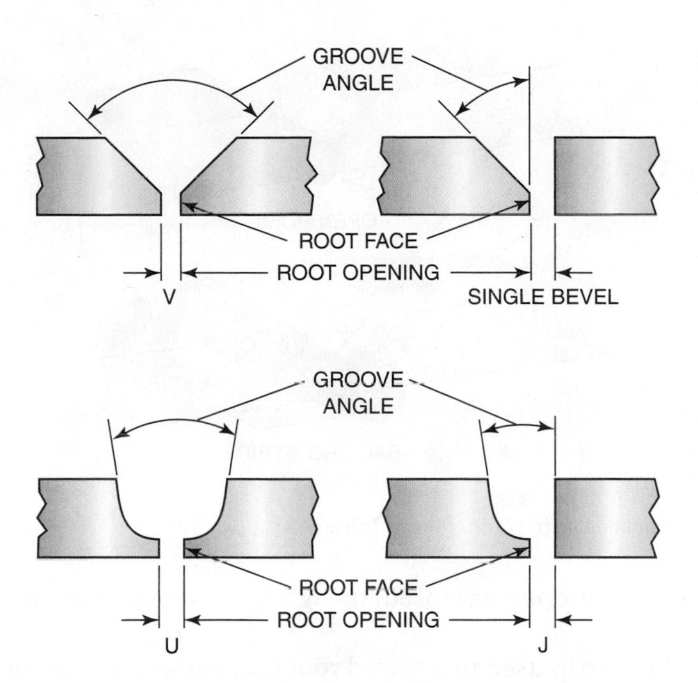

Figure 3.1
Standard grooves used to ensure satisfactory joint penetration

several feet thick can be welded with great success, **Figure 3.1.** The same welding techniques are used when making prepared welds of any thickness.

The root pass is used to fuse the parts together and seal off possible atmospheric contamination from the filler weld. Once the root pass is completed and cleaned, a hot pass may be used to improve the weld contour and burn out small spots of trapped slag. For high-quality welds, a grinder should be used on the root pass to clean it.

Filler welds and cover welds are often made with low hydrogen electrodes such as E7018. These passes are used to fill and cap the weld groove.

Welders are often qualified by passing a required qualification test using a groove weld. The type of joint, thickness of metal, type and size of electrode, and position are all specified by agencies issuing codes and standards. Except for the American Welding Society's Certified Welder program, taking a test according to one company or agency's specifications may not qualify a welder for another company or agency's testing procedures. But being able to pass one type of test will usually help the welder to pass other tests for the same type of joint, thickness of metal, type and size of electrode, and position. Information about the AWS Certified Welder program is available from the AWS's main office in Miami, Florida.

ROOT PASS

The **root pass** is the first weld bead of a **multiple pass weld.** The root pass fuses the two parts together and establishes the depth of weld metal penetration. A good root pass is needed in order to obtain a sound weld. The

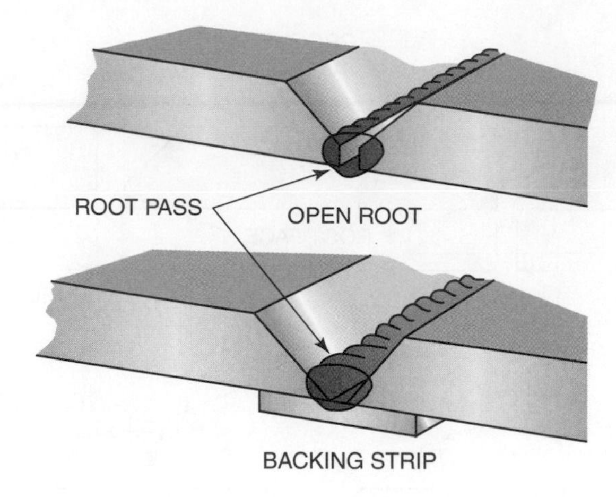

Figure 3.2
Root pass maximum deposit 1/4 in. (6 mm) thick

root may be either open or closed, using a backing strip or backing ring, **Figure 3.2.**

The backing strip used in a closed root may remain as part of the weld, or it may be removed. Because leaving the backing strip on a weld may cause it to fail due to concentrations of stresses along the backing strip, removable backup tapes have been developed. Backup tapes are made of high-temperature ceramics, **Figure 3.3,** that can be used to increase penetration and prevent **burn-through.** The tape can be peeled off after the weld is completed. Most welds do not use backing strips.

On plates that have the joints prepared on both sides, the root face may be ground or gouged clean before another pass is applied to both sides, **Figure 3.4.** This practice has been applied to some large diameter pipes. Welds, however, that can be reached from only one side must be produced adequately, without the benefit of being able to clean and repair the back side.

The open root weld is widely used in plate and pipe designs. The face side of an open root weld is not so important as the root surface on the back or inside, **Figure 3.5.** The face of a root weld may have some areas

(A) FIBERGLASS

(B) WELD ROOT PASS MADE USING CERAMIC BACKING TAPE

Figure 3.3
Welding backing tapes are available in different materials and shapes
Courtesy of ESAB Welding & Cutting Products

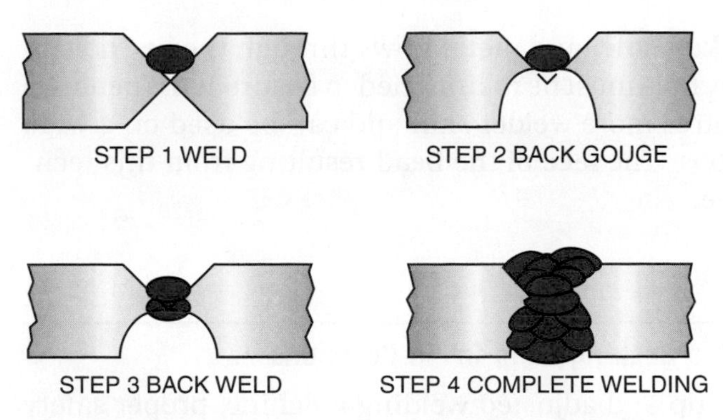

Figure 3.4
Using back gouging to ensure a sound weld root

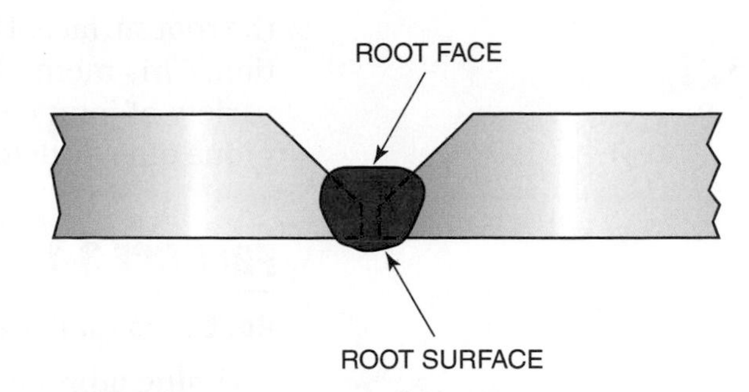

Figure 3.5
Ideal bead shape for the root pass

of poor uniformity in width, reinforcement, and buildup, or it may have other defects, such as undercut or overlap. As long as the root surface is correct, the front side can be ground, gouged, or burned out to produce a sound weld, **Figure 3.6.** For this reason, during the root pass practices, the weld will be evaluated from the root side only, as long as there are not too many defects on the face. To practice the open root welds, the welder will be using mild steel plate that is 1/8 in. (3 mm) thick. The root face for most grooved joints will be about the same size. This thin plate will help the welder build skill without taking too much time beveling the plate just to practice the root pass. Two different methods are used to make a root pass. One method is used only on joints with little or no root gap. This method requires a high amperage and short arc length. The arc length is so short that the electrode flux may drag along on the edges of the joint. The setup for this method must be correct in order for it to work.

The other method can be used on joints with wide, narrow, or varying root gaps. A stepping electrode manipulation and key hole control the penetration. The electrode is moved in and out of the **molten weld pool** as the weld progresses along the joint. The edge of the metal is burned back slightly by the electrode just ahead of the molten weld pool, **Figure 3.7.**

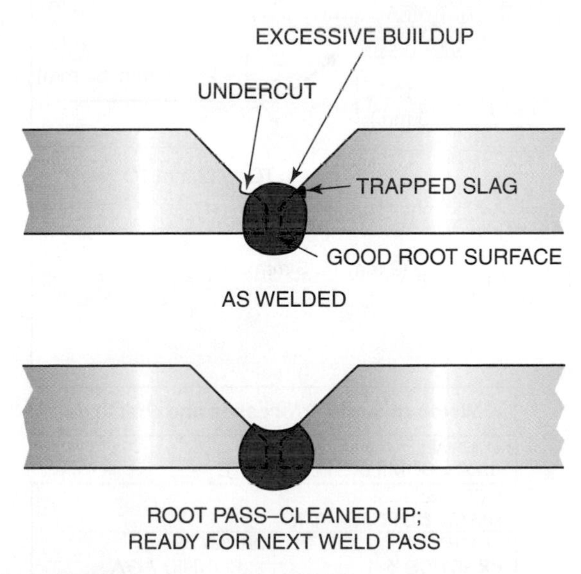

Figure 3.6
Grinding back the root pass to ensure a sound second pass

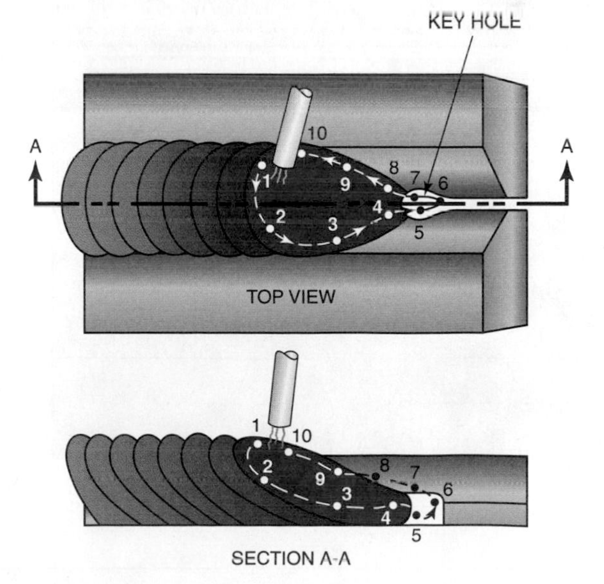

Figure 3.7
Electrode movement to open and use a keyhole

This is referred to as a **key hole,** and metal flows through the key hole to the root surface. The key hole must be maintained to ensure 100% penetration. This method requires more welder skill and can be used on a wide variety of joint conditions. The face of the bead resulting from this technique often is defect free.

PRACTICE 3-1

Root Pass on Plate with a Backing Strip in All Positions

Module 1
Key Indicator 1, 2, 4

Module 4
Key Indicator 6

Module 9
Key Indicator 1, 2

Using a properly set up and adjusted welding machine, proper safety protection, E6010 or E6011 arc welding electrodes having a 1/8-in. (3-mm) diameter, and one or more pieces of mild steel plate, 1/8 in. (3 mm) thick × 6 in. (152 mm) long, and one strip of mild steel, 1/8 in. (3 mm) thick × 1 in. (25 mm) wide × 6 in. (152 mm) long, you will make a root weld in all positions, **Figure 3.8.** Tack weld the plates together with a 1/16-in. (2-mm) to 1/8-in. (3-mm) root opening. Be sure there are no gaps between the backing strip and plates when the pieces are tacked together, **Figure 3.9.** If there is a small gap between the backing strip and the plates, it can be removed by placing the assembled test plates on an anvil and striking the tack weld with a hammer. This will close up the gap by compressing the tack welds, **Figure 3.10.**

Use a straight step or "T" pattern for this root weld. Push the electrode into the root opening so that there is good fusion with the backing strip and bottom edge of the plates. Failure to push the penetration deep into the joint will result in a cold lap at the root, **Figure 3.11.**

Watch the molten weld pool and keep its size as uniform as possible. As the molten weld pool increases in size, move the electrode out of the weld pool. When the weld pool begins to cool, bring the electrode back into the

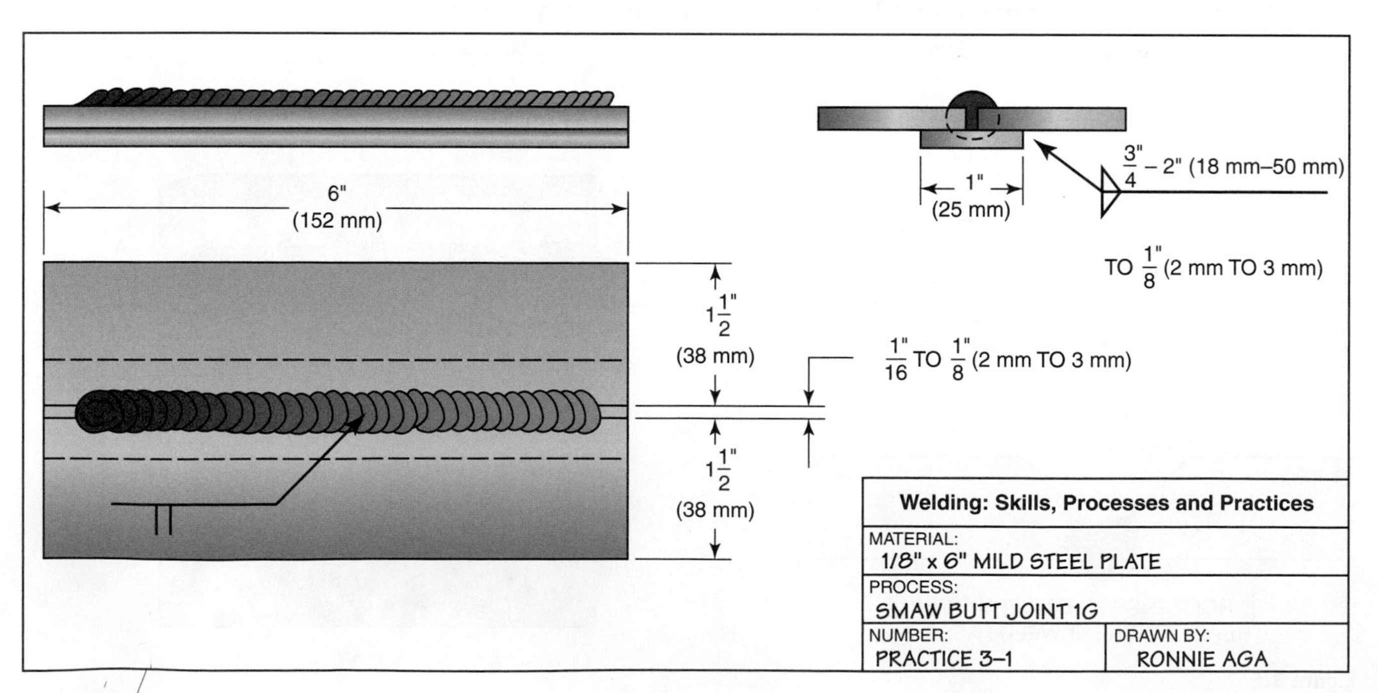

Figure 3.8
Square butt joint with a backing strip

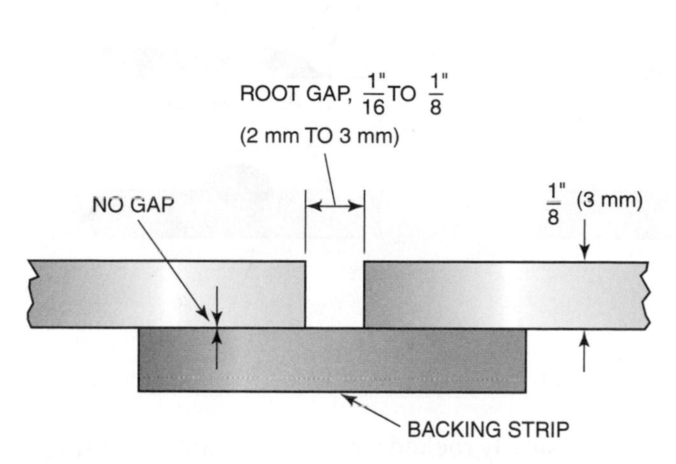

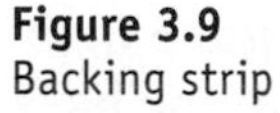

Figure 3.9
Backing strip

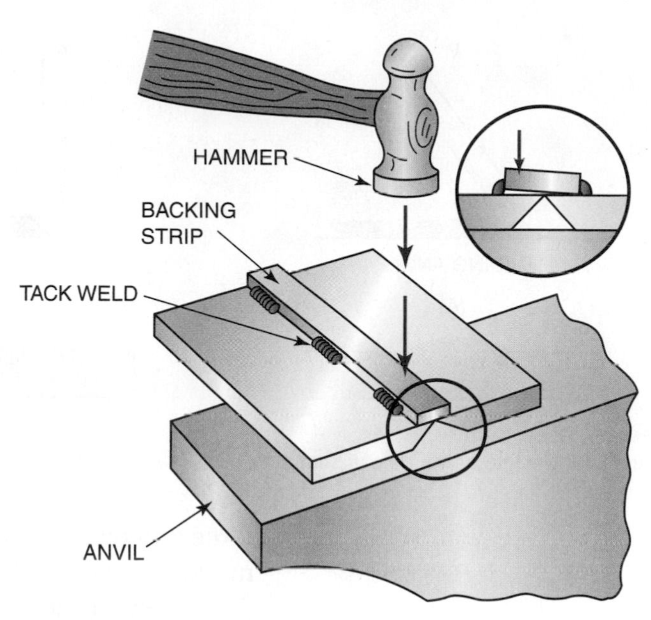

Figure 3.10
Using a hammer to align the backing strip and weld plates

molten weld pool. Use these weld pool indications to determine how far to move the electrode and when to return to the molten weld pool. After completing the weld, cut the plate and inspect the cross section of the weld for good fusion at the edges. Repeat the welds as necessary until you can consistently make welds free of defects. Turn off the welding machine and clean up your work area when you are finished welding.

Complete a copy of the "Student Welding Report" listed in Appendix I or provided by your instructor.

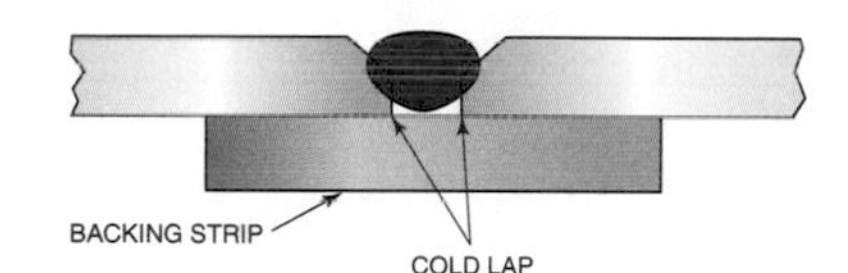

Figure 3.11
Incomplete root fusion

PRACTICE 3-2

Root Pass on Plate with an Open Root in All Positions

Using a properly set up and adjusted arc welding machine, proper safety protection, E6010 or E6011 arc welding electrodes with a 1/8-in. (3-mm) diameter, and two or more pieces of mild steel plate, 6 in. (152 mm) long × 1/8 in (3 mm) or 3/16 in. (4.7 mm.) thick, you will make a welded butt joint in all positions with 100% root penetration.

Module 1
Key Indicator 1, 2, 4

Module 4
Key Indicator 6

Module 9
Key Indicator 1, 2

- Tack weld the plates together with a root opening of 0 in. (0 mm) to 1/16 in. (2 mm).
- Using a short arc length and high amperage setting, make a weld along the joint.

You can change the electrode angle to control penetration and burn-through. As the trailing angle is decreased, making the electrode flatter to the plate, penetration, depth, and burn-through decrease, **Figure 3.12,** because both the arc force and heat are directed away from the bottom of the joint back toward the weld. Surface tension holds the metal in place, and the mass of the bead quickly cools the molten weld pool holding it in place. Increasing the electrode angle toward the perpendicular will increase penetration depth and possibly cause more burn-through. The arc

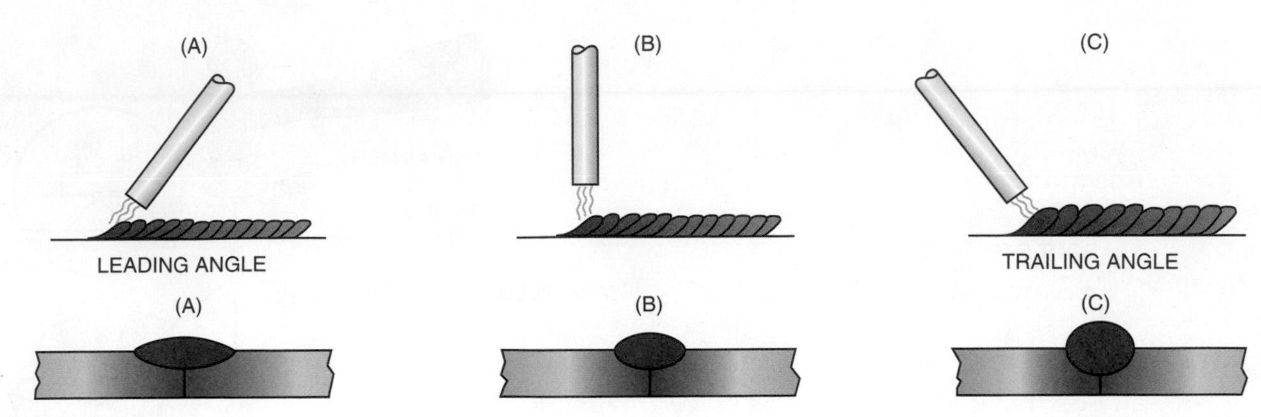

Figure 3.12
Effect of rod angle on weld bead shape

force and heat focused on the gap between the plates will push the molten metal through the joint.

The electrode holder can be slowly rocked from side to side while keeping the end of the electrode in the same spot on the joint, **Figure 3.13.** This will allow the arc force to better tie in the sides of the root to the base metal.

- When a burn-through occurs, rapidly move the electrode back to a point on the weld surface just before the burn-through.
- Lower the electrode angle and continue welding. If the burn-through does not close, stop the weld, chip, and wire brush the weld.
- Check the size of the burn-through. If it is larger than the diameter of the electrode, the root pass must be continued with the step method described in Practice 3-3. If the burn-through is not too large, lower the amperage slightly and continue welding.
- Watch the color of the slag behind the weld. If the weld metal is not fusing to one side, the slag will be brighter in color on one side. The

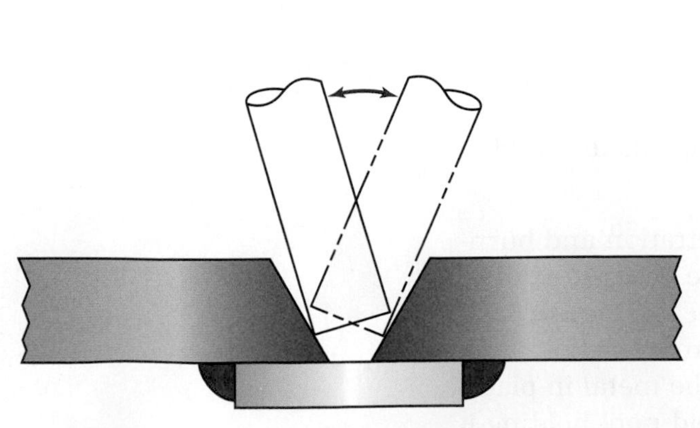

Figure 3.13
Rocking the top of the electrode while keeping the end in the same place helps control the bead shape

Figure 3.14
The weld toes appear uniform
Courtesy of Larry Jeffus

brighter color is caused by the slower cooling of the slag because there is less fused metal to conduct the heat away quickly.

- After the weld is completed, cooled, and chipped, check the back side of the plate for good root penetration. The root should have a small bead that will look as though it was welded from the back side, **Figure 3.14.** The penetration must be completely free of any drips of metal from the root face, called icicles.
- Repeat the welds as necessary until you can consistently make welds free of defects. Turn off the welding machine and clean up your work area when you are finished welding.

Complete a copy of the "Student Welding Report" listed in Appendix I or provided by your instructor.

PRACTICE 3-3

Open Root Weld on Plate Using the Step Technique in All Positions

Using the same setup, materials, and electrodes as described in Practice 3-2, you will make a welded butt joint in all positions with 100% root penetration.

Module 1
Key Indicator 1, 2, 4

Module 4
Key Indicator 6

Module 9
Key Indicator 1, 2

Tack weld the plates together with a root opening from 0 in. (0 mm) to 1/8 in. (3 mm). Using a medium amperage setting and a short stepping electrode motion, make a weld along the joint.

The electrode should be pushed deeply into the root to establish a key hole that will be used to ensure 100% root penetration. Once the key hole is established, the electrode is moved out and back in the molten weld pool at a steady, rhythmic rate. Watch the molten weld pool and key hole size to determine the rhythm and distance of electrode movement.

If the molten weld pool size decreases, the key hole will become smaller and may close completely. To increase the molten weld pool size and maintain the key hole, slow the rate of electrode movement and shorten the distance the electrode is moved away from the molten weld pool. This will increase the molten weld pool size and penetration because of increased localized heating.

If the molten weld pool becomes too large, metal may drip through the key hole, forming an icicle on the back side of the plate. Extremely large molten weld pool sizes can cause a large hole to be formed or cause burn-through. Repairing large holes can require much time and skill. To keep the molten weld pool from becoming too large, increase the travel speed, decrease the angle, shorten the arc length, or lower the amperage, **Table 3.1.**

The distance the electrode is moved from the molten weld pool and the length of time in the molten weld pool are found by watching the molten

Table 3.1 Changes Affecting Molten Weld Bead Size

	Amperage	Travel Speed	Electrode Size	Electrode Angle
To decrease puddle size	Decrease	Increase	Decrease	Leading
To increase puddle size	Increase	Decrease	Increase	Trailing

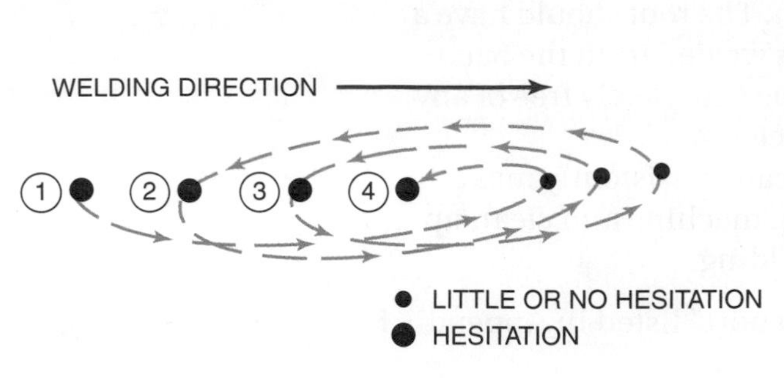

Figure 3.15
Weave pattern used to control the molten weld bead size

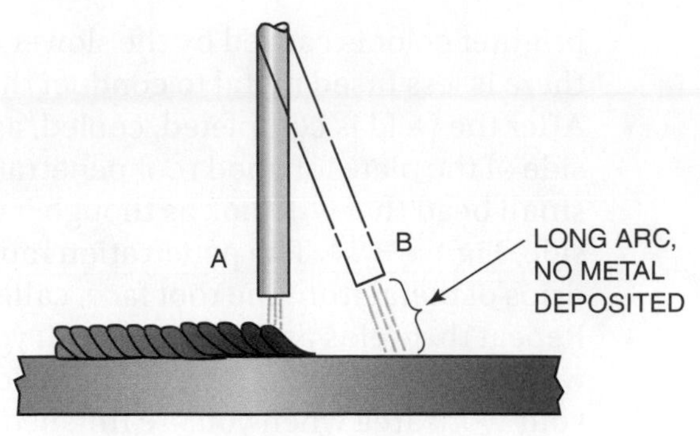

Figure 3.16
A long arc prevents metal or slag from being deposited ahead of the weld bead. A relatively quick movement from position "A" to position "B" and back to "A" is required.

weld pool. The molten weld pool size increases as you hold the arc in the molten weld pool until it reaches the desired size, about twice the electrode diameter, **Figure 3.15.** Move the electrode ahead of the molten weld pool, keeping the arc in the joint but being careful not to deposit any slag or metal ahead of the weld. To prevent metal and/or slag from transferring, raise the electrode to increase the arc length, **Figure 3.16.** Keep moving the electrode slowly forward as you watch the molten weld pool. The molten weld pool will suddenly start to solidify. At that time, move the electrode quickly back to the molten weld pool before it totally solidifies. Moving the electrode in a slight arc will raise the electrode ahead of the molten weld pool and automatically lower the electrode when it returns to the molten weld pool. Metal or slag deposited ahead of the molten weld pool may close the key hole, reduce penetration, and cause slag inclusions. Raising the end of the electrode too high or moving it too far ahead of the molten weld pool can cause all of the shielding gas to be blown away from the molten weld pool. If this happens, oxides can cause porosity. Keeping the electrode movement in balance takes concentration and practice.

Changing from one welding position to another requires an adjustment in timing, amperage, and electrode angle. The flat, horizontal, and overhead positions use about the same rhythm, but the vertical position may require a shorter time cycle for electrode movement. The amperage for the vertical position can be lower than that for the flat or horizontal, but the overhead position uses nearly the same amperage as flat and horizontal. The electrode angle for the flat and horizontal positions is about the same. For the vertical position, the electrode uses a sharper leading angle than does overhead, which is nearly perpendicular and may even be somewhat trailing.

Cool, chip, wire brush, and inspect both sides of the weld. The root surface should be slightly built up and look as though it was welded from that side (refer to Figure 3.14). Repeat the welds as necessary until you can consistently make welds free of defects. Turn off the welding machine and clean up your work area when you are finished welding.

Complete a copy of the "Student Welding Report" listed in Appendix I or provided by your instructor.

HOT PASS

The surface of a root pass may be irregular, have undercut, overlap, slag inclusions, or other defects, depending upon the type of weld, the code or standards, and the condition of the root pass. The surface of a root pass can be cleaned by grinding or by using a **hot pass.**

On critical, high-strength code welds, it is usually required that the root pass as well as each filler pass be ground (refer to Figure 3.6). This grinding eliminates weld discontinuities caused by slag entrapments. It also can be used to remove most of the E60 series weld metal so that the stronger weld metal can make up most of the weld. When high-strength, low-alloy welding electrodes are used, this grinding is important to remove most of the low-strength weld deposit. This will leave the weld made up of nearly 100% of the high-strength weld metal, **Figure 3.17.**

The fastest way to clean out trapped slag and make the root pass more uniform is to use a hot pass. The hot pass uses a higher than normal amperage setting and a fast travel rate to reshape the bead and burn out trapped slag. After chipping and wire brushing the root pass to remove all the slag possible, a welder is ready to make the hot pass. The ideal way to apply a hot pass is to rapidly melt a large surface area, **Figure 3.18,** so that the trapped slag can float to the surface. The slag, mostly silicon dioxide (SiO_2), may not melt itself, so the surrounding steel must be melted to enable it to float free. The silicon dioxide may not melt because it melts at about 3100°F (1705°C), which is more than 500°F (270°C) hotter than the temperature at which the surrounding steel melts, around 2600°F (1440°C).

A very small amount of metal should be deposited during the hot pass so that the resulting weld is concave. A concave weld, compared to a

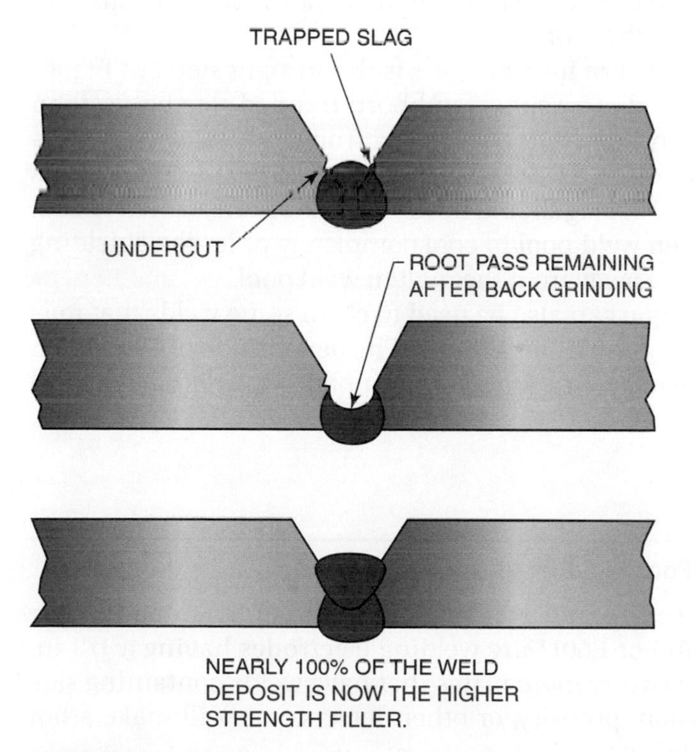

Figure 3.17
Back grinding to remove both discontinuities; filler metal used for the root pass

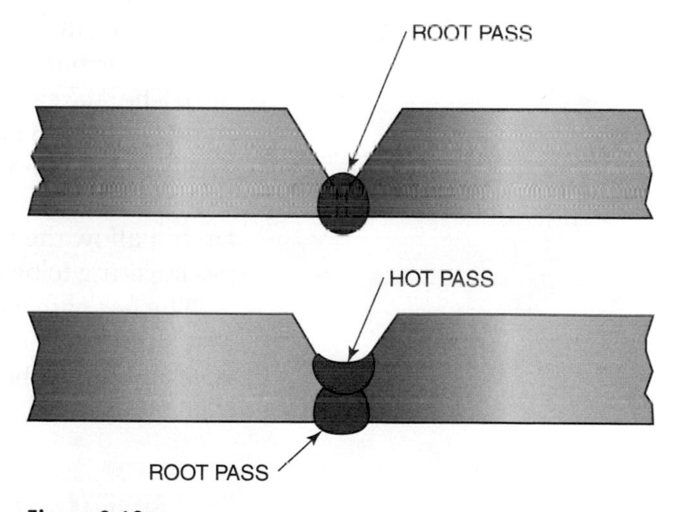

Figure 3.18
Using the hot pass to clean up the face of the root pass

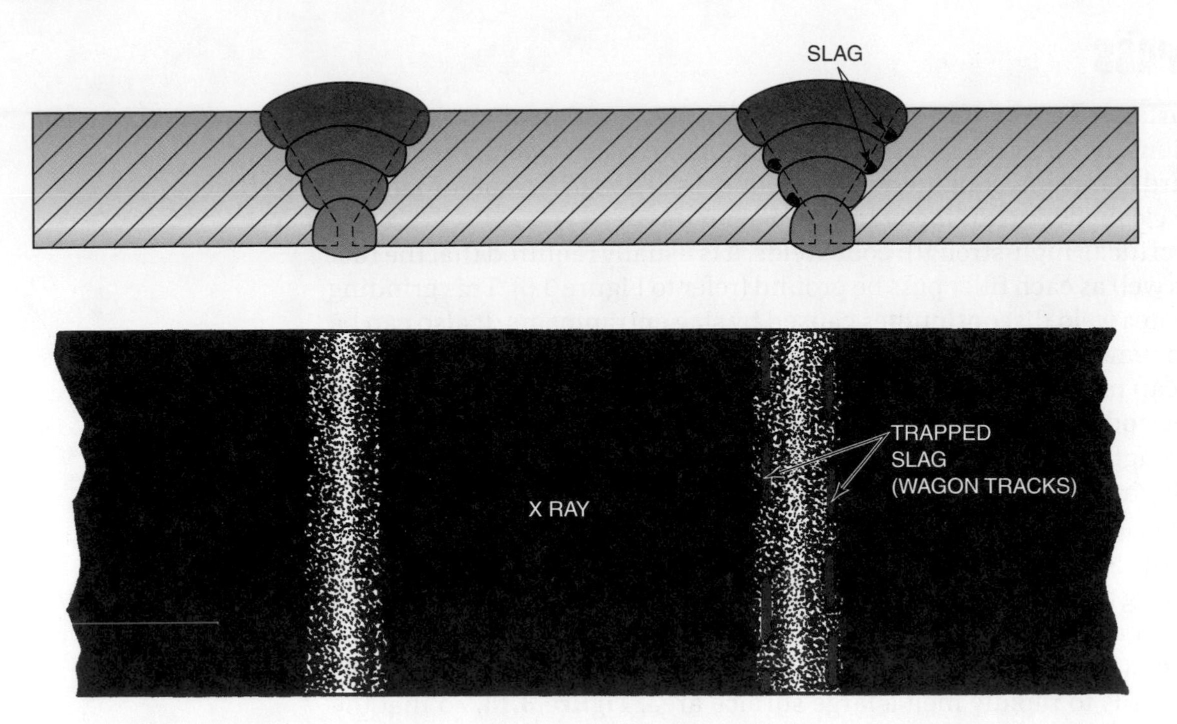

Figure 3.19
Slag trapped between passes will show on an X ray

convex weld, is more easily cleaned by chipping, wire brushing, or grinding. Failure to clean the convex root weld will result in a discontinuity showing up on an X ray. Such discontinuities are called **wagon tracks, Figure 3.19.**

The hot pass can also be used to repair or fill small spots of incomplete fusion or pinholes left in the root pass.

The normal weave pattern for a hot pass is the straight step or "T" pattern. The "T" can be used to wash out stubborn trapped slag better than the straight step pattern. The frequency of electrode movement is dependent upon the time required for the molten weld pool to start cooling. As with the root pass, metal or slag should not be deposited ahead of the bead. Do not allow the molten weld pool to cool completely or let the shielding gas covering to be blown away from the molten weld pool.

The hot pass technique can also be used to clean some welds that may first require grinding or gouging for a repair. The penetration of the molten weld pool must be deep enough to free all trapped slag and burn out all porosity.

EXPERIMENT 3-1

Hot Pass to Repair a Poor Weld Bead

Module 1
Key Indicator 1, 2, 4

Module 4
Key Indicator 4

Module 9
Key Indicator 1, 2

Using a properly set up and adjusted arc welding machine, proper safety protection, E6010 or E6011 arc welding electrodes having a 1/8-in. (3-mm) diameter, and two or more plates that have welds containing slag inclusions, lack of fusion, porosity, or other defects, you will make a hot pass to burn out the defects.

Chip and wire brush the weld bead. If necessary, use a punch to break apart large trapped slag deposits. The poorer the condition of the weld,

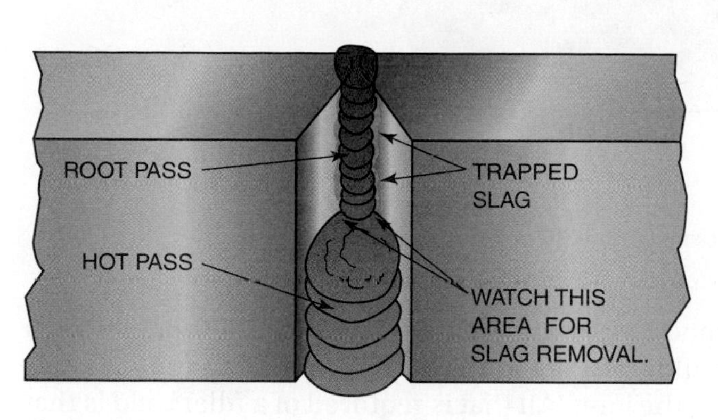

Figure 3.20
Burning out trapped slag by using a hot pass

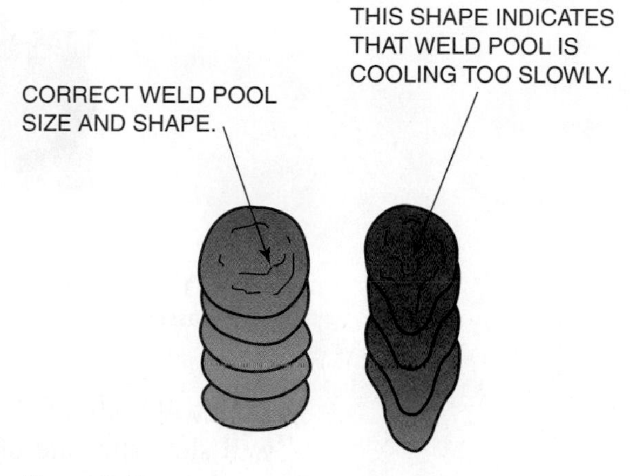

Figure 3.21
The shape of the weld pool can indicate the temperature of the surrounding base metal

the more vertical the joint should be for the hot pass. Large slag deposits tend to float around the molten weld pool and stay trapped in deep pockets in the flat position. With the weld in the vertical position, the slag can run out of the joint and down the face of the weld. Set the amperage as high as possible without overheating and burning up the electrode. Start at the bottom and weld upward using a combination of straight-step and "T" patterns to keep the weld deposit uniform. Watch the back edge of the molten weld pool for size and the weld crater for the complete burning out of impurities, **Figure 3.20.**

The plate may start to become overheated because of the high heat input. If you notice that the weld bead is starting to cool too slowly and is growing in length, you should stop welding, **Figure 3.21.** Allow the plate to cool before continuing the weld.

After the weld is completed, cool, chip, and inspect it for uniformity. The plate can be cut at places where you know large discontinuities existed before to see if they were repaired or only covered up. If you wish, this experiment can be repeated on other defects and joints.

Welds that have large defects in addition to excessive buildup may require some grinding to remove the buildup. Turn off the welding machine and clean up your work area when you are finished welding.

Complete a copy of the "Student Welding Report" listed in Appendix I or provided by your instructor.

CAUTION

This hot pass technique is designed to be used on noncritical, noncode welds only. It should not be used to cover bad welds or as a means of repairing the work of a welder who is less skilled.

FILLER PASS

After the root pass is completed and it has been cleaned, the groove is filled with weld metal. These weld beads make up the **filler passes.** More than one pass is often required.

Filler passes are made with stringer beads or weave beads. For multiple pass welds, the weld beads must overlap along the edges. They should overlap enough so that the finished bead is smooth, **Figure 3.22.** Stringer beads usually overlap about 50%, and weave beads overlap approximately 25%.

Each weld bead must be cleaned before the next bead is started. Slag left on the plate between welds cannot be completely burned out because

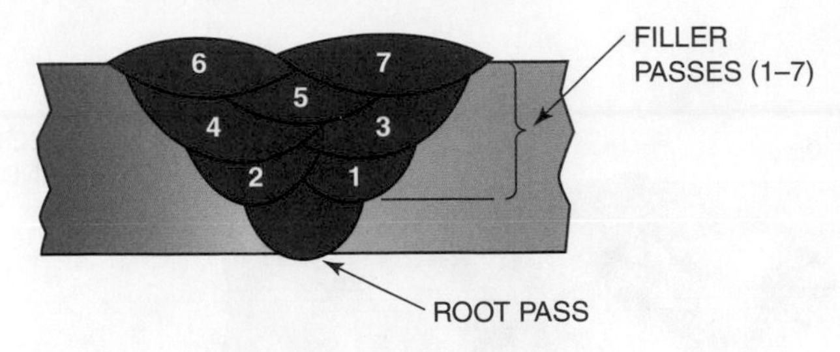

Figure 3.22
Filler passes—maximum thickness 1/8 in. (3 mm) each pass

filler welds should be made with a low amperage setting. Deep penetration will slow the rate of buildup in the joint. Deeply remelting the previous weld metal may weaken the joint. All that is required of a filler weld is that it be completely fused to the base metal.

Chipping, wire brushing, and grinding are the best ways to remove slag between filler weld passes. After the weld is completed, it can be checked by ultrasonic or radiographic nondestructive testing. Most schools are not equipped to do this testing. Therefore, a quick check for soundness can be made by destructive testing. One method of testing the deposited weld metal is by cutting and cross-sectioning the weld with an abrasive wheel and inspecting the weld. Another fast way to inspect filler passes is to cut a groove through the weld with a gouging tip. Watch the hot metal as it is washed away. The black spots that appear in the cut are slag inclusions. If only a few small spots appear, the weld probably will pass most tests. But, if a long string or large pieces of inclusions appear, the weld will most likely fail.

PRACTICE 3-4

Multiple Pass Filler Weld on a V-joint in All Positions

Module 1
Key Indicator 1, 2, 4

Module 4
Key Indicator 5

Module 9
Key Indicator 1, 2

Using a properly set up and adjusted arc welding machine, proper safety protection, E6010 or E6011 arc welding electrodes having a 1/8-in. (3-mm) diameter, and two or more pieces of mild steel plate, 6 in. (152 mm) long × 3 in. (76 mm) wide × 3/8 in. (10 mm) thick, you will make a multiple pass filler weld on a V-joint.

Tack weld the plates together at the corner so that they form a V, **Figure 3.23.** Starting at one end, make a stringer bead along the entire length using the straight step or "T" weave pattern. Thoroughly clean off the slag from the weld before making the next bead. **Figure 3.24** shows the suggested sequence for locating the beads. Continue making welds and cleaning them until the weld is 1 in. (25 mm) or more thick. Both ends of the weld may taper down. If it is important that the ends be square, metal tabs are welded on the ends of the plate for starting and stopping, **Figure 3.25.** The tabs are removed after the weld is completed.

After the weld is completed, it can be visually inspected for uniformity. If nondestructive testing is available, it may be checked for discontinuities. The weld also may be inspected by sectioning it with an abrasive wheel or gouging out with a torch. Repeat these welds until they are mastered. Turn off the welding machine and clean up your work area when you are finished welding.

Complete a copy of the "Student Welding Report" listed in Appendix I or provided by your instructor.

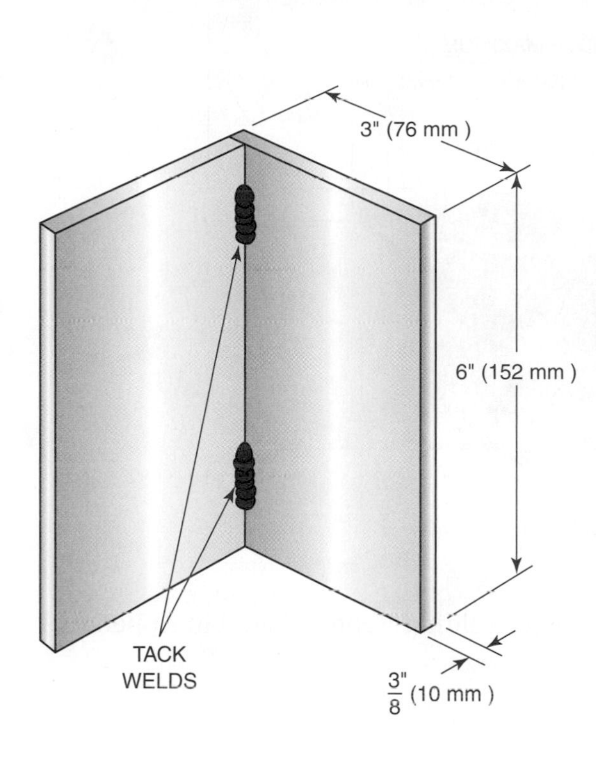

Figure 3.23
Tack weld the plates for the filler weld practice

Figure 3.24
Filler pass buildup sequence

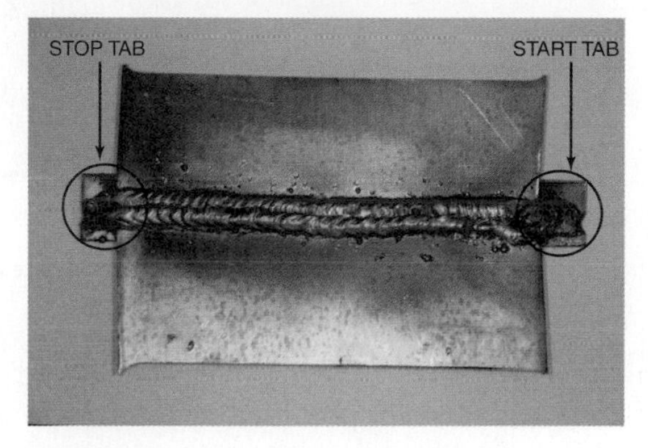

Figure 3.25
Run-off tabs help control possible under-fill or burn back at the starting and stopping points of a groove weld. The tabs can be removed.
Courtesy of Larry Jeffus

PRACTICE 3-5

Multiple Pass Filler Weld on a V-joint in All Positions Using E7018 Electrodes

Using the same setup and materials as in Practice 3-4, and E7018 arc welding electrodes in place of E6010 or E6011 electrodes, you will make a multiple pass filler weld on a V-joint.

Module 1
Key Indicator 1, 2, 4

Module 4
Key Indicator 5

Module 9
Key Indicator 1, 2

- Tack weld the two plates together at the corners so that they form a V.
- Using a slow, straightforward motion, with little or no stepping, "T" or inverted "V" motion, make a stringer bead along the root of the joint.
- Chip the slag and repeat the weld until there is a buildup of 1 in. (25 mm) or more.

If during the weld the buildup should become uneven or large slag entrapments occur, they should be ground out. In industry, groove welds in plate 1 in. (25 mm) or more are normally repaired. Making these repairs now is good experience for the welder. All welders will at some time make a weld that may need repairing.

- After the weld is completed, visually test and nondestructive test the weld for external and internal discontinuities.
- Repeat the welds as necessary until you can consistently make welds free of defects. Turn off the welding machine and clean up your work area when you are finished welding.

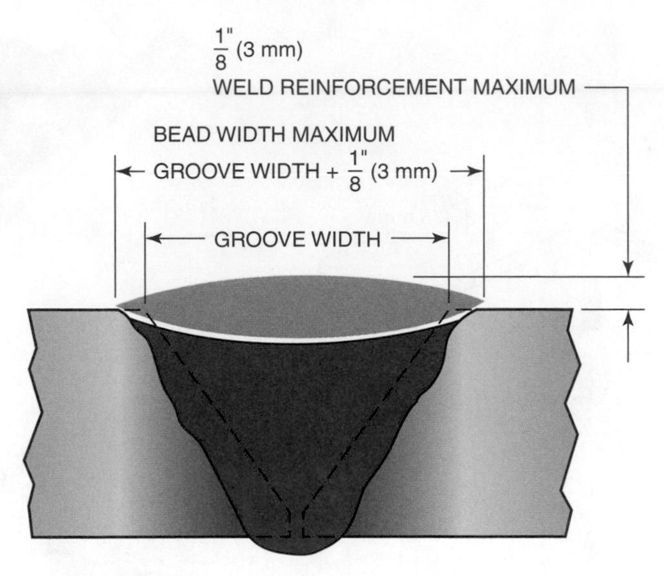

Figure 3.26
The cover pass should not be excessively large

Complete a copy of the "Student Welding Report" listed in Appendix I or provided by your instructor.

COVER PASS

The last weld bead on a multiple pass weld is known as the **cover pass.** The cover pass may use a different electrode weave, or it may be the same as the filler beads. Keeping the cover pass uniform and neat looking is important. Most welds are not tested, and often the inspection program is only visual. Thus, the appearance might be the only factor used for accepting or rejecting welds.

The cover pass should be free of any visual defects such as undercut, overlap, porosity, or slag inclusions. It should be uniform in width and reinforcement. A cover pass should not be more than 1/8 in. (3 mm) wider than the groove opening, **Figure 3.26.** Cover passes that are too wide do not add to the weld strength.

PRACTICE 3-6

Cover Bead in All Positions

Module 1
Key Indicator 1, 2, 4

Module 4
Key Indicator 4

Module 9
Key Indicator 1, 2

Using a properly set up and adjusted arc welding machine, proper safety protection, E7018 arc welding electrodes having a 1/8-in. (3-mm) diameter, and one or more pieces of mild steel plate, 6 in. (152 mm) long × 1/4 in. (6 mm) thick, you will make a cover bead in each position.

Remember, any time an E7018 low-hydrogen-type electrode is to be used, the weave pattern, if used, must not be any larger than 2-1/2 times the diameter of the electrode. This weave cannot be any larger than 5/16 in. (7 mm) wide. Start welding at one end of the plate and weld to the other end. The weld bead should be about 5/16 in. (7 mm) wide having no more than 1/8 in. (3 mm) of uniform buildup, **Figure 3.27.** The weld buildup should have a smooth transition at the toe, with the plate and the face somewhat

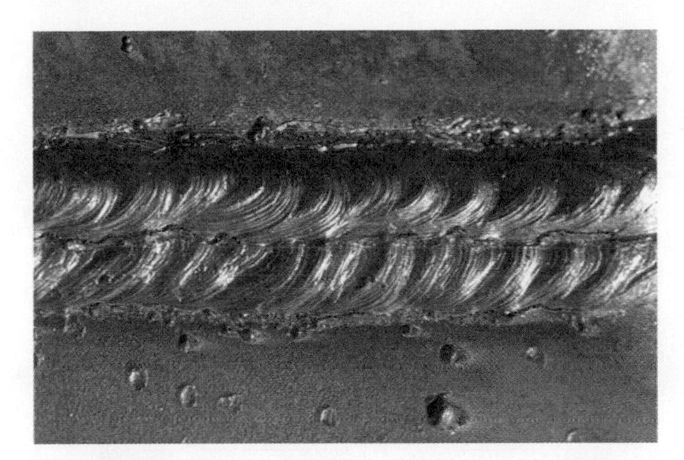

Figure 3.27
Practice cover pass
Courtesy of Larry Jeffus

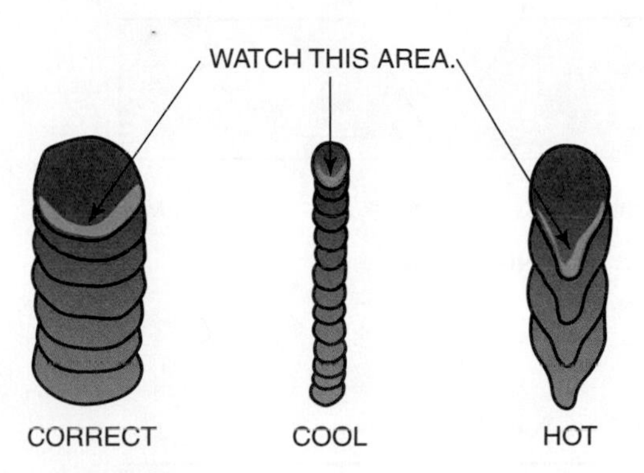

Figure 3.28
Watch the back edge of the weld pool to determine the correct current

convex. Undercut at the toe and a concave or excessively built-up face are the most common problems. Watch the sides of the bead for undercut. When undercut occurs, keep the electrode just ahead of the spot until it is filled in. There should be a smooth transition between the weld and the plate (refer to Figure 3.26). The shape of the bead face can be controlled by watching the trailing edge of the molten weld pool. That trailing edge is the same as the finished bead, **Figure 3.28.**

Deep penetration is not required with this weld and may even result in some weakening. After the weld is completely cooled, chip and inspect it for uniformity and defects. Repeat the welds as necessary until you can consistently make welds free of defects. Turn off the welding machine and clean up your work area when you are finished welding.

Complete a copy of the "Student Welding Report" listed in Appendix I or provided by your instructor.

PLATE PREPARATION

When welding on thick plate, it is impossible or impractical for the welder to try to get 100% penetration without preparing the plate for welding. The preparation of the plate is usually in the form of a **weld groove.** The groove can be cut into one side or both sides of the plate, and it may be cut into either just one plate or both plates of the joint, **Figure 3.29.** The type, depth, angle, and location of the groove are usually determined by a code standard that has been qualified for the specific job.

For SMA welds on plate 1/4 in. (6 mm) or thicker that need to have a weld with 100% joint penetration, the plate must be grooved. The groove may be ground, flame-cut, gouged, or machined on the edge of the plate before or after the assembly. Bevels and V-grooves are best if they are cut before the parts are assembled. J-grooves and U-grooves can be cut either before or after assembly, **Figure 3.30.** The lap joint is seldom prepared with a groove because little or no strength can be gained by grooving this joint. The only advantage to grooving the lap joint design is to give additional clearance.

Plates that are thicker than 3/8 in. (10 mm) can be grooved on both sides but may be prepared on only one side. The choice to groove one

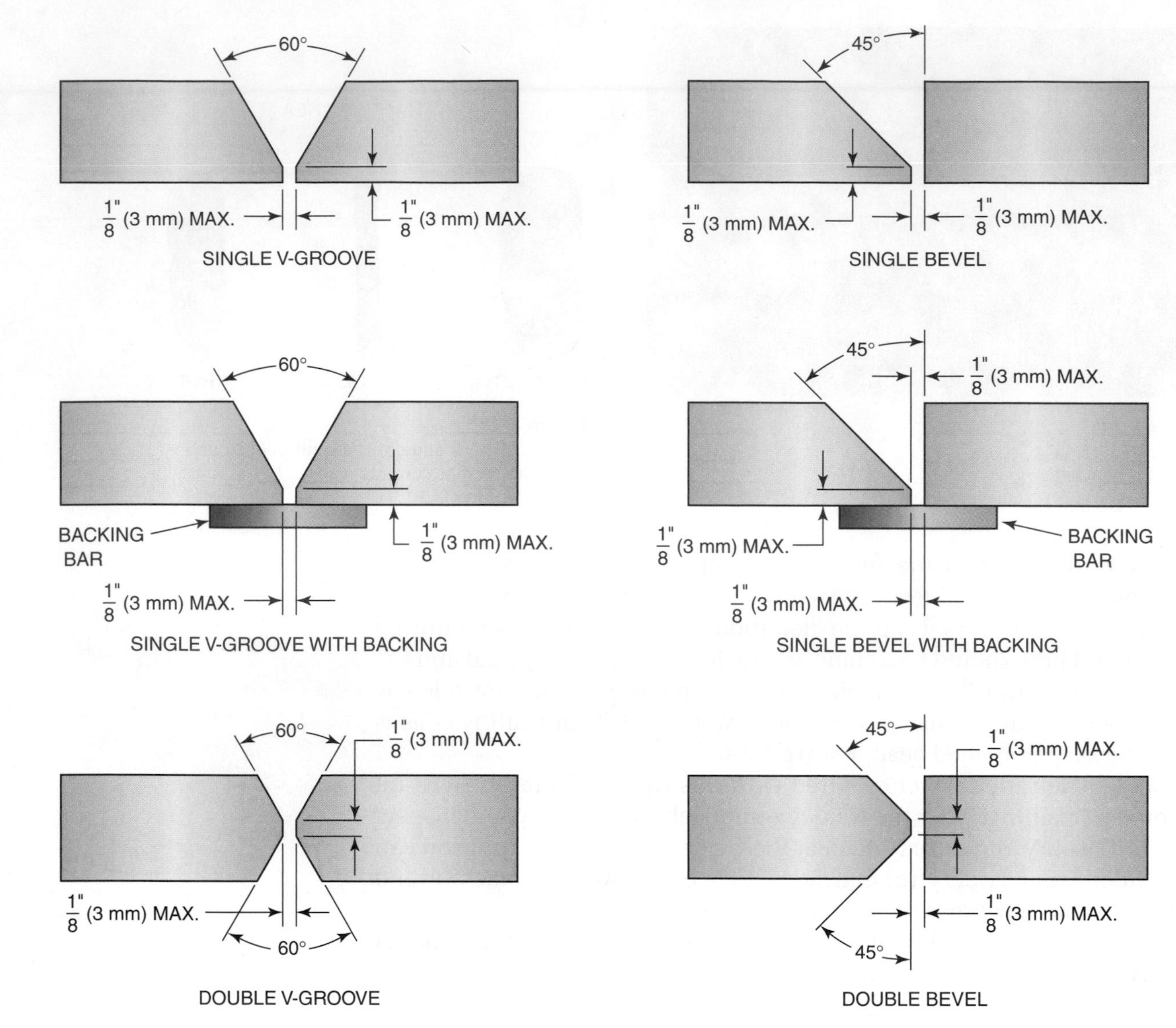

Figure 3.29
Typical butt joint preparations

or both sides is most times determined by joint design, position, and application. A tee joint in thick plate is easier to weld and will have less distortion if it is grooved on both sides. Plate in the flat position is usually grooved on only one side unless it can be repositioned. Welds that must have little distortion or that are going to be loaded equally from both sides are usually grooved on both sides. Sometimes plates are either grooved and welded or just welded on one side, and then back gouged and welded, **Figure 3.31. Back gouging** is a process of cutting a groove in the back side of a joint that has been welded. Back gouging can ensure 100% fusion at the root and remove discontinuities of the root pass. This process can also remove the root pass metal if the properties of the metal are not desirable to the finished weld, **Figure 3.32.** After back gouging, the groove is then welded. See Chapters 6, 7, and 8 in *Welding: Skills, Processes and Practices for Entry-Level Welders: Book One* for more information on the various methods of gouging.

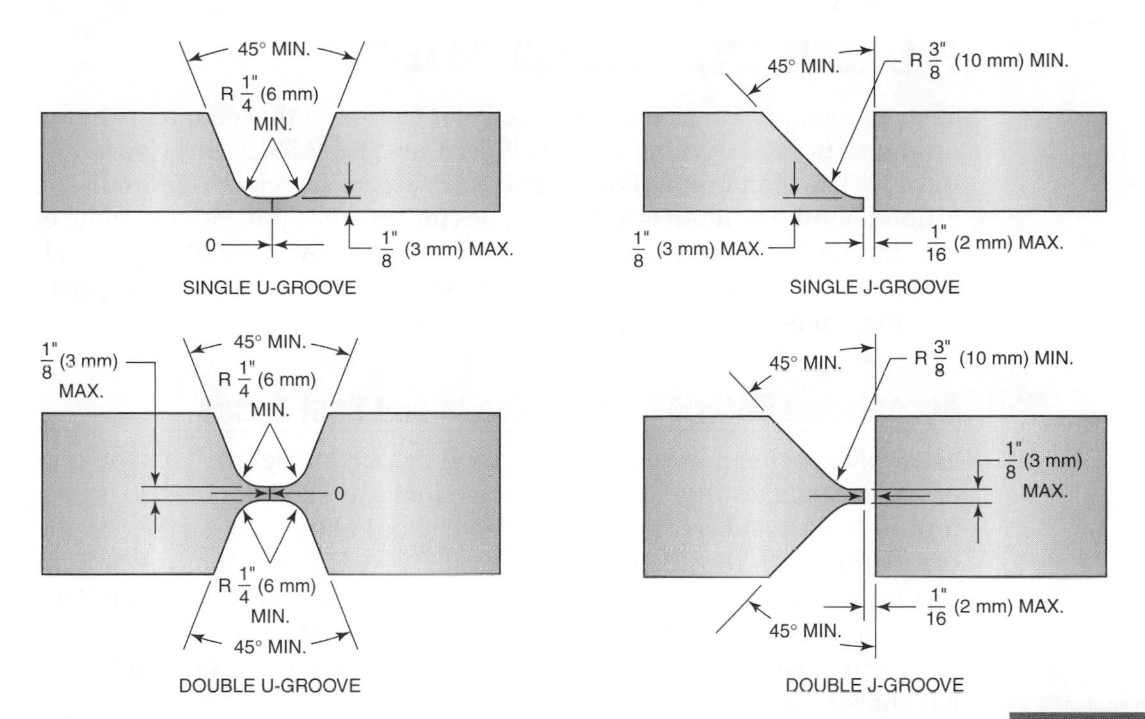

Figure 3.30
Typical butt joint preparations

Note

Root faces on bevels and grooves with backing bars show a 1/8-in. root face. The 1/8-in. root face is a maximum allowable dimension and many welders prefer root faces prepared thinner, or all the way to a "knife edge" (see Figure 3.2).

Heavy plate and pipe sections requiring preparations are often used in products manufactured under a code or standard. The American Welding Society, American Society of Mechanical Engineers, and American Bureau of Ships are a few of the agencies that issue codes and specifications. The AWS D1.1 and the ASME Boiler and Pressure Vessel (BPV) Section IX standards will be used in this chapter as the standards for multiple pass groove welds that will be tested. The groove depth and angle are determined by the plate or pipe thickness and process.

After the plate is beveled, a grinder can be used to clean off oxides and improve the fitup.

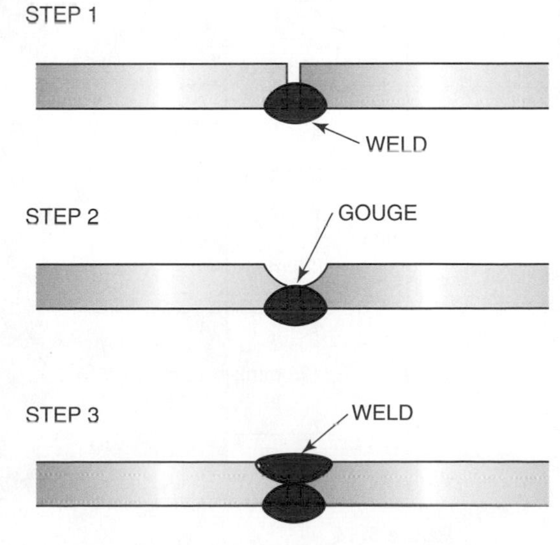

Figure 3.31
Back gouging sequence for a weld to ensure 100% joint penetration

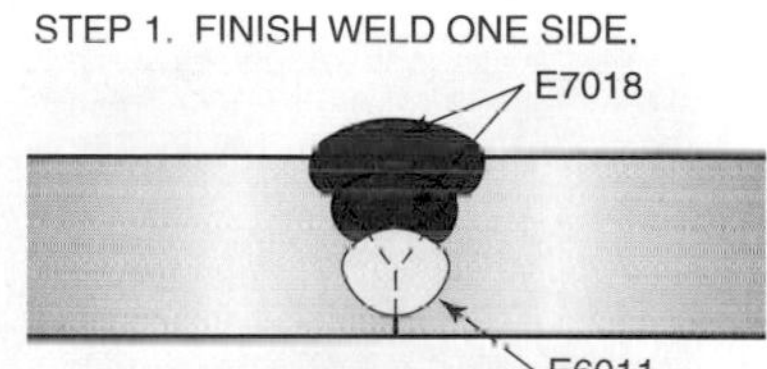

STEP 2. BACK GOUGE TO REMOVE ALL E6011 DEPOSITED IN FIRST WELD.

E7018

STEP 3. WELD U-GROOVE.

100% E7018

Figure 3.32
Back gouging to remove all weld metal used for the root pass or tacking

PREPARING SPECIMENS FOR TESTING

The detailed preparation of specimens for testing in this chapter is based on the structural welding code AWS D1.1 and the ASME BPV Code, Section IX. The maximum allowable size of fissures (crack or opening) in a **guided bend specimen** is given in codes for specific applications. Some of the standards are listed in ASTM E190 or AWS B4.0, AWS QC10, AWS QC11, and others. Copies of these publications are available from the appropriate organizations.

Acceptance Criteria for Face Bends and Root Bends

The weld specimen must first pass visual inspection before it can be prepared for bend testing. Visual inspection looks to see that the weld is uniform in width and reinforcement. There should be no arc strikes on the plate other than those on the weld itself. The weld must be free of both incomplete fusion and cracks. The joint penetration must be either 100% or as specified by the specifications. The weld must be free of overlap and undercut must not exceed either 10% of the base metal or 1/32 in. (0.8 mm), whichever is less.

Correct **weld specimen** preparation is essential for reliable results. The weld must be uniform in width and reinforcement and have no undercut or overlap. The weld reinforcement and backing strip, if used, must be removed flush to the surface, **Figure 3.33.** They can be machined or ground off. The plate thickness after removal must be a minimum of 3/8 in. (10 mm), and the pipe thickness must be equal to the pipe's original wall thickness. The specimens may be cut out of the test weldment by using an abrasive disc, by sawing, or by cutting with a torch. Flame-cut specimens must have the edges ground or machined smooth after cutting.

Figure 3.33
Plate ground in preparation for removing test specimens
Courtesy of Larry Jeffus

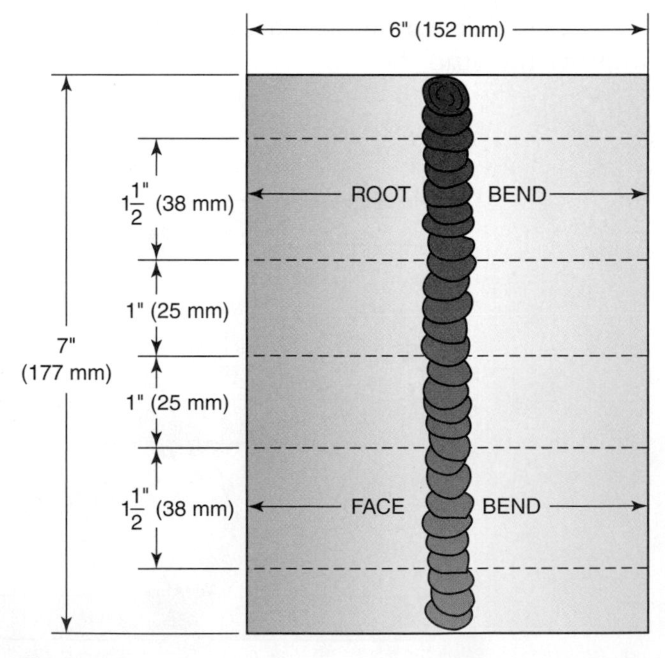

Figure 3.34
Sequence for removing guided bend specimens from the plate once welding is complete

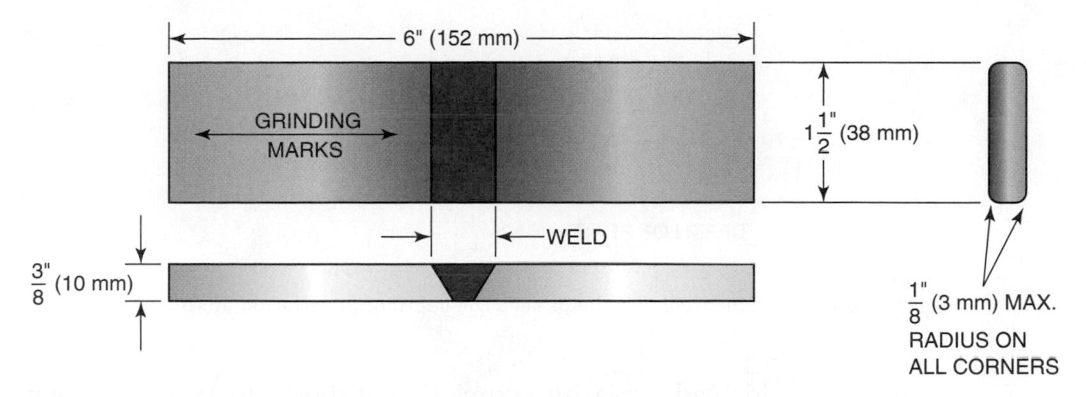

Figure 3.35
Guided bend specimen

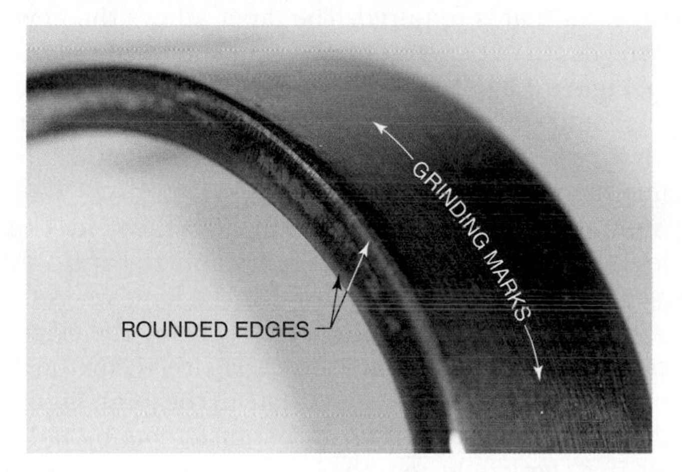

Figure 3.36
Guided bend test specimen
Courtesy of Larry Jeffus

This procedure is done to remove the heat-affected zone caused by the cut, **Figure 3.34.**

All corners must be rounded to a radius of 1/8 in. (3 mm) maximum, and all grinding or machining marks must run lengthwise on the specimen, **Figure 3.35** and **Figure 3.36.** Rounding the corners and keeping all marks running lengthwise reduce the chance of good weld specimen failure due to poor surface preparation.

The weld must pass both the face and root bends in order to be acceptable. After bending there can be no single defects larger than 1/8 in. (3 mm), and the sum of all defects larger than 1/32 in. (0.8 mm) but less than 1/8 in. (3 mm) must not exceed a total of 3/8 in. (10 mm) for each bend specimen. An exception is made for cracks that start at the edge of the specimen and do not start at a defect in the specimen.

RESTARTING A WELD BEAD

On all but short welds, the welding bead will need to be restarted after a welder stops to change electrodes. Because the metal cools as a welder changes electrodes and chips slag when restarting, the penetration and buildup may be adversely affected.

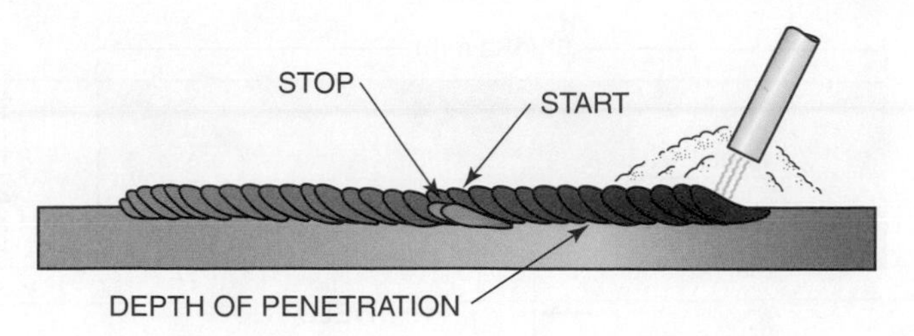

Figure 3.37
Tapering the size of the weld bead helps keep the depth of penetration uniform

When a weld bead is nearing completion, it should be tapered so that when it is restarted the buildup will be more uniform. To taper a weld bead, the travel rate should be increased just before welding stops. A 1/4-in. (6-mm) taper is all that is required. The taper allows the new weld to be started and the depth of penetration reestablished without having excessive buildup, **Figure 3.37.**

The slag should always be chipped and the weld crater should be cleaned each time before restarting the weld. This is important to prevent slag inclusions at the start of the weld.

The arc should be restarted in the joint ahead of the weld. The electrodes must be allowed to heat up so that the arc is stabilized and a shielding gas cloud is reestablished to protect the weld. Hold a long arc as the electrode heats up so that metal is not deposited. Slowly bring the electrode downward and toward the weld bead until the arc is directly on the deepest part of the crater where the crater meets the plate in the joint, **Figure 3.38.** The electrode should be low enough to start transferring metal. Next, move the electrode in a semicircle near the back edge of the weld crater. Watch

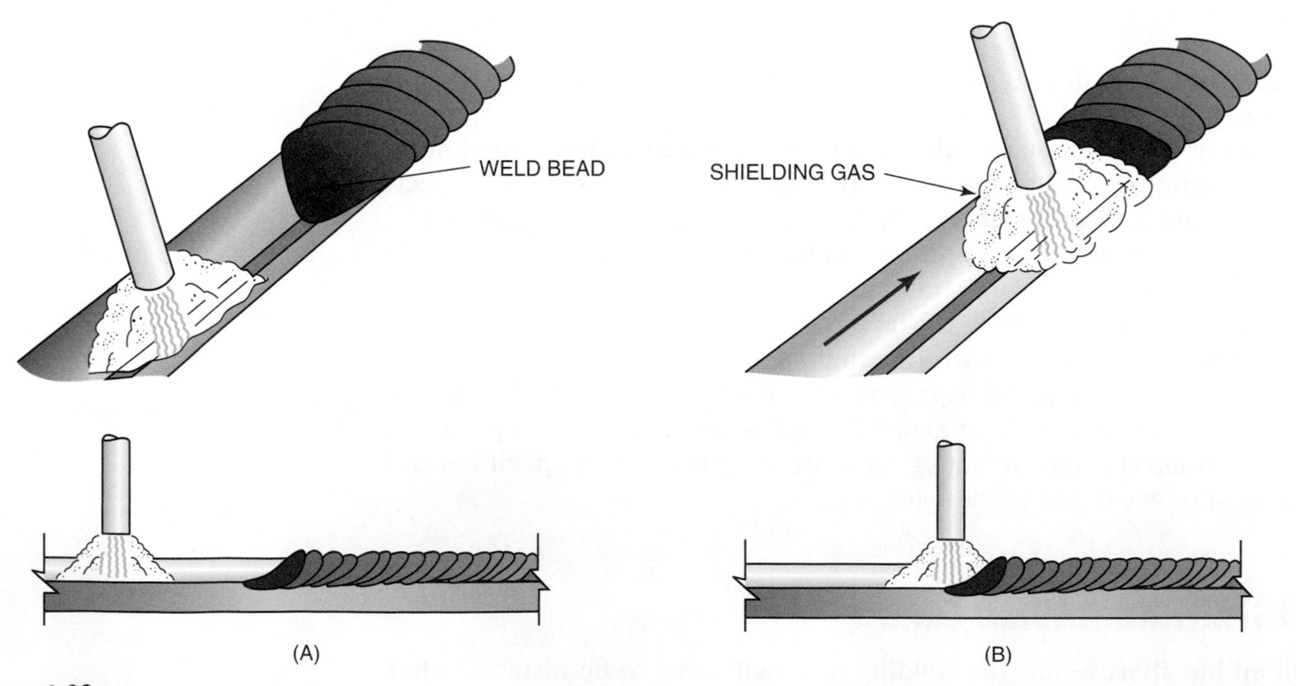

Figure 3.38
When restarting the arc, strike the arc ahead of the weld in the joint (A). Hold a long arc and allow time for the electrode to heat up, forming the protective gas envelope. Move the electrode so that the arc is focused directly on the leading edge (root) of the previous weld crater (B).

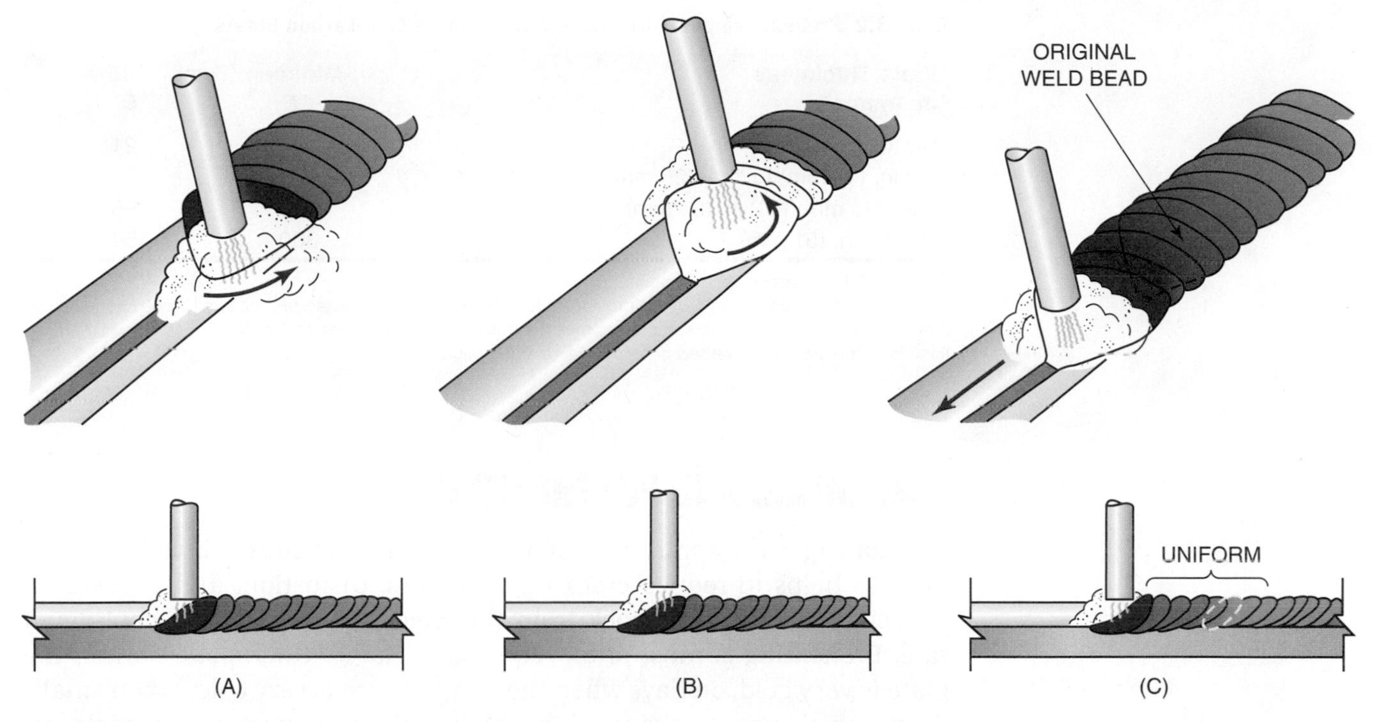

Figure 3.39
When restarting the weld pool after the root has been heated to the melting temperature, move the electrode upward along one side of the crater (A). Move the electrode along the top edge, depositing new weld metal (B). When the weld is built up uniformly with the previous weld, continue along the joint (C).

the buildup and match your speed in the semicircle to the deposit rate so that the weld is built up evenly, **Figure 3.39.** Move the electrode ahead and continue with the same weave pattern that was being used previously.

The movement to the root of the weld and back up on the bead serves both to build up the weld and reheat the metal so that the depth of penetration will remain the same. If the weld bead is started too quickly, penetration is reduced and buildup is high and narrow.

Starting and stopping weld beads in corners should be avoided. Tapering and restarting are especially difficult in corners, and this often results in defects, **Figure 3.40.**

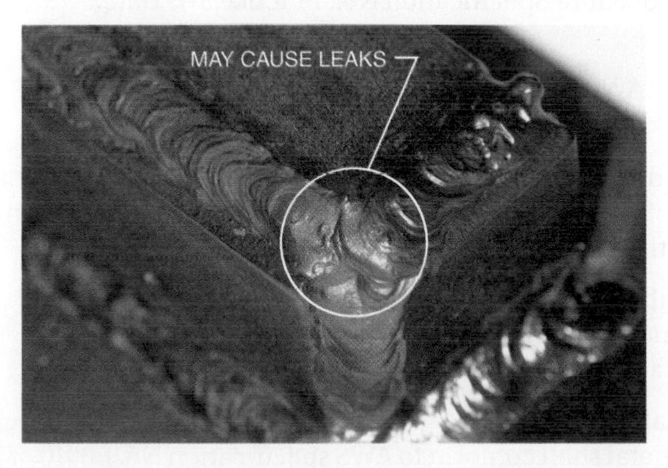

Figure 3.40
Correct method of welding through a corner. Stopping on a corner may cause leaks.
Courtesy of Larry Jeffus

Table 3.2 Preheat Temperatures for Arc Welding on Low-carbon Steels*

Plate Thickness in. (mm)	Minimum Temperature °F	°C
Up to 1/2 in. (13 mm)	70	21
1/2 in. (13 mm) to 1 in. (25 mm)	100	38
1 in. (25 mm) to 2 in. (51 mm)	200	95
Over 2 in. (51 mm)	300	150

*Metal should be above the dew point.
Allow 1 hour for each inch in order to provide uniform heating or for localized preheating. Check to ensure that preheat temperatures are sufficient to melt thermal crayons a minimum of 3 inches (76 mm) in all directions from the area to be welded prior to welding arc application.

PREHEATING AND POSTHEATING

Preheating is the application of heat to the metal before it is welded. This process helps to reduce cracking, hardness, distortion, and stresses by reducing the thermal shock from the weld and slowing the cool-down rate. Preheating is most often required on large, thick plates, when the plate is very cold, on days when the temperature is very cold, when small-diameter electrodes are used, on high-carbon or manganese steels, on complex shapes, or with fast welding speeds.

With the practices that are to be tested in this chapter, preheating should be used if the base metal to be welded is very cold. It may also be used to reduce distortion on thick sections and to reduce hardness caused by the rapid cooling of the weld, which may result in weld failure. Preheating the metal will slow the weld cooling rate, which results in a more ductile weld. **Table 3.2** lists the recommended preheat temperatures for plain carbon steels.

PRACTICE 3-7

Limited Thickness Welder Performance Qualification Test Plate with Backing and F3 and F4 Electrodes

Module 1
Key Indicator 1, 2, 4

Module 4
Key Indicator 1, 2, 3, 4, 6

Module 9
Key Indicator 1, 2

Welding Procedure Specification (WPS)

Welding Procedure Specification No.: Practice 3-7. Date:

Title:

Welding SMAW of plate to plate.

Scope:

This procedure is applicable for V-groove plate with a backing strip within the range of 3/8 in. (10 mm) through 3/4 in. (20 mm). Welding may be performed in the following positions: 1G, 2G, 3G, and 4G.

Base Metal:

The base metal shall conform to M1020 or A36.
Backing material specification M1020 or A36.

Filler Metal:

The filler metal shall conform to AWS specification No. E6010 or E6011 root pass and E7018 for the cover pass from AWS specification A5.1. This filler metal falls into F-number F3 and F4 and A-number A-1.

Shielding Gas:

The shielding gas, or gases, shall conform to the following compositions and purity: N/A.

Joint Design and Tolerances:

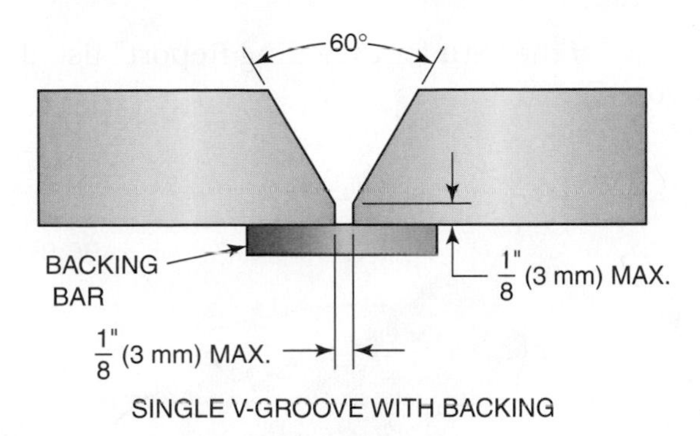

Preparation of Base Metal:

The V-groove is to be ground, flame-cut, or machined on the edge of the plate before the parts are assembled. Root faces may be prepared from a knife edge to the maximum allowable thickness of 1/8 in. All parts must be cleaned prior to welding of all contaminants, such as paints, oils, grease, or primers. Both inside and outside surfaces within 1 in. (25 mm) of the joint must be mechanically cleaned using a wire brush or grinder.

Electrical Characteristics:

The current shall be AC or DCRP.

The base metal shall be on the work lead or negative side of the line.

Preheat:

The parts must be heated to a temperature higher than 70°F (21°C) before any welding is started.

Backing Gas:

N/A

Welding Technique:

Tack weld the plates together with the backing strip. There should be about a 1/8-in. (3-mm) root gap between the plates. Use the E6010 or E6011 arc welding electrodes to make a root pass to fuse the plates and backing strip together. Clean the slag from the root pass and use either a hot pass or grinder to remove any trapped slag.

Using the E7018 arc welding electrodes, make a series of filler welds in the groove until the joint is filled.

Interpass Temperature:

The plate should not be heated to a temperature higher than 400°F (205°C) during the welding process. After each weld pass is completed, allow it to cool; the weldment must not be quenched in water.

Cleaning:

The slag can be chipped and/or ground off between passes but can only be chipped off of the cover pass.

Inspection:

Visually inspect the weld for uniformity and other discontinuities with criteria found in practice 3-8. If the weld passes the visual inspection, then it is to be prepared and guided bend tested according to criteria found in practice 3-8. Repeat each of the welds as needed until you can pass this test.

Complete a copy of the "Student Welding Report" listed in Appendix I or provided by your instructor.

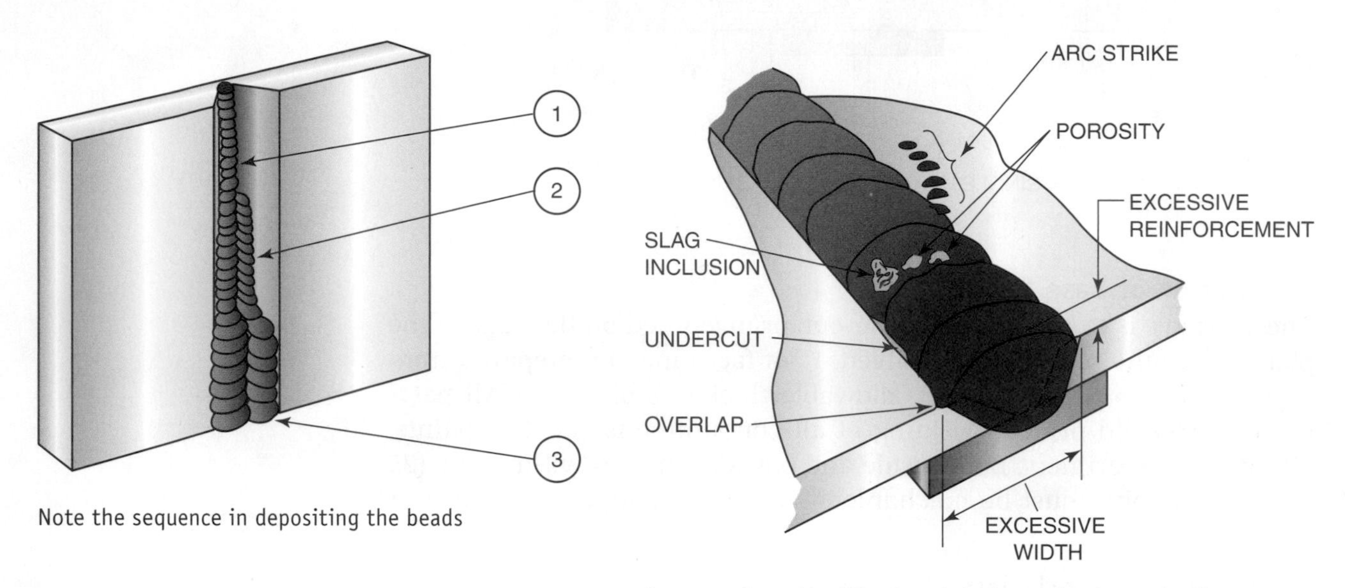

Note the sequence in depositing the beads

Common discontinuities found during a visual examination

Sketches:

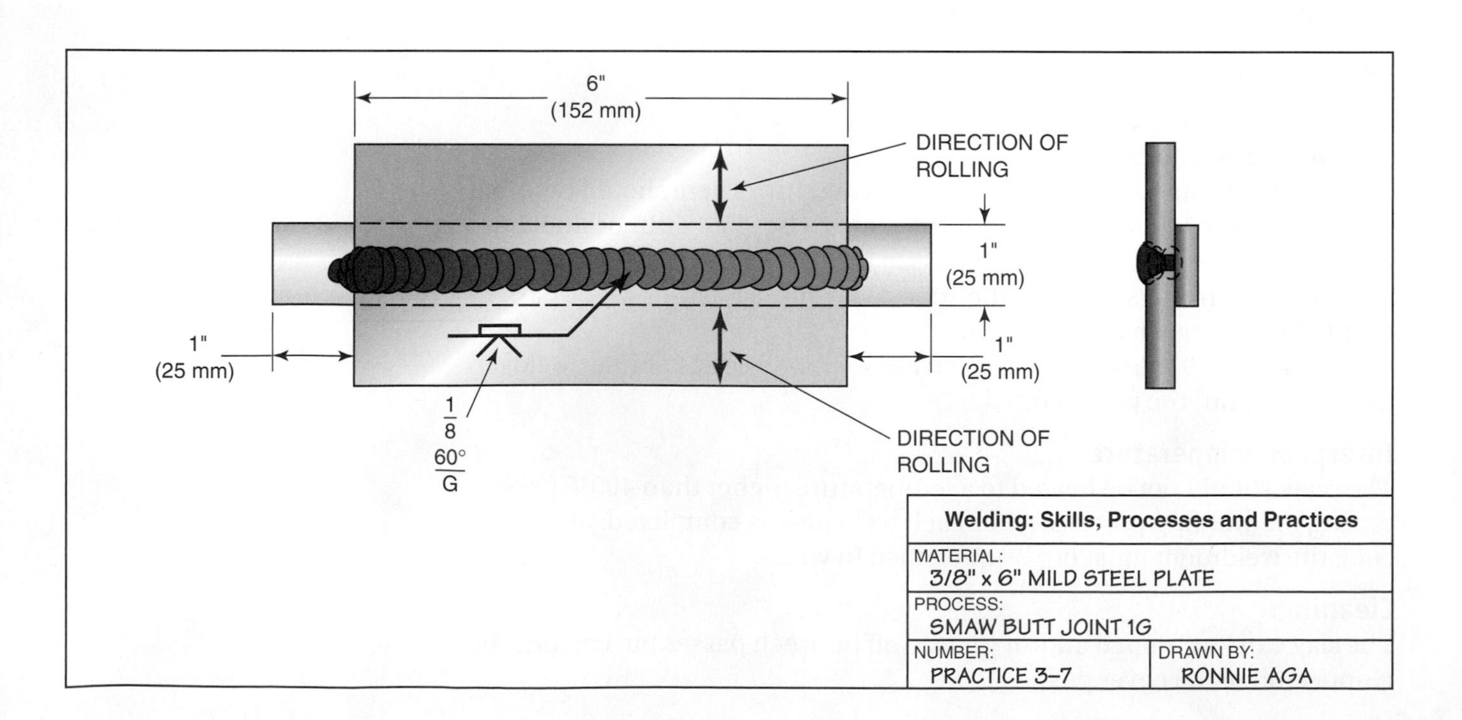

PRACTICE 3-8

SMAW Workmanship Sample and Welder Qualification Test Plate for Limited Thickness Horizontal 2G and 3G Positions with E7018 Electrodes

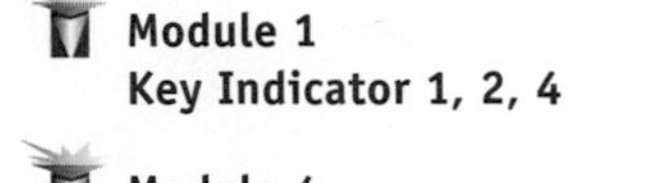
Module 1
Key Indicator 1, 2, 4

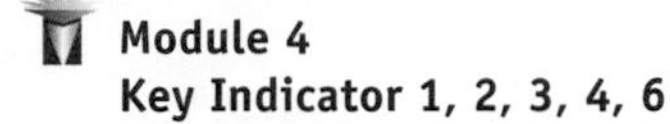
Module 4
Key Indicator 1, 2, 3, 4, 6

Module 9
Key Indicator 1, 2

Dimensional Tolerances:

2G Test plates: two (2); each 3/8 in. (10 mm) thick, 3 in. (75 mm) wide, and 7 in. (175 mm) long, one having a 45° bevel along one edge.
3G Test plates: two (2); each 3/8 in. (10 mm) thick, 3 in. (75 mm) wide, and 7 in. (175 mm) long, one having a 45° included bevel.
Backing strip: one (1); each either 1/4 in. (6 mm) or 3/8 in. (10 mm) thick, 1 in. (25 mm) wide, and 9 in. (228 mm) long.

Welding Procedure Specification (WPS)

Welding Procedure Specification No.: Practice 3-8. Date:

Title:

Welding SMAW of plate to plate.

Scope:

This procedure is applicable for single bevel or V-groove plate with a backing strip within the range of 3/16 in. (5 mm) through 3/4 in. (20 mm).
Welding may be performed in the following positions: 2G and 3G.

Base Metal:

The base metal shall conform to carbon steel M-1 or P-1, Group 1 or 2.
Backing material specification carbon steel M-1 or P-1, Group 1, 2, or 3.

Filler Metal:

The filler metal shall conform to AWS specification no. E7018 from AWS specification A5.1. This filler metal falls into F-number F-4 and A-number A-1.

Shielding Gas:

The shielding gas, or gases, shall conform to the following compositions and purity: N/A.

Joint Design and Tolerances:

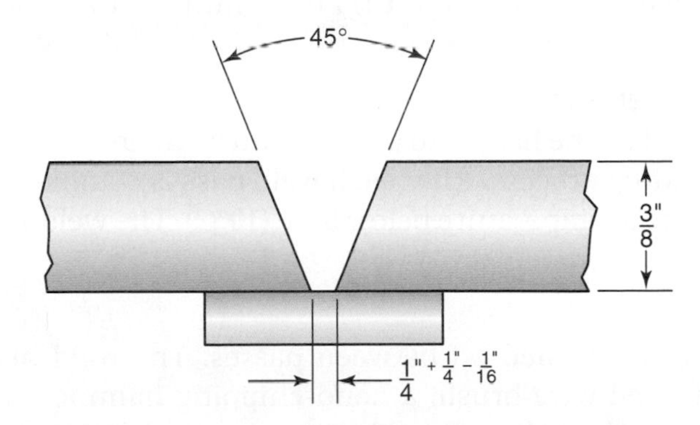

Preparation of Base Metal:

The bevel is to be flame-cut on the edge of the plate before the parts are assembled. The beveled surface must be smooth and free of notches. Any roughness or notches that are deeper than 1/64 in. (0.4 mm) must be ground smooth.

All hydrocarbons and other contaminations, such as cutting fluids, grease, oil, and primers, must be cleaned off all parts and filler metals before welding. This cleaning can be done with any suitable solvents or detergents. The backing strip, groove face, and inside and outside plate surface within 1 in. (25 mm) of the joint must be mechanically cleaned of slag, rust, and mill scale. Cleaning must be done with a wire brush or grinder down to bright metal.

Electrical Characteristics:

The current shall be direct current electrode positive (DCEP). The base metal shall be on the negative side of the line.

Welds	Filler Metal Diameter	Current	Amperage Range
Tack	3/32 in. (2.4 mm)	DCEP	70 to 115
Root and Filler	1/8 in. (3 mm)	DCEP	115 to 165
Filler	1/8 in. (3 mm)	DCEP	115 to 140

Preheat:

The parts must be heated to a temperature higher than 50°F (10°C) before any welding is started.

Backing Gas:

N/A

Safety:

Proper protective clothing and equipment must be used. The area must be free of all hazards that may affect the welder or others in the area. The welding machine, welding leads, work clamp, electrode holder, and other equipment must be in safe working order.

Welding Technique:

Tack weld the plates together with the backing strip. There should be about a 1/4-in. (6-mm) root gap between the plates. Use the E7018 arc welding electrodes to make a root pass to fuse the plates and backing strip together. Clean the slag from the root pass, being sure to remove any trapped slag along the sides of the weld.

Using the E7018 arc welding electrodes, make a series of stringer or weave filler welds, no thicker than 1/4 in. (6 mm), in the groove until the joint is filled.

Interpass Temperature:

The plate should not be heated to a temperature higher than 350°F (175°C) during the welding process. After each weld pass is completed, allow it to cool but never to a temperature below 50°F (10°C). The weldment must not be quenched in water.

Cleaning:

The slag must be cleaned off between passes. The weld beads may be cleaned by a hand wire brush, a hand-chipping hammer, a punch and hammer, or a needle scaler. All weld cleaning must be performed with the test plate in the welding position.

Visual Inspection Criteria for Entry Welders*:

There shall be no cracks, no incomplete fusion. There shall be no incomplete joint penetration in groove welds except as permitted for partial joint penetration welds.

*Courtesy of American Welding Society

The Test Supervisor shall examine the weld for acceptable appearance, and shall be satisfied that the welder is skilled in using the process and procedure specified for the text.

Undercut shall not exceed the lesser of 10% of the base metal thickness or 1/32 in. (0.8 mm).

Where visual examination is the only criterion for acceptance, all weld passes are subject to visual examination, at the discretion of the Test Supervisor.

The frequency of porosity shall not exceed one in each 4 in. (100 mm) of weld length and the maximum diameter shall not exceed 3/32 in. (2.4 mm).

Welds shall be free from overlap.

Sketches:

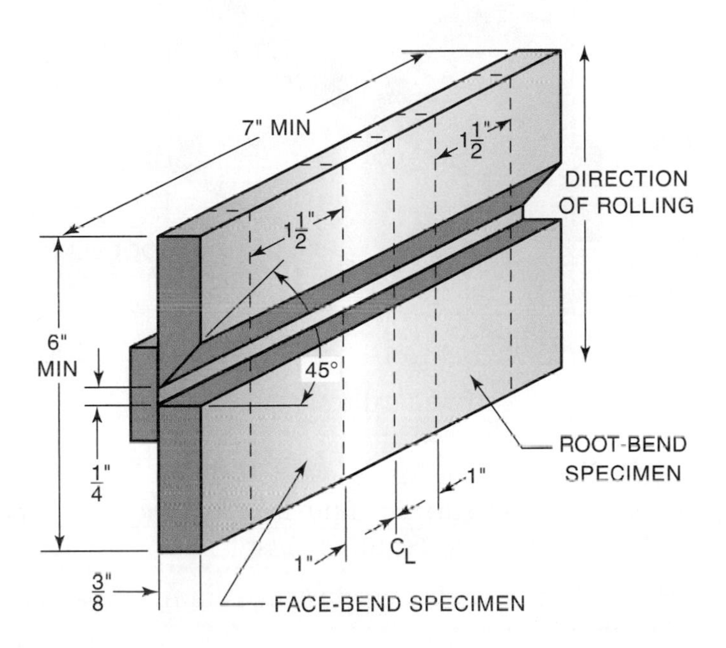

Bend Test:

The weld is to be mechanically tested only after it has passed the visual inspection. Be sure that the test specimens are properly marked to identify the welder, the position, and the process.

Specimen Preparation

For 3/8-in. (10-mm) test plates, two specimens are to be located in accordance with the requirements below. One is to be prepared for a transverse face bend, and the other is to be prepared for a transverse root bend.

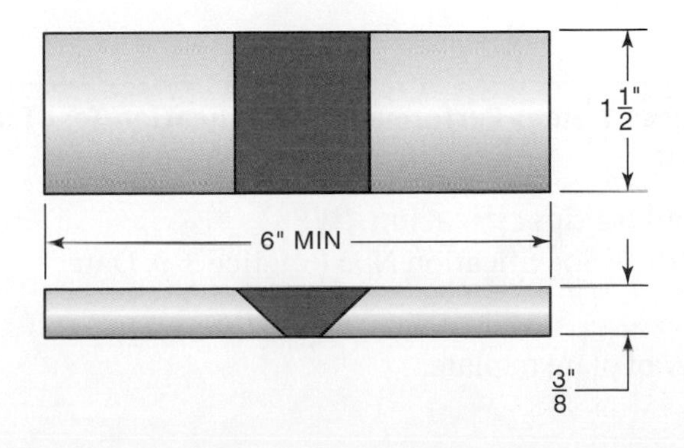

- *Transverse face bend.* The weld is perpendicular to the longitudinal axis of the specimen and is bent so that the weld face becomes the tension surface of the specimen. Transverse face bend specimens shall comply with the requirements of below.
- *Transverse root bend.* The weld is perpendicular to the longitudinal axis of the specimen and is bent so that the weld root becomes the tension surface of the specimen. Transverse root bend specimens shall comply with the requirements of above.

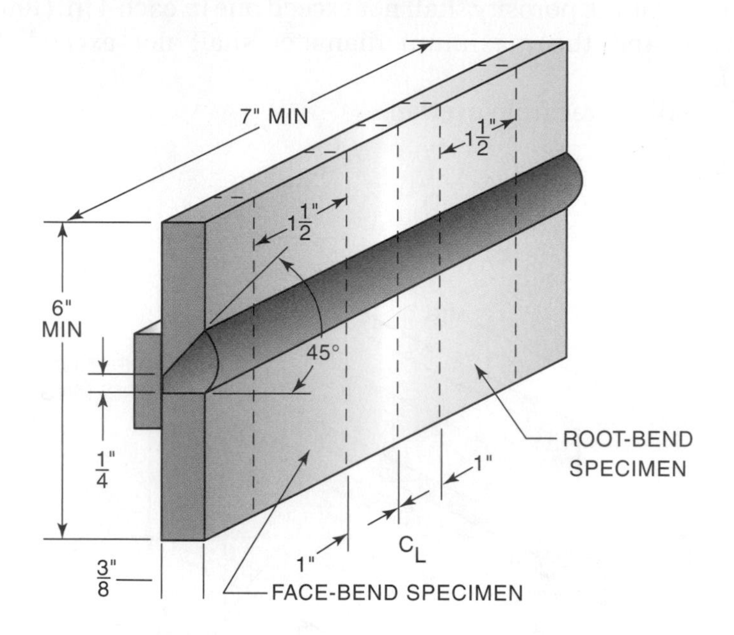

Acceptance Criteria for Bend Test*:

For acceptance, the convex surface of the face and root bend specimens shall meet both of the following requirements:

1. No single indication shall exceed 1/8 in. (3 mm), measured in any direction on the surface.
2. The sum of the greatest dimensions of all indications on the surface, which exceed 1/32 in. (0.8 mm), but are less than or equal to 1/8 in. (3 mm), shall not exceed 3/8 in. (10 mm).

*Courtesy of American Welding Society

Cracks occurring at the corner of the specimens shall not be considered unless there is definite evidence that they result from slag or inclusions or other internal discontinuities.

Complete a copy of the "Student Welding Report" listed in Appendix I or provided by your instructor.

PRACTICE 3-9

Module 1
Key Indicator 1, 2, 4

Module 4
Key Indicator 1, 2, 3, 6

Module 9
Key Indicator 1, 2

Limited Thickness Welder Performance Qualification Test Plate Without Backing

Welding Procedure Specification (WPS)

Welding Procedure Specification No.: Practice 3-9. Date:

Title:

Welding SMAW of plate to plate.

Scope:
This procedure is applicable for V-groove plate without a backing strip within the range of 3/8 in. (10 mm) through 3/4 in. (20 mm).

Welding may be performed in the following positions: 1G, 2G, 3G, and 4G.

Base Metal:
The base metal shall conform to M1020 or A36.
Backing material specification: M1020 or A36.

Filler Metal:
The filler metal shall conform to AWS specification no. E6010 or E6011 root pass and E7018 for the cover pass from AWS specification A5.1. This filler metal falls into F-number F3 and F4 and A-number A-1.

Shielding Gas:
The shielding gas, or gases, shall conform to the following compositions and purity: N/A.

Joint Design and Tolerances:

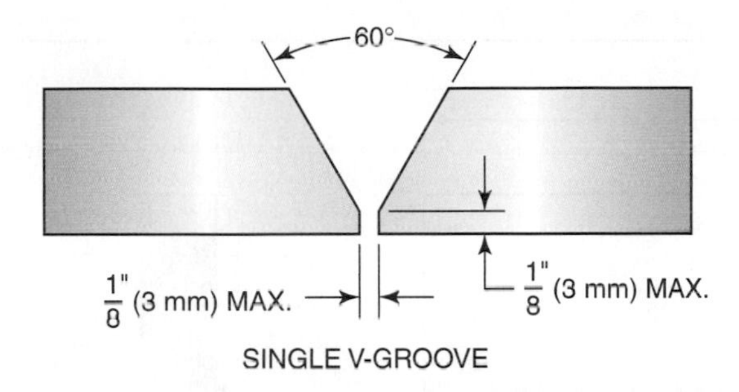

Preparation of Base Metal:
The V-groove is to be ground, flame-cut, or machined on the edge of the plate before the parts are assembled. All parts must be cleaned prior to welding of all contaminants, such as paints, oils, grease, or primers. Both inside and outside surfaces within 1 in. (25 mm) of the joint must be mechanically cleaned using a wire brush or grinder.

Electrical Characteristics:
The current shall be DCRP. The base metal shall be on the work lead or negative side of the line.

Preheat:
The parts must be heated to a temperature higher than 70°F (21°C) before any welding is started.

Backing Gas:
N/A

Welding Technique:
Tack weld the plates together; there should be about a 1/8-in. (3-mm) root gap between the plates. Use the E6010 or E6011 arc welding electrodes to make a root pass to fuse the plates together. Clean the slag from the root pass and use either a hot pass or grinder to remove any trapped slag.

Using the E7018 arc welding electrodes, make a series of filler welds in the groove until the joint is filled.

Interpass Temperature:
The plate, outside of the heat affected zone, should not be heated to a temperature higher than 400°F (205°C) during the welding process. After each weld pass is completed, allow it to cool; the weldment must not be quenched in water.

Cleaning:
The slag can be chipped and/or ground off between passes but can only be chipped off of the cover pass.

Inspection:
Visually inspect the weld for uniformity and other discontinuities by using the criteria found in Practice 3-8. If the weld passes the visual inspection, then it is to be prepared and guided bend tested according to the "guided bend test" criteria found in Practice 3-8. Repeat each of the welds as needed until you can pass this test.

Complete a copy of the "Student Welding Report" listed in Appendix I or provided by your instructor.

Sketches:

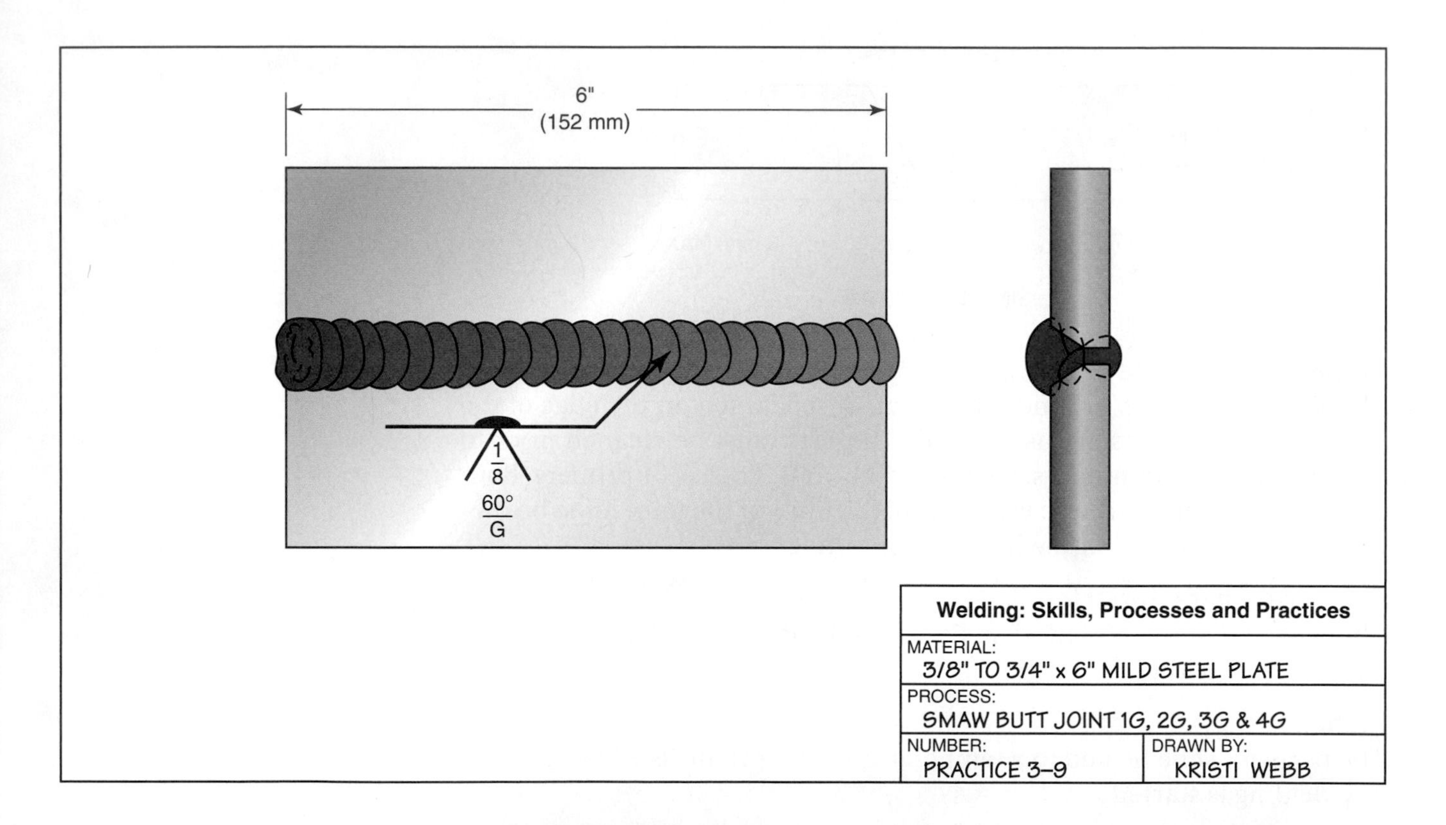

Postheating is the application of heat to the metal after welding. Postheating is used to slow the cooling rate and reduce hardening.

Interpass temperature is the temperature of the metal during welding. The interpass temperature is given as a minimum and maximum. The minimum temperature is usually the same as the preheat temperature. If the plate cools below this temperature during welding, it should be reheated. The maximum temperature may be specified to keep the plate below a certain phase change temperature for the mild steel used in these practices. The maximum interpass temperature occurs when the weld bead cannot be controlled because of a slow cooling rate. When this happens, the plate should be allowed to cool down, but not below the minimum interpass temperature.

If, during the welding process, a welder must allow a practice weldment to cool so that the weld can be completed later, the weldment should be cooled slowly and then reheated before starting to weld again. A weld that is to be tested or that is done on any parts other than scrap should not be quenched in water.

POOR FITUP

Ideally, all welding will be performed on joints that are properly fitted. Most welds produced to a code or standard are properly fitted. Repair, prototype, and job shop welding, however, may not be cut and fitted properly. These welds must be performed under less than ideal conditions, but they still must be strong and have a good appearance.

To make a good weld on a poorly fitted joint requires some special skills. These welds also require a good welder, one whose skill is developed. A skilled welder can watch the molten weld pool and knows how to correct for problems before they develop into disasters. The welder must be able to read the molten weld pool correctly to make needed changes in amperage, current, electrode movement, electrode angle, and timing.

The amperage setting may have to be adjusted up or down by only a few amperes to make the necessary changes in molten weld pool size. Adjusting the machine is often preferable to lengthening the weave pattern excessively. The current may be changed from AC to DCSP or DCRP to vary the amount of heat input to the molten weld pool. Some electrodes can operate better than other electrodes with lower amperages on some currents. The current also will alter the forcefulness of some electrodes.

The "U," "J," and straight step patterns are usually the best to use, but they should not be moved more than required to close the gap or opening. On some poor-fitting joints, it is necessary to break and restart the arc in order to keep the molten weld pool under control. This will result in a weld with porosity, slag inclusions, and other defects. But it is often better to have a poor weld than to have no weld. A poor root weld can be capped with a sound weld to improve joint strength.

Changing the electrode angle from leading to trailing improves poor fit. Sometimes a very flat angle will also help. The time interval that the electrode is moved into and out of the molten weld pool is critical in maintaining weld control. Returning to a molten weld pool too often or too soon can cause the molten weld pool to drop out of the joint. In most cases, a welder should return to the molten weld pool only after it has started to cool.

On some joints, it is possible to make stringer beads on both sides of the joint until the gap is closed, **Figure 3.41.** Note that the beads are made alternately from the edges of the joint to the center. Welds made in this manner can have good weld soundness and strength, but they require more time to complete.

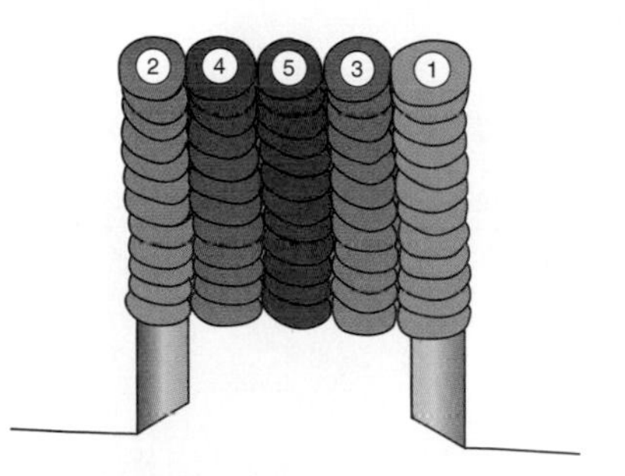

Figure 3.41
Multiple stringer beads used to close a large gap

PRACTICE 3-10

Single V-groove Open Root Butt Joint with an Increasing Root Opening

Module 1
Key Indicator 1, 2, 4

Module 4
Key Indicator 6

Module 9
Key Indicator 1, 2

For this practice, you will need a properly set up and adjusted arc welding machine, proper safety protection, E6010 or E6011 and E7018 arc welding electrodes having a 1/8-in. (3-mm) diameter, and two or more pieces of mild steel plate, 3/8 in. (10 mm) thick × 4 in. (102 mm) wide × 12 in. (305 mm) long. You will weld a single V-groove open root butt joint that has a poor fitup, starting from the close end.

Tack weld the plates together with a root opening of 1/8 in. (3 mm) at one end and 1/2 in. (13 mm) at the other end, **Figure 3.42.** Using the E6010 or the E6011 electrode, start the root pass at the narrow end and weld to the other end. As the root pass progresses along the widening root gap, care must be taken to maintain molten weld pool control. The "J" or "U" weave pattern works best with low current settings. Long time intervals for electrode movements will give the best weld control.

When the root pass is completed, clean the weld and make a hot pass to burn out any trapped slag. Finish the weld with filler passes using the E7018 electrode. Cool, chip, and inspect the weld for uniformity and defects. Repeat these welds until you can consistently make welds free of defects. Turn off the welding machine and clean up your work area when you are finished welding.

Complete a copy of the "Student Welding Report" listed in Appendix I or provided by your instructor.

PRACTICE 3-11

Single V-groove Open Root Butt Joint with a Decreasing Root Opening

Module 1
Key Indicator 1, 2, 4

Module 4
Key Indicator 6

Module 9
Key Indicator 1, 2

Using the same setup, equipment, and materials as described in Practice 3-10, you will weld a single V-groove open root butt joint that has a poor fitup, starting from the wide end.

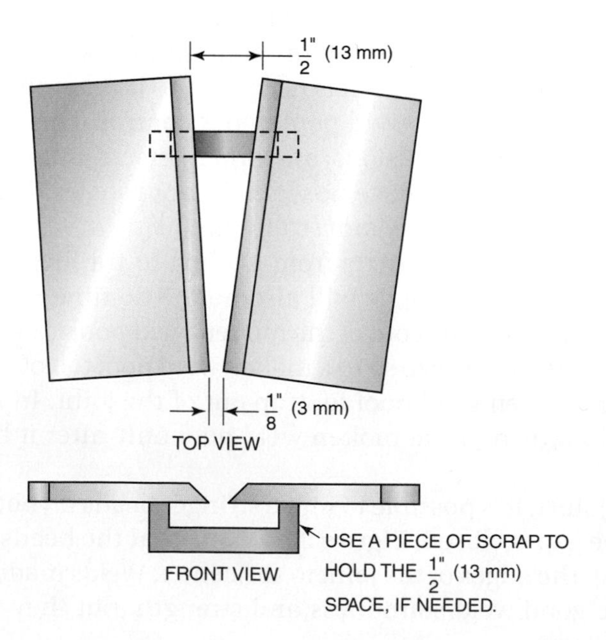

Figure 3.42
Welding specimen with poor fitup

As in Practice 3-10, tack weld the plates together with a root opening of 1/8 in. (3 mm) at one end and 1/2 in. (13 mm) at the other end. Using the E6010 or E6011 electrode, start welding the root pass at the wide end. Both sides of the joint must be built up until it is possible to get the metal to flow together. The "J" or "U" weave pattern works best to control the bead.

When the root pass is completed, make a hot pass to clean out any trapped slag before making the filler passes with the E7018 electrode. After the weld is completed, cool, chip, and inspect it for uniformity and defects. Repeat these welds until you can consistently make welds free of defects. Turn off the welding machine and clean up your work area when you are finished welding.

Complete a copy of the "Student Welding Report" listed in Appendix I or provided by your instructor.

SUMMARY

Grooved welds on approximately 1/2-in. (13-mm)-thick plate are the most common test plates given to new welding applicants. Groove welds are used by many companies as the base welding skills performance test requirement for employment. The vertical and overhead positions are the most commonly used for the test. It is often assumed in the welding industry that a uniform, visually defect-free weld will successfully pass destructive testing. This assumption has great basis in fact because, in most cases, such a weld reflects the welder's skills required to produce quality welds, so, in many cases, applicant test welds are evaluated only by the weld shop foreman or supervisor for visual defects. For this reason, you should always attempt to make your welds as uniform in appearance as possible. Learning how to make a "pretty" groove weld can often mean the difference between successfully earning the job and losing out to another applicant.

REVIEW

1. Why are some weld joints grooved?
2. Sketch four of the standard grooves used for welded joints.
3. Why are some backing strips removed from the finished weld?
4. Why are backing tapes used on some joints?
5. Why is it very important to make a weld with a good root surface?
6. What are the two common methods of making a root pass on an open root joint?
7. How can small gaps between the weld plate and backing strip be closed?
8. What effect does changing from a trailing angle to a leading angle have on a weld?
9. What benefit would there be to the root pass if the electrode holder were rocked back and forth while keeping the electrode tip in the joint?
10. What might cause the bright color on the flux as a weld cools?

11. What can happen if the molten weld pool becomes too large on a root weld?
12. What can be done to increase the amount of high-strength welding electrode in the final weld if the root weld was made with a low-alloy electrode?
13. What is the purpose of the hot pass?
14. Why should a filler weld pass not have deep penetration?
15. Why is it important to have a good cover pass?
16. What can watching the back edge of a weld pool help you determine?
17. Other than penetration, why would thick butt joints be grooved on both sides?
18. List the things that a weld must be inspected for before it is ground for bend testing.
19. What determines the acceptance or rejection of a bend specimen?
20. What technique can be used to make restarting weld beads easier and more uniform?
21. Why should some weldments be preheated before welding starts?
22. What is postheating used for?
23. How can a wide gap in a joint be closed by welding?

CHAPTER 4

Gas Tungsten Arc Welding Equipment, Setup, Operation, and Filler Metals

OBJECTIVES

After completing this chapter, the student should be able to

- describe the gas tungsten arc welding process and list the other terms used to identify the process
- list four attributes of a quality tungsten electrode
- list four precautions taken to limit tungsten erosion
- contrast the various types of tungsten electrodes and how they are used
- shape the end of the tungsten electrode and clean it
- grind a point on a tungsten electrode using an electric grinder
- remove a contaminated tungsten end
- melt the end of the tungsten electrode into the desired shape
- list two advantages and two disadvantages of water-cooled GTA welding torches and air-cooled torches
- describe the purposes of the three hoses connecting a water-cooled torch to the welding machine
- choose an appropriate nozzle for the job
- accurately read a flowmeter
- compare and contrast the three types of welding current used for GTA welding
- list three advantages and three disadvantages of the shielding gases used in the GTA welding process
- set preflow time and postflow time on a GTA welding workstation
- list three problems that can occur as a result of an incorrect gas flow rate
- set up a GTAW workstation for carbon steel, stainless steel, and aluminum operations
- strike a GTA welding arc on carbon steel, stainless steel, and aluminum

KEY TERMS

cleaning action
collet
flowmeter
frequency
inert gas
postflow time
preflow time
rectification
spark gap oscillator
tungsten

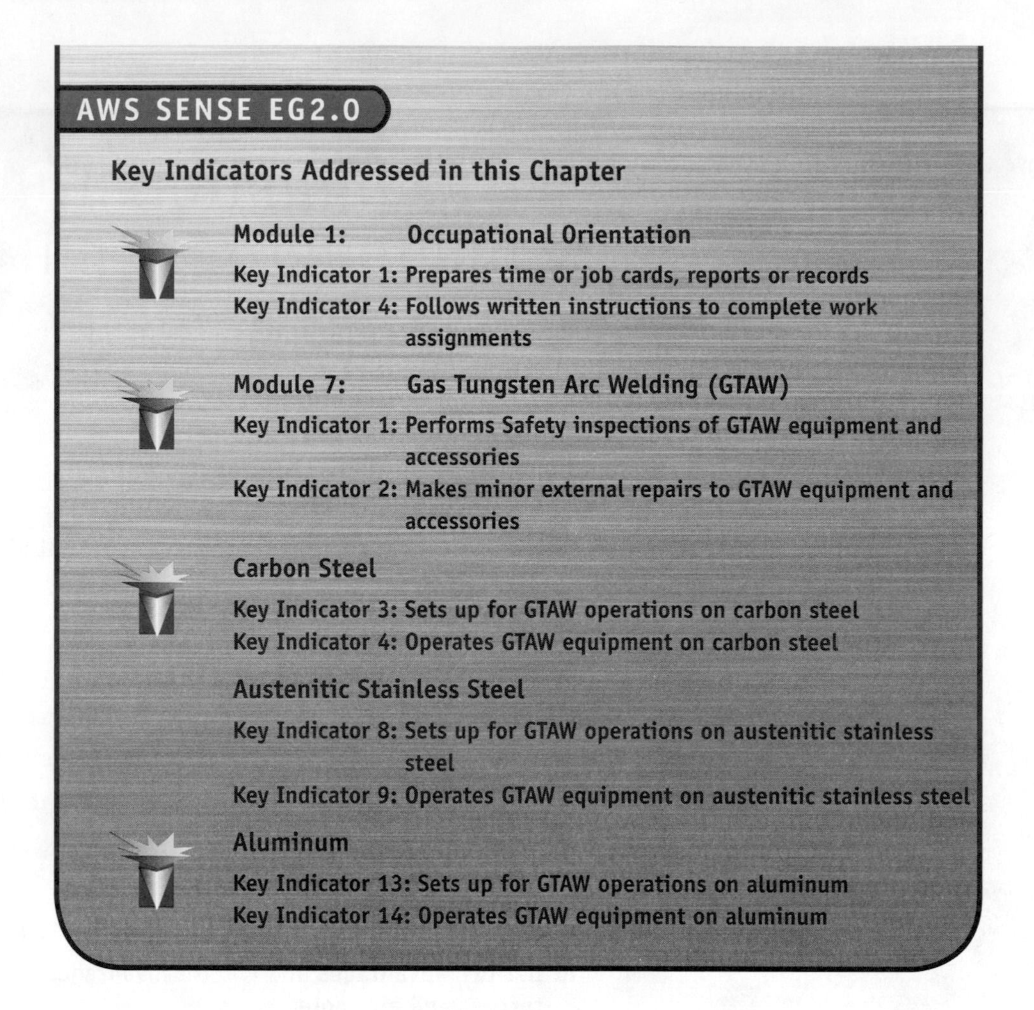

AWS SENSE EG2.0

Key Indicators Addressed in this Chapter

Module 1: Occupational Orientation

Key Indicator 1: Prepares time or job cards, reports or records

Key Indicator 4: Follows written instructions to complete work assignments

Module 7: Gas Tungsten Arc Welding (GTAW)

Key Indicator 1: Performs Safety inspections of GTAW equipment and accessories

Key Indicator 2: Makes minor external repairs to GTAW equipment and accessories

Carbon Steel

Key Indicator 3: Sets up for GTAW operations on carbon steel

Key Indicator 4: Operates GTAW equipment on carbon steel

Austenitic Stainless Steel

Key Indicator 8: Sets up for GTAW operations on austenitic stainless steel

Key Indicator 9: Operates GTAW equipment on austenitic stainless steel

Aluminum

Key Indicator 13: Sets up for GTAW operations on aluminum

Key Indicator 14: Operates GTAW equipment on aluminum

INTRODUCTION

The gas tungsten arc welding (GTAW) process is sometimes referred to as TIG, or heliarc. The term *TIG* is short for tungsten inert gas welding, and the much older term *heliarc* was used because helium was the first gas used for the process. The aircraft industry developed the GTAW process for welding magnesium during the late 1930s and the early 1940s. During that time, helium was the primary shielding gas used, along with DCEP welding current. These caused many of the problems that limited application of the GTA welding process. But improvements in gas composition and a better understanding of the importance of polarity improved the process's effectiveness and reduced its cost.

To use this process, an arc is established between a nonconsumable tungsten electrode and the base metal, which is called the work. Under the correct welding conditions, the electrode does not melt, although the work does at the spot where the arc impacts its surface and produces a molten weld pool. The filler metal is thin wire that is fed directly into the molten weld pool, where it melts. Since hot tungsten is sensitive to oxygen contamination, a good inert shielding gas is required to keep the air away from the hot tungsten and molten weld metal. The inert gas provides the needed arc characteristics and protects the molten weld pool. Because fluxes are not used, the welds produced are sound, free of slags, and as corrosion-resistant as the parent metal.

Before development of the GTAW process, welding aluminum and magnesium was difficult. The welds produced were porous and prone to corrosion.

When argon became plentiful and DCEN was recognized as more suitable than DCEP, the GTA process became more common. Until the development of Gas Metal Arc Welding (GMAW) in the late 1940s, GTAW was the only acceptable process for welding such reactive materials as aluminum, magnesium, titanium, and some grades of stainless steel, regardless of thickness. Reactive metals are ones that are easily contaminated when heated to their molten welding temperatures. Contamination can come from the air or can be picked up from surfaces containing oxides, oils, paints, or similar materials. Although economical for welding sheet metal, the process proved tedious and expensive for joining sections much thicker than 1/4 in. (6 mm). The eye-hand coordination required to make GTA welds is very similar to the coordination required for oxyfuel gas welding. GTA welding is often easier to learn when a person can oxyfuel gas weld.

TUNGSTEN

Tungsten, atomic symbol W, has the following properties:

- high tensile strength: 500,000 lb/in.^{2} (3447 kg/mm^2)
- hardness: Rockwell C45
- high melting temperature: 6170°F (3410°C)
- high boiling temperature: 10,700°F (5630°C)
- good electrical conductivity

Tungsten is produced mainly by reduction of its trioxide with hydrogen. Powdered tungsten is then purified to 99.95+%, compressed, and sintered (heated to a temperature below melting where grain growth can occur) to make an ingot. The ingot is heated to increase ductility and then is swaged and drawn through dies to produce electrodes. These electrodes are available in sizes varying from 0.01 in. to 0.25 in. (0.25 mm to 6 mm) in diameter. The tungsten electrode, after drawing, has a heavy black oxide that is later chemically cleaned or ground off.

The high melting temperature and good electrical conductivity make tungsten the best choice for a nonconsumable electrode. Its arc temperature, around 11,000°F (6000°C), is much higher than its melting temperature but not much higher than its boiling temperature of 10,700°F (5630°C).

As the tungsten electrode becomes hot, the arc between the electrode and the work will stabilize. Because electrons are more freely emitted from hot tungsten, the very highest temperature possible at the tungsten electrode tip is desired. Maintaining a balance between the heat necessary to have a stable arc and heat that is high enough to melt the tungsten requires an understanding of the GTA torch and electrode.

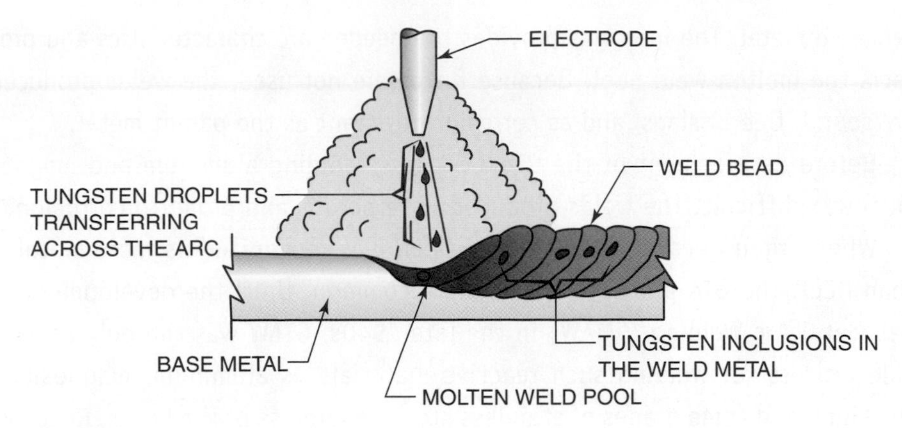

Figure 4.1
Some tungsten will erode from the electrode, be transferred across the arc, and become trapped in the weld deposit

The thermal conductivity of tungsten and the heat input are prime factors in the use of tungsten as an electrode. In general, tungsten is a good conductor of heat. This conductive property is what allows the tungsten electrode to withstand the arc temperature well above its melting temperature. The heat of the arc is conducted away from the electrode's end so fast that it does not reach its melting temperature. For example, a wooden match burns at approximately 3000°F (1647°C). Because aluminum melts at 1220°F (971°C), a match should easily melt an aluminum wire. However, a match will not even melt a 1/16-in. (2-mm) aluminum wire. The aluminum, like a tungsten electrode, conducts the heat away so quickly that it will not melt.

Because of the intense heat of the arc, some erosion of the electrode will occur. This eroded metal is transferred across the arc, **Figure 4.1.** Slow erosion of the electrode results in limited tungsten inclusions in the weld, which are acceptable. Standard codes give the size and amount of tungsten inclusions that are allowable in various types of welds. The tungsten inclusions are hard spots that cause stresses to concentrate, possibly resulting in weld failure. Although tungsten erosion cannot be completely eliminated, it can be controlled. A few ways of limiting erosion include:

- having good mechanical and electrical contact between the electrode and the collet
- using as low a current as possible
- using a water-cooled torch
- using as large a size of tungsten electrode as possible
- using DCEN current
- using as short an electrode extension from the collet as possible
- using the proper electrode end shape
- using an alloyed tungsten electrode

The torch end of the electrode is tightly clamped in a collet. The collet inside the torch is cooled by air or water. The **collet** is the cone-shaped sleeve that holds the electrode in the torch. Heat from both the arc and the tungsten electrode's resistance to the flow of current must be absorbed by the collet and torch. To ensure that the electrode is being cooled properly, be sure the collet connection is clean and tight. And for water-cooled torches, make sure water flow is adequate.

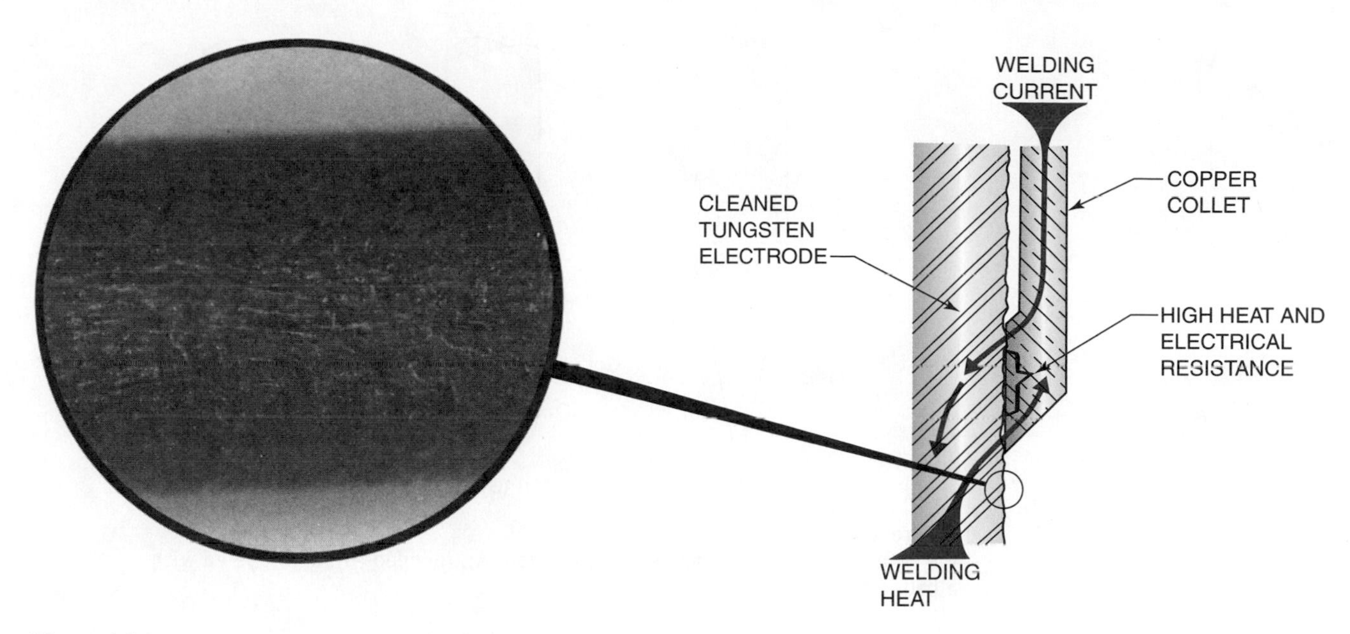

Figure 4.2
Irregular surface of a cleaned tungsten electrode (poor heat transfer to collet)
Photo Courtesy of Larry Jeffus

Collet-tungsten connection efficiency is shown in **Figure 4.2** and **Figure 4.3.**

Large-diameter electrodes conduct more current because the resistance heating effects are reduced. However, excessively large sizes may result in too low a temperature for a stable arc.

In general, the current-carrying capacity at DCEN is about ten times greater than that at DCEP.

The particular electrode tip shape impacts the temperature and erosion of the tungsten. With DCEN, a pointed tip concentrates the arc as much as possible and improves arc, starting with either a high-voltage electrical

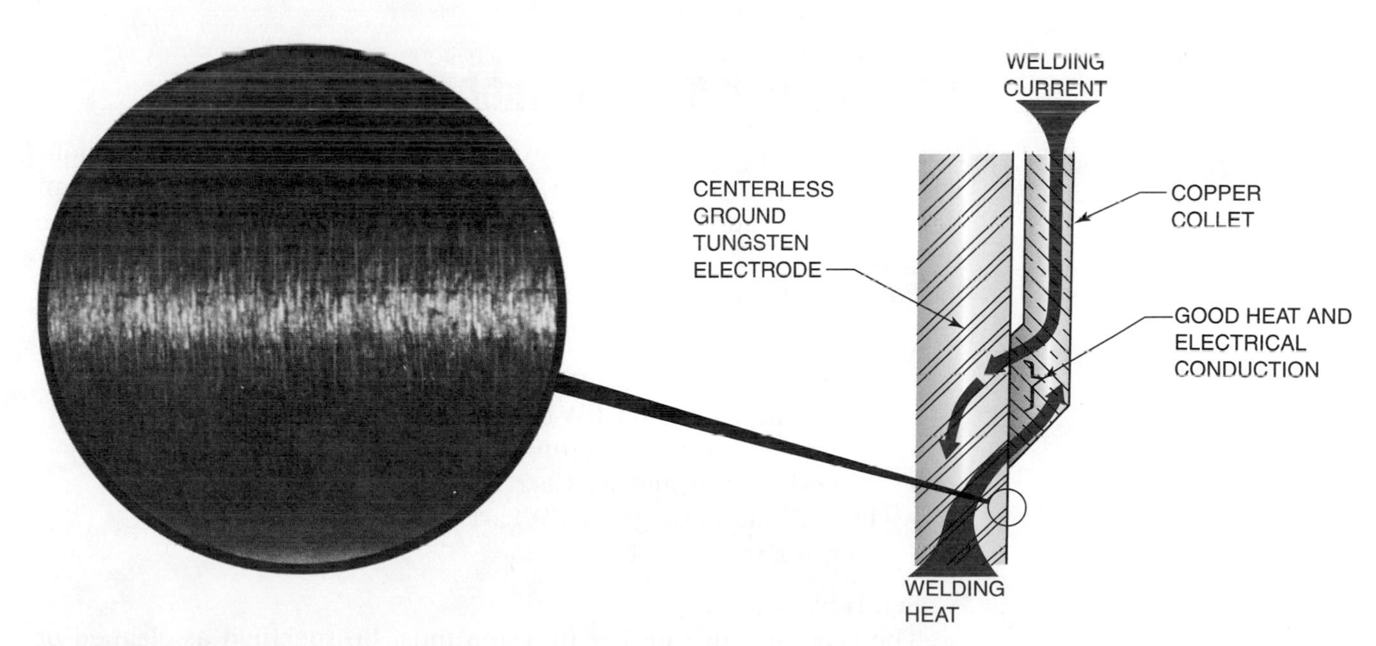

Figure 4.3
Smooth surface of a centerless ground tungsten electrode (good heat transfer to collet)
Photo Courtesy of Larry Jeffus

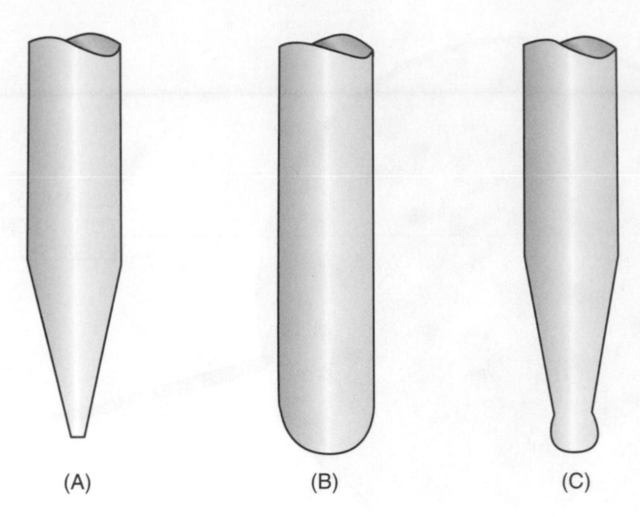

Figure 4.4
Basic tungsten electrode and shapes: pointed (A), rounded (B), and tapered with a balled end (C)

discharge or a touch start. Because DCEN does not put much heat on the tip, it is relatively cool, the point is stable, and it can survive extensive use without damage, **Figure 4.4A.**

With alternating current, the tip is subjected to more heat than with DCEN. To allow a larger mass at the tip to withstand the higher heat, the tip is rounded. The melted end must be small to ensure the best arc stability, **Figure 4.4B.**

DCEP has the highest heat concentration on the electrode tip. For this reason, a slight ball of molten tungsten is suspended at the end of a tapered electrode tip. The liquid tungsten surface tension, with the larger mass above the molten ball, holds it in place like a drop of water on your fingertip, **Figure 4.4C.**

TYPES OF TUNGSTEN ELECTRODES

Pure **tungsten** has a number of properties that make it an excellent nonconsumable electrode for the GTA welding process. These properties can be improved by adding cerium, lanthanum, thorium, or zirconium to the tungsten.

For GTA welding, tungsten electrodes are classified as the following:

- pure tungsten, EWP
- 1% thorium tungsten, EWTh-1
- 2% thorium tungsten, EWTh-2
- 1/4% to 1/2% zirconium tungsten, EWZr
- 2% cerium tungsten, EWCe-2
- 1% lanthanum tungsten, EWLa-1
- alloy not specified, EWG

See **Table 4.1.**

The type of finish on the tungsten must be specified as cleaned or ground. More information on composition and other requirements for tungsten welding electrodes is available in the AWS publication A5.12,

Table 4.1 Tungsten Electrode Types and Identification

AWS Classification	Tungsten Composition	Tip Color
EWP	Pure tungsten	Green
EWTh-1	1% thorium added	Yellow
EWTh-2	2% thorium added	Red
EWZr	1/4% to 1/2% zirconium added	Brown
EWCe-2	2% cerium added	Orange
EWLa-1	1% lanthanum added	Black
EWLa-1.5	1.5% lanthanum added	Gold
EWLa-2	2% lanthanum added	Blue
EWG	Alloy not specified	Grey

Specifications for Tungsten and Tungsten Alloy Electrodes for Arc Welding and Cutting.

Pure Tungsten Electrodes, EWP

Pure tungsten has the poorest heat resistance and electron emission characteristic of all the tungsten electrodes. It has a limited use with AC welding of metals, such as aluminum and magnesium, and is not recommended for use with inverter type welding power supplies. New alloying elements are now available with higher current carrying capacity; better emission characteristics and greater durability have replaced pure tungsten electrodes for AC welding in many shops.

Thoriated Tungsten Electrodes, EWTh-1 and EWTh-2

Thorium oxide (ThO_2), when added to tungsten in percentages of up to 0.6%, improves tungsten's current-carrying capacity. The addition of 1% to 2% of thorium oxide does not further improve current-carrying capacity. It does, however, help with electron emission. This can be observed by a reduction in the electron force (voltage) required to maintain an arc of a specific length. Thorium also increases the serviceable life of the tungsten. The improved electron emission of the thoriated tungsten allows it to carry approximately 20% more current. This also results in a corresponding reduction in electrode tip temperature, resulting in less tungsten erosion and subsequent weld contamination.

Thoriated tungsten also provides a much easier arc starting characteristic than pure or zirconiated electrodes. Thoriated tungsten works well with direct current electrode negative (DCEN). It can maintain a sharpened point well. It is very well suited for making welds on steel, steel alloys (including stainless), nickel alloys, and most other metals, other than aluminum or magnesium.

Thoriated electrodes do not work well with alternating current (AC). It is difficult to maintain a balled end, which used to be required and is still commonly used for AC welding. A thorium spike, **Figure 4.5,** may also develop on the end of a thorium electrode when it is overheated, disrupting a smooth arc. Furthermore, whether balled or sharpened to a point, when a thoriated tungsten electrode is overheated by exceeding its maximum amperage rating, small vaporized droplets of tungsten will transfer across the arc and into the work, possibly making the weld rejectable (see Figure 4.1).

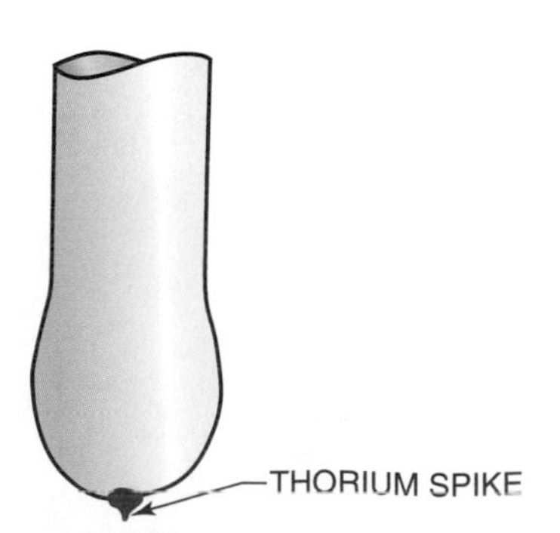

Figure 4.5
Thorium spike on a balled end tungsten electrode

When a thorium spike becomes visible on an electrode tip, it should be assumed that the electrode was overheated and that the weld may be contaminated by vaporized droplets.

CAUTION

Thorium is a very low-level radioactive oxide, but the level of radioactive contamination from a thorium electrode has not been found to be a health hazard during welding. It is, however, recommended that grinding dust be contained. Because of concern in other countries regarding radioactive contamination to the welder and welding environment, thoriated tungsten has been replaced with other alloys.

Zirconium Tungsten Electrodes, EWZr

Zirconium oxide (ZrO2) also helps tungsten emit electrons freely. The addition of zirconium to the tungsten has the same effect on the electrode characteristic as thorium, but to a lesser degree. Because zirconium tungsten is more easily melted than thorium tungsten, ZrO2 electrodes are recommended to be used with AC current. Because of the ease in forming the desired balled end on thorium versus zirconium tungsten, zirconium tungsten is normally the electrode chosen for AC welding of aluminum and magnesium alloys. Zirconiated tungsten is more resistant to weld pool contamination than pure tungsten, thus providing excellent weld qualities with minimal contamination.

Zirconiated tungsten also has the advantage over thoriated tungsten in that it is not radioactive.

Cerium Tungsten Electrodes, EWCe-2

Cerium oxide (CeO2) is added to tungsten to improve the current-carrying capacity in the same manner as does thorium. These electrodes were developed to replace thoriated tungsten electrodes because they are not made of a radioactive material. Cerium oxide electrodes have a current-carrying capacity similar to that of thoriated tungsten; however, they have an improved arc starting and arc stability characteristic, similar to that of thoriated tungsten. They can also provide a longer life than do most other electrodes, including thorium.

These electrodes have strong welder appeal for DC welding at low current settings; however, they are also used more and more with AC processes. In fact, ceriated electrodes along with lanthanum oxide electrodes are being recommended as a multipurpose electrode by the manufacturers of inverter based welding power supplies. With excellent arc starts at low amperages, ceriated tungsten has become more and more common in orbital tube and pipe manufacturing operations as well as with thin sheet and delicate work. Cerium electrodes contain approximately 2% cerium oxide.

Lanthanum Tungsten Electrodes, EWLa-1, EWLa-1.5, EWLa-2

Lanthanated tungsten is available as 1%, 1.5%, and 2%. Lanthanum trioxide has the lowest work function of any of the materials, thus it usually starts easiest and has the lowest temperature at the tip, which resists grain growth and promotes longer service life. Tests have proven that lanthanum electrodes last longer than do thorium electrodes when they are not overheated, and lanthanum electrodes are also resistant to thermal shock, which makes them a good choice for pulsed GTAW operations. Lanthanum electrodes require about 15% less power to initiate and maintain low current arcs. Lanthanum tungsten is a "rare earth" material and is not radioactive. Whereas lanthanated electrodes have been used successfully in Europe and Japan since 1993, lanthanum tungsten is a relative newcomer to the United States. This electrode has been primarily used for DC welding, but it is now being recommended, along with cerium electrodes, for AC welding, given its high current-carrying capacity and resistance to spitting during AC operations. The higher current-carrying capacity and resistance to spitting and spikes allow lanthanum and cerium electrodes to be used with a pointed tip prep on AC operations with aluminum alloys. The pointed tip allows for a more controllable arc and the ability to make smaller weld profiles in fillet joints. While slightly more expensive than thoriated electrodes, lanthanum can be used for just about any shop operation, thus eliminating confusion about electrode selection when multiple alloys are welded.

Alloy Not Specified, EWG

The EWG classification is for electrodes whose alloys have been modified by manufacturers. Such alloys have been developed and tested by manufacturers to meet specific welding criteria. The blend of alloying oxides are proprietary, meaning that specific alloy compositions are not normally available from manufacturers; however, they do provide welding characteristics for these electrodes.

SHAPING THE TUNGSTEN ELECTRODE

The desired end shape of tungsten can be obtained by grinding, breaking, remelting the end, or using chemical compounds. Tungsten is brittle and easily broken. Welders must be sure to make a smooth, square break where they want it to be located.

Grinding

A grinder or belt sander is often used to clean a contaminated tungsten electrode or to put a point at the end of the tungsten. The grinder or sander used to sharpen tungsten should have a fine, hard stone or a fine grit media for the belt (80-120 grit). It should be used for grinding tungsten only. Because of the hardness of the tungsten and its brittleness, the grinding stone chips off small particles of the electrode. A coarse grinding stone will result in more tungsten breakage and a poorer finish. If the grinder is used for metals other than tungsten, particles of these metals may become trapped on the tungsten as it is ground. The metal particles will quickly break free when the arc is started, resulting in contamination.

EXPERIMENT 4-1

Grinding the Tungsten to the Desired Shape

Module 1
Key Indicator 1, 4

Module 7
Key Indicator 2
Carbon Steel
Key Indicator 3
Austenitic Stainless Steel
Key Indicator 8
Aluminum
Key Indicator 13

Using an electric grinder with a fine grinding stone or a belt sander with 80-120 grit abrasive media, one piece of tungsten 2 in. (51 mm) long or longer, and safety glasses, you will grind a point on tungsten electrodes.

Because of the hardness of the tungsten, it will become hot. Its high thermal conductivity means that the heat will be transmitted quickly to your fingers. To prevent overheating, only light pressure should be applied

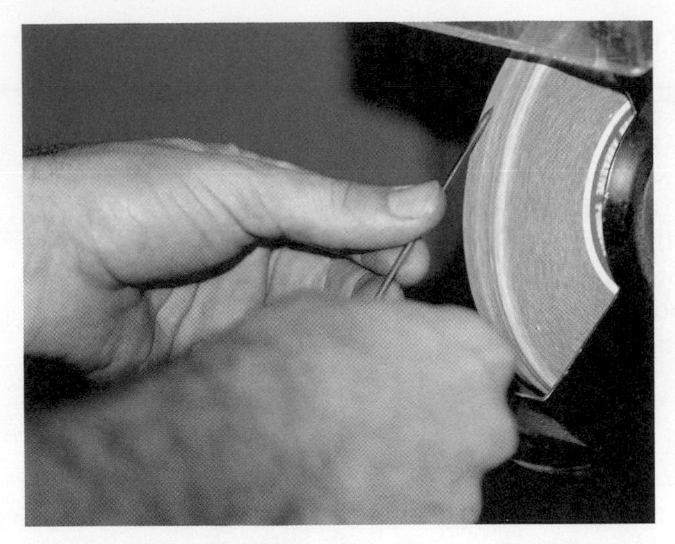

Figure 4.6
Correct method of grinding a tungsten electrode
Courtesy of Larry Jeffus

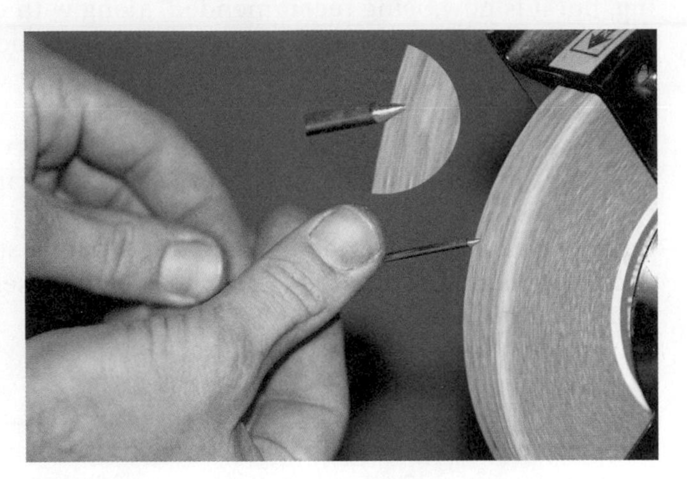

Figure 4.7
Incorrect method of grinding a tungsten electrode
Courtesy of Larry Jeffus

Note

If an electrode has a large aluminum deposit at the end from contact with the work or filler metal, the deposit should be broken off before grinding on a stone wheel because nonferrous materials like aluminum, brass, and copper have a tendency to smear onto the wheel, which may cause a dangerous out-of-balance condition in the wheel.

CAUTION

When holding one end of the tungsten against the grinding wheel, the other end of the tungsten must not be directed toward the palm of your hand, **Figure 4.8**. This will prevent the tungsten from being stuck into your hand if the grinding wheel catches it and suddenly pushes it downward.

against the grinding wheel or belt. This will also reduce the possibility of accidentally breaking the tungsten.

Grind the tungsten so that the grinding marks run lengthwise, **Figure 4.6** and **Figure 4.7.** Lengthwise grinding reduces the amount of small particles of tungsten contaminating the weld. Move the tungsten up and down as it is twisted during grinding. This will prevent the tungsten from becoming hollow-ground.

Complete a copy of the "Student Welding Report" listed in Appendix I or provided by your instructor.

Breaking and Remelting

Tungsten is hard but brittle, resulting in a low impact strength. If tungsten is struck sharply, it will break without bending. When it is held against a sharp corner and hit, a fairly square break will result. **Figure 4.9, Figure 4.10,**

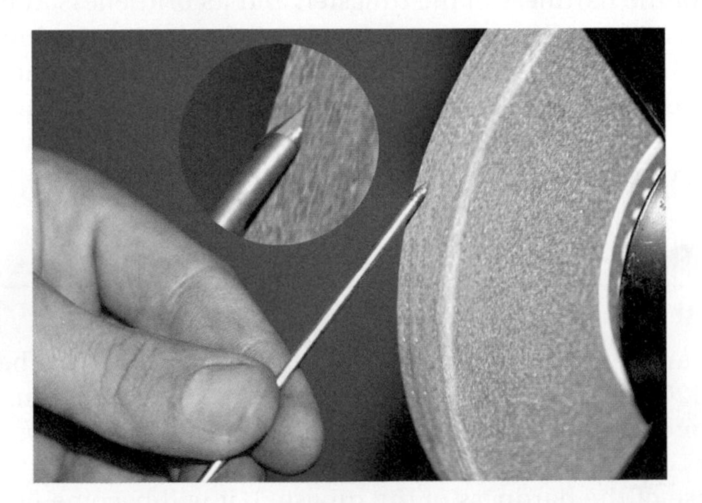

Figure 4.8
Correct way of holding a tungsten when grinding
Courtesy of Larry Jeffus

Figure 4.9
Breaking the contaminated end from a tungsten by striking it with a hammer
Courtesy of Larry Jeffus

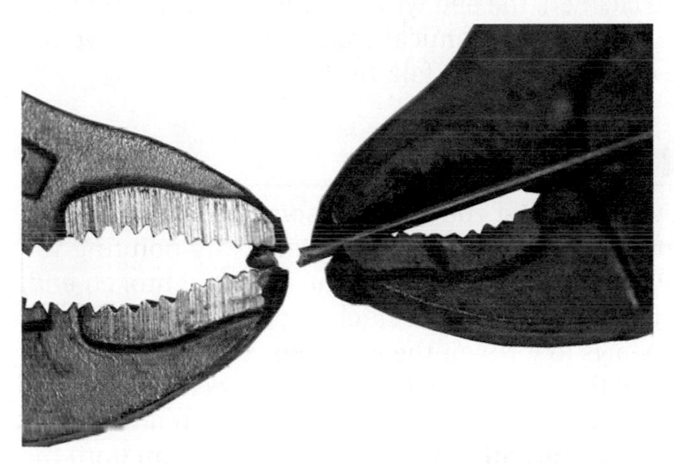

Figure 4.10
Correctly breaking the tungsten using two pairs of pliers
Courtesy of Larry Jeffus

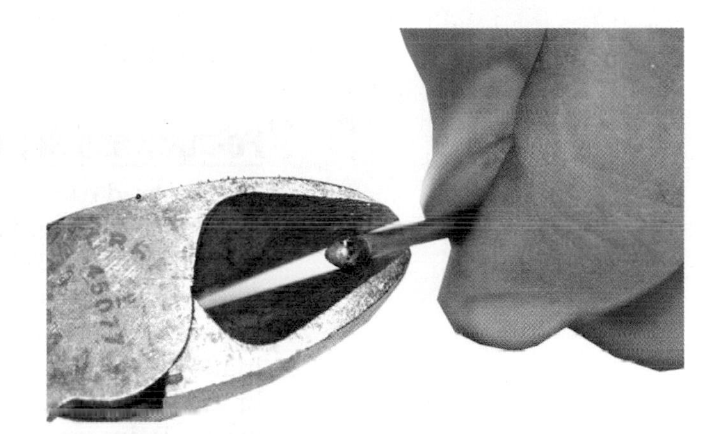

Figure 4.11
Using wire cutters to correctly break the tungsten
Courtesy of Larry Jeffus

and **Figure 4.11** show ways to break the tungsten correctly on a sharp corner using a hammer, with two pliers, and wire cutters.

Once the tungsten has been broken squarely, the end may be melted back so that it becomes somewhat rounded. The breaking and remelting technique is an older technique appropriate for pure tungsten and zirconiated tungsten electrodes due to their lower current-carrying capacity and inability to hold a pointed end. Cerium and lanthanum electrodes will maintain a pointed end much longer; therefore, this technique is rarely used with them unless the wider bead profile produced by a hemispherical tip prep is desired. This is accomplished by switching the welding current to DCEP and striking an arc under argon shielding on a piece of copper. If copper is not available, another piece of clean metal can be used. Do not use carbon, as it will contaminate the tungsten.

Note

Always wear eye protection as the end of the electrode can fly a good distance when struck.

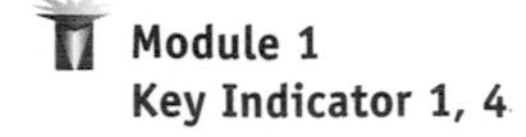

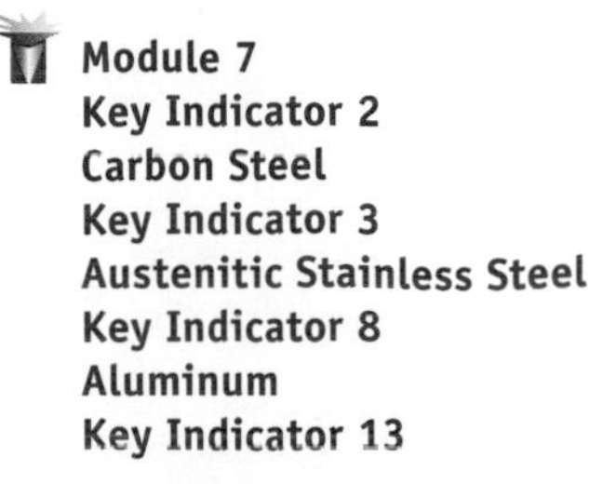

Module 1
Key Indicator 1, 4

Module 7
Key Indicator 2
Carbon Steel
Key Indicator 3
Austenitic Stainless Steel
Key Indicator 8
Aluminum
Key Indicator 13

EXPERIMENT 4-2

Removing a Contaminated Tungsten End by Breaking

Using short scrap pieces of tungsten, pliers or wire cutters, and a light machinist's hammer, you will break the end from the tungsten.

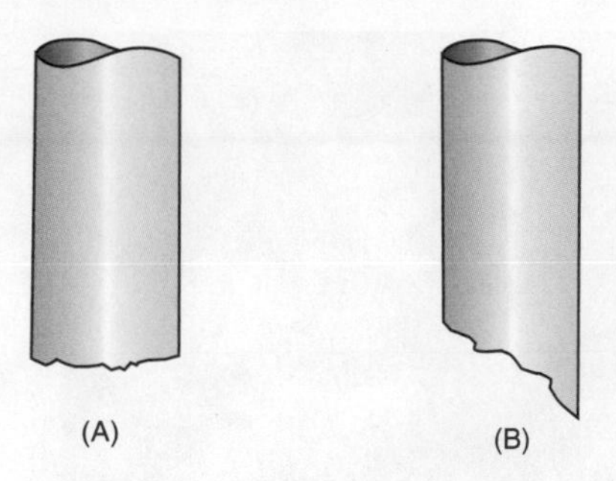

Figure 4.12
(A) Correctly broken tungsten electrode; (B) incorrectly broken tungsten electrode

Break about 1/4 in. (6 mm) from the end of the tungsten using the appropriate method, depending upon the diameter of the tungsten. Observe the break; it should be square and relatively smooth, **Figure 4.12.** The end of the tungsten should be broken only if the tungsten is badly contaminated.

Complete a copy of the "Student Welding Report" listed in Appendix I or provided by your instructor.

Chemical Cleaning and Pointing

The tungsten can be cleaned and pointed using one of several compounds. The tungsten is heated by shorting it against the work. The tungsten is then dipped in the compound, a strong alkaline, which rapidly dissolves the hot tungsten. The chemical reaction occurs so quickly that enough additional heat is produced to keep the tungsten hot, **Figure 4.13.** When the tungsten is removed, cooled, and cleaned, the end will be tapered to a fine point. If the electrode is contaminated, the chemical compound will dissolve the tungsten, allowing the contamination to fall free.

Pointing and Remelting

The tapered tungsten with a balled end, a shape sometimes used for AC or DCEP welding, is made by first grinding or chemically pointing the electrode. Using DCEP, as in the procedure for the remelted broken end, strike an arc on some copper under argon shielding and slowly increase the current until a ball starts to form on the tungsten. The ball should be made large enough so that the color of the end stays between dull red and bright red. If the color turns white, the ball is too small and should be made larger. To increase the size of the ball, simply apply more current until the end begins to melt. Surface tension will pull the molten tungsten up onto the tapered end. Lower the current and continue welding. DCEP is seldom used for welding. If the tip is still too hot, it may be necessary to increase the size of the tungsten.

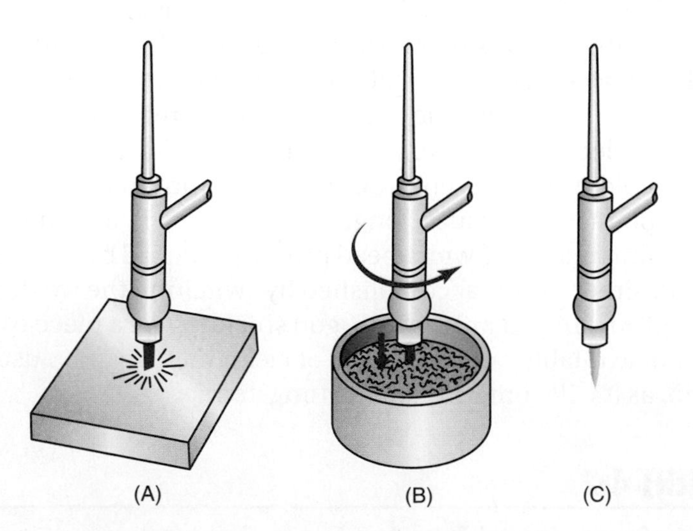

Figure 4.13
Chemically cleaning and pointing tungsten: (A) shorting the tungsten against the work to heat it to red hot, (B) inserting the tungsten into the compound and moving it around, and (C) cleaned and pointed tungsten ready for use

EXPERIMENT 4-3

Melting the Tungsten End Shape

Using a properly set up GTA welding machine, proper safety protection, one piece of copper or other clean piece of metal, and the tungsten that was sharpened and broken in Experiments 4-1 and 4-2, you will melt the end of the tungsten into the desired shape.

Properly install the tungsten, set the argon gas flow, switch the current to DCEP, and turn on the machine. Strike an arc on the copper and slowly increase the amperage. Watch the tungsten as it begins to melt and stop the current when the desired shape has been obtained, **Figure 4.14.**

Complete a copy of the "Student Welding Report" listed in Appendix I or provided by your instructor.

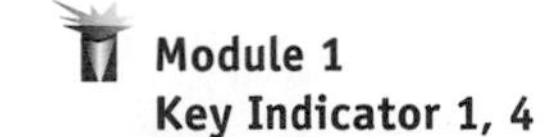

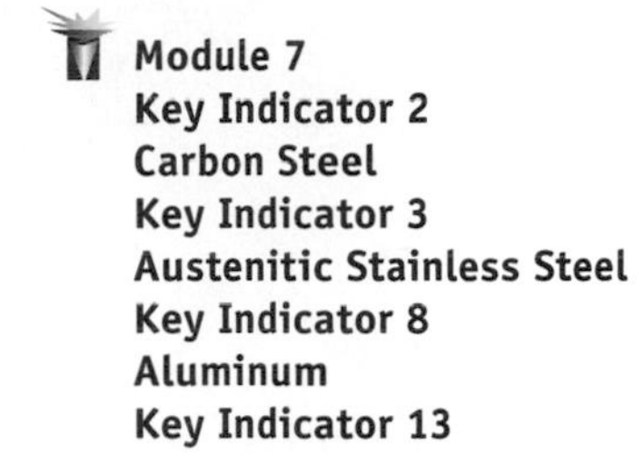

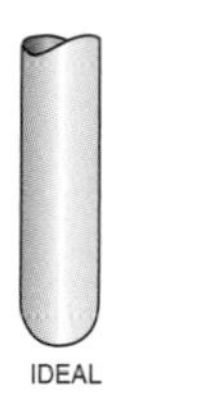
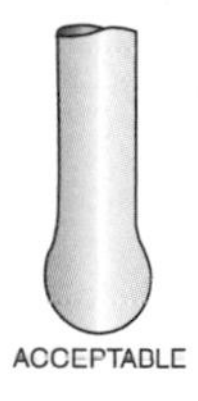
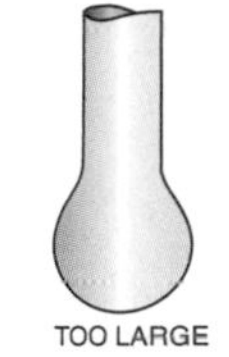

Figure 4.14
Melting the tungsten end shape

GTA WELDING EQUIPMENT

Figure 4.15 and **Figure 4.16** show two industrial applications of gas tungsten arc welding.

Torches

GTA welding torches are available water-cooled or air-cooled. The heat transfer efficiency for GTA welding may be as low as 20%. This means that 80% of the heat generated does not enter the weld. Much of this

Figure 4.15
Semiautomatic operation allows a stainless steel part to be GTA welded as it is turned past the torch
Courtesy of Lincoln Electric Company

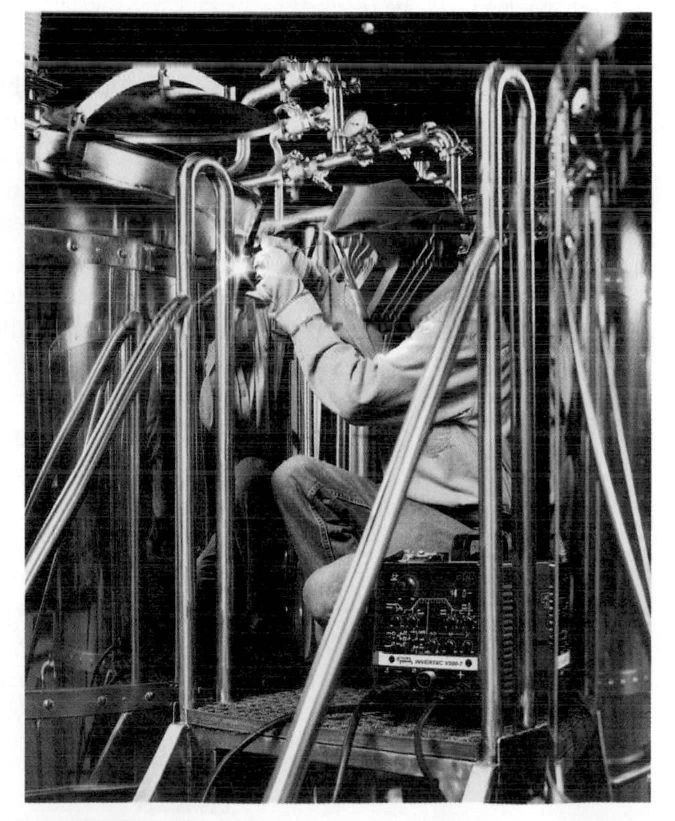

Figure 4.16
An operator GTA welds a cap ring on a pneumatic tank
Courtesy of Lincoln Electric Company

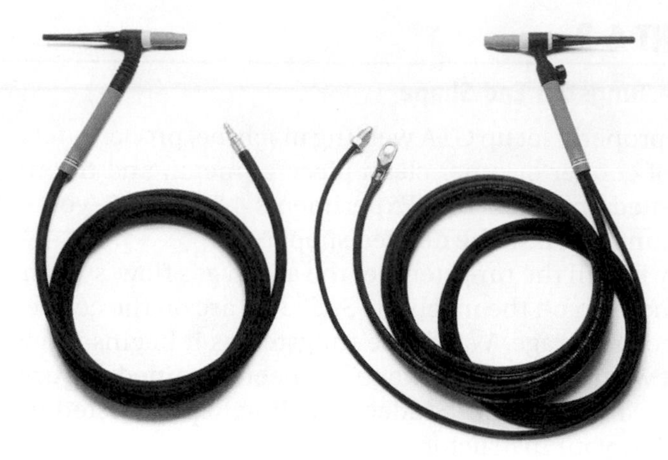

Figure 4.17
Power cable safety fuse
Courtesy of ESAB Welding & Cutting Products

> **Note**
>
> With the introduction of multipurpose electrode types, such as cerium and lanthanum, it is often no longer necessary to produce a balled end on an electrode when welding aluminum with alternating current. However, when making surfacing welds or buildups on aluminum castings, a hemispherical end preparation on an electrode will produce a wider arc and weld pool, which is desirable in these types of operations.

heat stays in the torch and must be removed by some type of cooling method.

The water-cooled GTA welding torch is more efficient at removing waste heat than is an air-cooled torch. The water-cooled torch, as compared to the air-cooled torch, operates at a lower temperature, resulting in a lower tungsten temperature and less erosion.

The air-cooled torch is more portable because it has fewer hoses, and it may be easier to manipulate than the water-cooled torch. Also, the water-cooled torch requires a water reservoir or other system to give the needed cooling. The cooling water system should contain some type of safety device, **Figure 4.17,** to make it possible to shut off the power if the water flow is interrupted. The power cable is surrounded by the return water to keep it cool so that a smaller-size cable can be used. Without the cooling water, the cable quickly overheats and melts through the hose.

The water can become stopped or restricted for a number of reasons, such as a kink in the hose, a heavy object set on the hose, or failure to turn on the system. Water pressures higher than 35 psi (241 kg/mm^2) may cause the water hoses to burst. When an open system is used, a pressure regulator must be installed to prevent pressures that are too high from damaging the hoses.

GTA welding torch heads are available in a variety of amperage ranges and designs, **Figure 4.18.** The amperage listed on a torch is the maximum rating and cannot be exceeded without possible damage to the torch. The various head angles allow better access in tight places. Some of the heads can be swiveled easily to new angles. The back cap that both protects and tightens the tungsten can be short or long. Short caps are beneficial for getting the torch into restricted areas. Long caps can accommodate a full-length electrode, which will carry more heat away from the tip and allow slightly more current to be used before the electrode overheats. **Figure 4.19** and **Figure 4.20.**

Hoses

A water-cooled torch has three hoses connecting it to the welding machine. The hoses are for shielding gas to the torch, cooling water to the

Figure 4.18
GTA welding torches
Courtesy of American Torch Tip

torch and cooling water return, and housing the power cables to the torch, **Figure 4.21.** Air-cooled torches may have one hose for shielding gas attached to the power cable, **Figure 4.22.**

The shielding gas hose must be plastic to prevent the gas from being contaminated. Rubber hoses contain oils that can be picked up by the gas, resulting in weld contamination.

The water-in hose may be made of any sturdy material. Water hose fittings have left-hand threads, and gas hose fittings have right-hand threads.

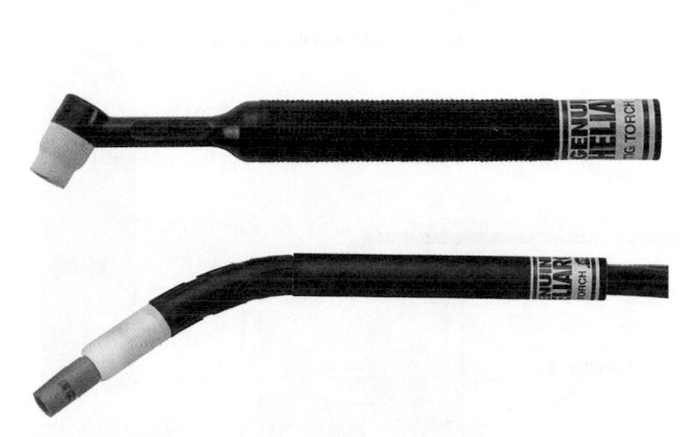

Figure 4.19
Short back caps are available for torches when space is a problem

Figure 4.20
Long back caps allow tungstens that are a full 7 in. (177 mm) long to be used
Courtesy of ESAB Welding & Cutting Products

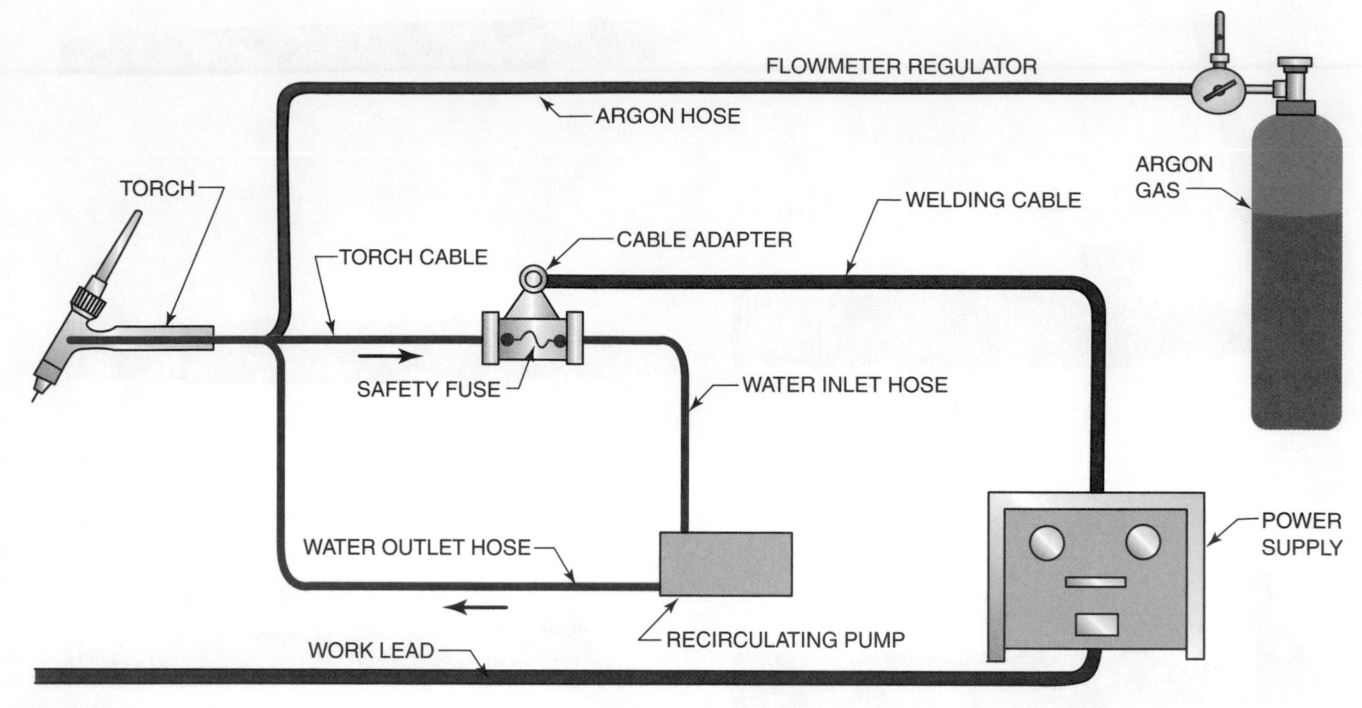

Figure 4.21
Schematic of a GTA welding setup with a water-cooled torch

This prevents the water and gas hoses from accidentally being reversed when they are attached to the welder. The return water hose also contains the welding power cable. This permits a much smaller-size cable to be used because the water keeps it cool.

The water must be supplied to the torch head and return around the cable. This allows the head to receive the maximum cooling from the water before the power cable warms it. Running the water through the torch first has another advantage. That is, when the water solenoid is closed, there is no water pressure in the hoses, which is particularly important. This fea-

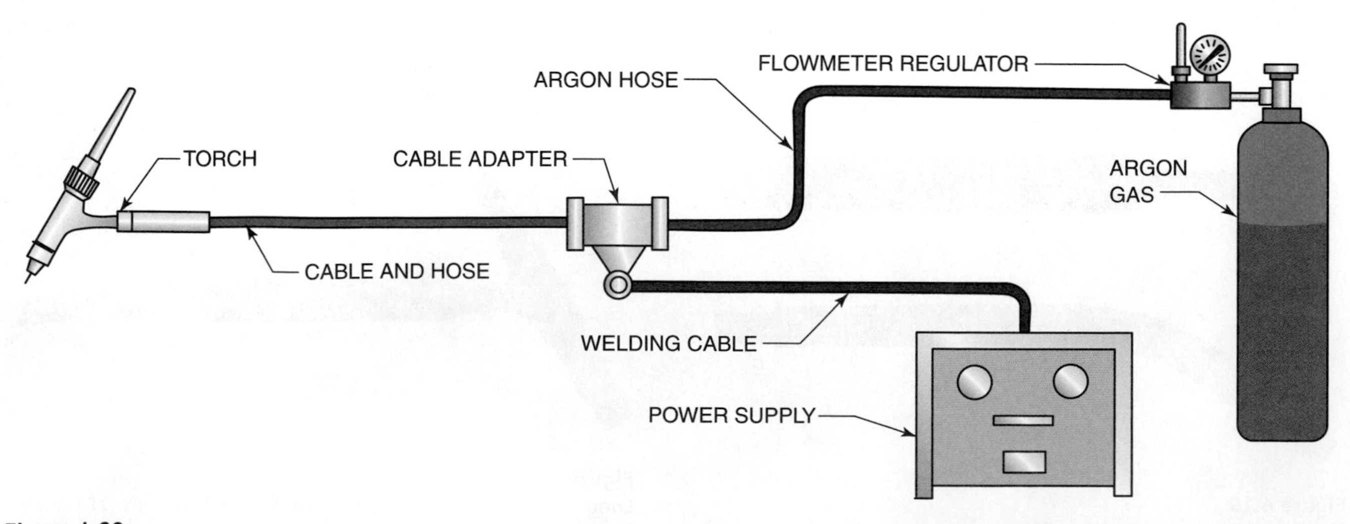

Figure 4.22
Schematic of a GTA welding setup with an air-cooled torch

Figure 4.23
Zip-on protective covering also helps keep the hoses neat
Courtesy of ESAB Welding & Cutting Products

Figure 4.24
A bracket holds the leads off the floor
Courtesy of Larry Jeffus

ture also prevents condensation in the torch. If a water leak should occur during welding, the welding power is stopped, closing the water solenoid and thus stopping the leak.

A protective covering can be used to prevent the hoses from becoming damaged by hot metal, **Figure 4.23.** Even with this protection, the hoses should be supported, **Figure 4.24,** so that they are not underfoot on the floor. By supporting the hoses, the chance of their being damaged by hot sparks is reduced.

Nozzles

The nozzle, or cup, is used to direct the shielding gas directly on the welding zone. The nozzle size is determined by the diameter of the opening and its length, **Table 4.2.** Nozzles may be made from ceramic, such as alumina or silicon nitride (opaque), or from fused quartz (clear). The nozzle may also have a gas lens to improve the gas flow pattern.

The nozzle size, both length and diameter, is often the welder's personal preference. Occasionally, a specific choice must be made based upon joint design or location. Small nozzle diameters allow the welder to better see the molten weld pool and can be operated with lower gas flow rates. Larger nozzle diameters can give better gas coverage, even in drafty places.

Ceramic nozzles are heat resistant and offer a relatively long life. The useful life of a ceramic nozzle is affected by the current level and prox-

Table 4.2 Recommended Cup Sizes

Tungsten Electrode Diameter		Nozzle Orifice Diameter	
in.	(mm)	in.	(mm)
1/16	(2)	1/4 to 3/8	(6 to 10)
3/32	(2.4)	3/8 to 7/16	(10 to 11)
1/8	(3)	7/16 to 1/2	(11 to 13)
3/16	(5)	1/2 to 3/4	(13 to 20)

imity to the work. Silicon nitride nozzles will withstand much more heat, resulting in a longer useful life.

The fused quartz (glass) used in a nozzle is a special type that can withstand the welding heat. These nozzles are no more easily broken than ceramic ones but are more expensive. The added visibility with glass nozzles in tight, hard-to-reach places is often worth the added expense.

The longer a nozzle, the longer the tungsten must be extended from the collet. This can cause higher tungsten temperatures, resulting in greater tungsten erosion. When using long nozzles, it is better to use low amperages or a larger-size tungsten.

Flowmeter

The **flowmeter** may be merely a flow regulator used on a manifold system or it may be a combination flow and pressure regulator used on an individual cylinder, **Figure 4.25** and **Figure 4.26.**

The flow is metered or controlled by opening a small valve at the base of the flowmeter. The rate of flow is then read in units of cfh (cubic feet per hour), or L/min (liters per minute). The reading is taken from a fixed scale that is compared to a small ball floating on the stream of gas. Meters from various manufacturers may be read differently. For example, they may read from the top, center, or bottom of the ball, **Figure 4.27.** The ball floats on top of the stream of gas inside a tube that gradually increases in diameter in the upward direction. The increased size allows

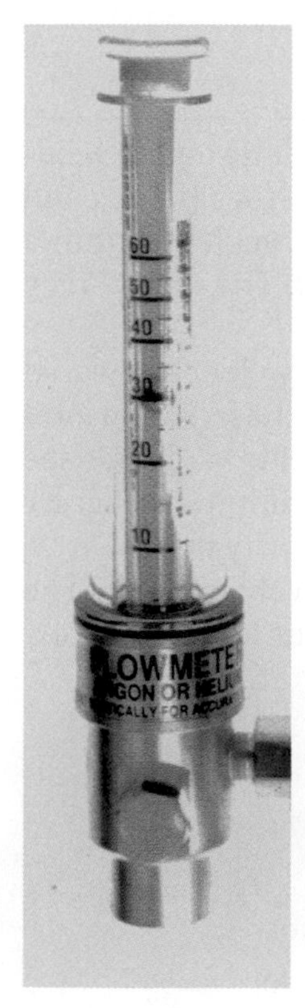

Figure 4.25
Flowmeter
Courtesy of Controls Corporation of America

Figure 4.26
Flowmeter regulator
Courtesy of Controls Corporation of America

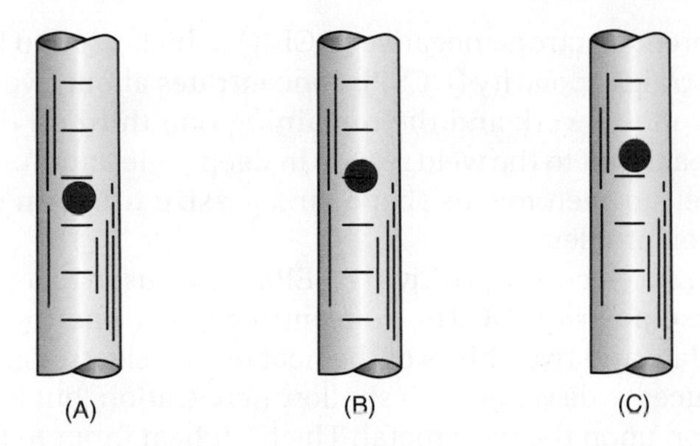

Figure 4.27
Three methods of reading a flowmeter (A) top of ball, (B) center of ball, and (C) bottom of ball

more room for the gas flow to pass by the ball. If the tube is not vertical, the reading is not accurate, but the flow is unchanged. Also, when using a line flowmeter, it is important to have the correct pressure. Changes in pressure will affect the accuracy of the flowmeter reading. In order to get accurate readings, be sure the gas being used is read on the proper flow scale. Less dense gases, such as helium and hydrogen, will not support the ball on as high a column with the same flow rate as a denser gas, such as argon.

TYPES OF WELDING CURRENT

All three types of welding current, or polarities, can be used for GTA welding. Each current has individual features that make it more desirable for specific conditions or with certain types of metals.

The major differences among the currents are in their heat distributions and the presence or degree of arc cleaning. **Figure 4.28** shows the heat distribution for each of the three types of currents.

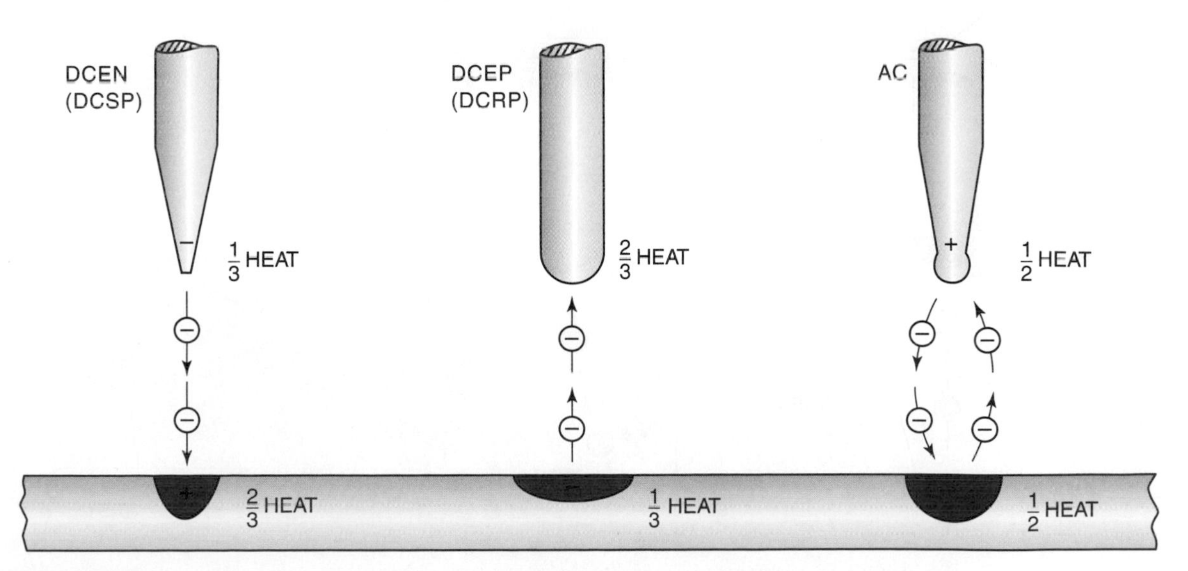

Figure 4.28
Heat distribution between the tungsten electrode and the work with each type of welding current

Direct current electrode negative (DCEN), which used to be called direct current straight polarity (DCSP), concentrates about two-thirds of its welding heat on the work and the remaining one-third on the tungsten. The higher heat input to the weld results in deep penetration. The low heat input into the tungsten means that a smaller-size tungsten can be used without erosion problems.

Direct current electrode positive (DCEP), which used to be called direct current reverse polarity (DCRP), concentrates only one-third of the arc heat on the plate and two-thirds of the heat on the electrode. This type of current produces wide welds with shallow penetration, but it has a strong cleaning action upon the base metal. The high heat input to the tungsten indicates that a large-size tungsten is required, and the end shape with a ball must be used. The low heat input to the metal and the strong cleaning action on the metal make this a good current choice for thin, heavily oxidized metals and magnesium. The metal being welded will not emit electrons as freely as does tungsten, so the arc may wander or be more erratic than DCEN. Because of the near molten state of the electrode tip during DCEP operations, it is almost always done in the flat position.

There are many theories as to why DCEP has a **cleaning action.** The most probable explanation is that the electrons accelerated from the cathode surface lift the oxides that interfere with their movement. The positive ions accelerated to the metal's surface provide additional energy. In combination, the electrons and ions cause the surface erosion needed to produce the cleaning. Although this theory is disputed, it is important to note that cleaning does occur, that it requires argon-rich shield gases and DCEP polarity, and that it can be used to advantage, **Figure 4.29.**

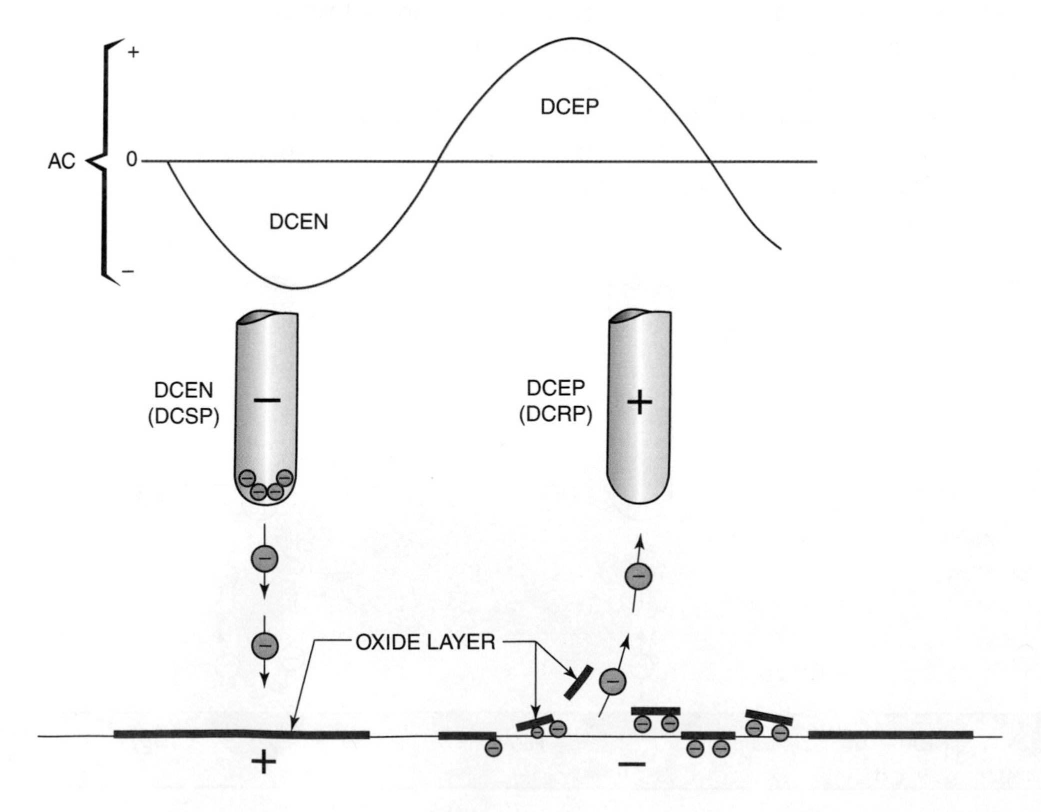

Figure 4.29
Electrons collect under the oxide layer during the DCEP portion of the cycle and lift the oxides from the surface

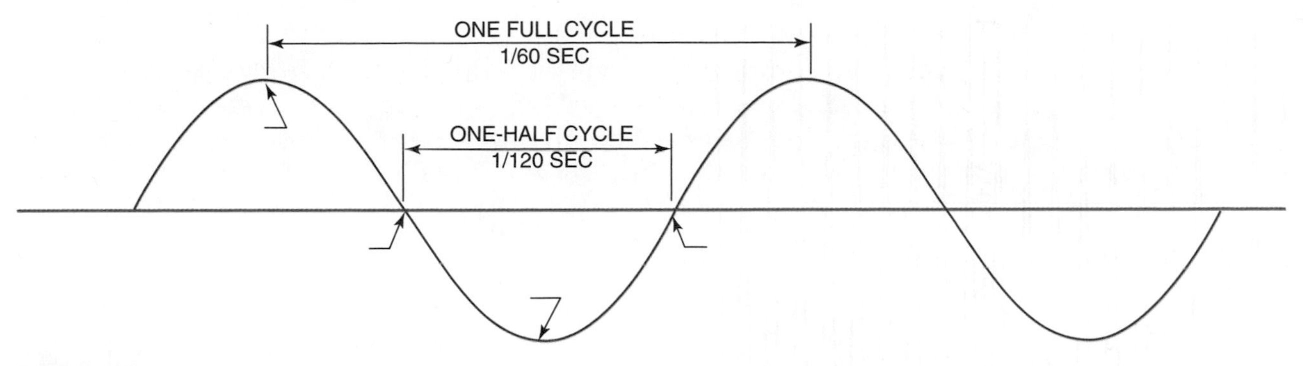

Figure 4.30
Sine wave of alternating current at 60 cycle

Alternating current (AC) concentrates about half of its heat on the work and the other half on the tungsten. Alternating current is DCEN half of the time and DCEP the other half of the time. The **frequency** at which the current cycles is the rate at which it makes a full change in direction, **Figure 4.30.** In the United States, the electric current cycles at the rate of 60 times per second, or 60 hertz (60 Hz). Referring again to Figure 4.30, the current is at its maximum peak at points A and B. The rate gradually decreases until it stops at points C and D. The arc at these points is extinguished and, as the current reversal begins, must be reestablished. This event requires the emission of electrons from the cathode to ionize the shielding gas. When the hot, emissive electrode becomes the cathode, reestablishing the arc is easy. However, it is often quite difficult to reestablish the arc when the colder and less emissive workpiece becomes the cathode. Because voltage from the power supply is designed to support a relatively low voltage arc, it may be insufficient to initiate electron flow. When the arc does not reignite consistently, it becomes destabilized and can cause poor welding performance. This phenomenon is called **rectification.** Thus, a voltage assist from another source is needed. A high-voltage but low-current **spark gap oscillator** commonly provides the assist at a relatively low cost. The high frequency ensures that a voltage peak will occur reasonably close to the current reversal in the welding arc, creating a low-resistance ionized path for the welding current to follow, **Figure 4.31A** and **Figure 4.31B.** This same device is often used to initiate direct current arcs, a particularly useful technique for mechanized welding.

The high-frequency current is established by capacitors discharging across a gap set on points inside the machine. Changing the point gap setting will change the frequency of the current. The closer the points are, the higher the frequency; the wider the spacing between the points, the lower the frequency. The voltage is stepped up with a transformer from the primary voltage supplied to the machine. The available amperage to the high-frequency circuit is very low. Thus, when the circuit is complete, the voltage quickly drops to a safe level. The high frequency is induced on the primary welding current in a coil.

The high frequency may be set so that it automatically cuts off after the arc is established, when welding with DC. It is kept on continuously with AC and transformer rectifier power supplies. When used in this manner, it is referred to as alternating current, high-frequency stabilized, or ACHF. Most of the newer inverter power supplies have advanced circuitry that

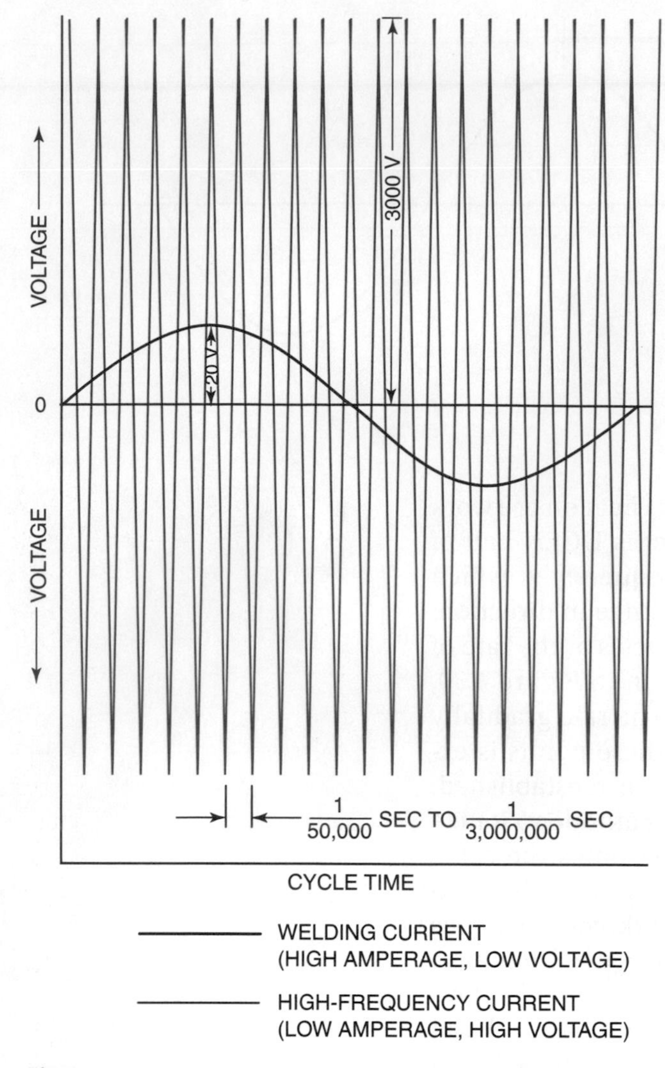

Figure 4.31A
High-frequency arc starting current shown over the low-frequency welding current

Figure 4.31B
The high frequency first appears as a blue glow around the tungsten before the welding current starts its arc
Courtesy of Larry Jeffus

switches between DCEN and DCEP so quickly that rectification cannot occur and high frequency is not used.

AC BALANCE CONTROL

While older transformer rectifier welding power supplies can only produce AC power in the form of a sine wave (Figure 4.30), newer transformer rectifiers and inverter power supplies can provide enhanced AC output by controlling the dwell time spent on each side of the AC cycle.

The two additional AC waveforms are called square wave and advanced square wave. **Figure 4.32** shows a comparison of the three waveforms. The waveforms represent how the current is measured in amplitude and time as the current moves to a high point, changes direction and polarity, and crosses over the zero point while moving to the maximum amperage on the other side. When the current has moved up and down, hitting both maximum amperages, this represents one cycle. AC electrical power in the United States is delivered at 60 cycles per second.

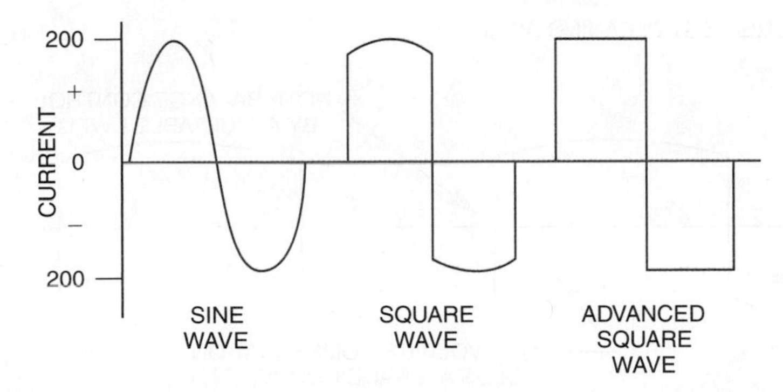

Figure 4.32
Comparison of the three different AC waveforms all representing a time balanced condition and operating at 200 amperes

Welding machines with square wave capabilities have a knob or digital readout that controls the amount of time the current spends on the EN and EP half of the AC cycle. If the machine is adjusted to have more dwell time spent on the EN side, the result will be deeper penetration and reduced cleaning action. The dial usually says "max penetration," **Figure 4.33.** An additional benefit of an AC arc balanced to favor EN is that the extra current delivered to the work will reduce the heat at the tip of the tungsten, which may allow a smaller diameter electrode, or even a pointed end, to be used.

> **Note**
>
> Increased EP balance is not a substitute for excellent surface prep, **Figure 4.34.**

If the machine is adjusted to have more dwell time spent on the EP side of the half cycle, the result will be a wider, shallower bead profile with greater cleaning action. A machine setting favoring more dwell time on EP (called "max cleaning") will be a benefit when it is impractical or impossible to completely remove small amounts of surface contaminants or when welding on aluminum castings. The extra EP will help to etch away more or thicker oxides. When balance controls are set toward "max cleaning," a balled end will be required on the electrode, and if the ball becomes larger than 1-1/2 times the electrode diameter, the next larger size should be selected (Figure 4.14).

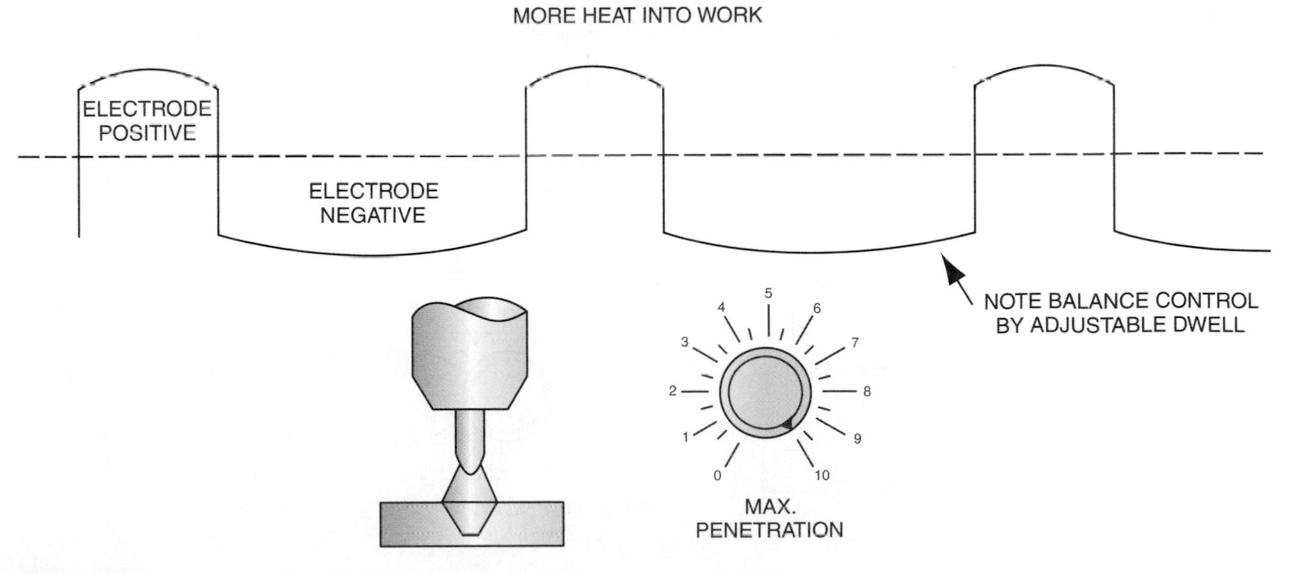

Figure 4.33
Maximum penetration balance control setting

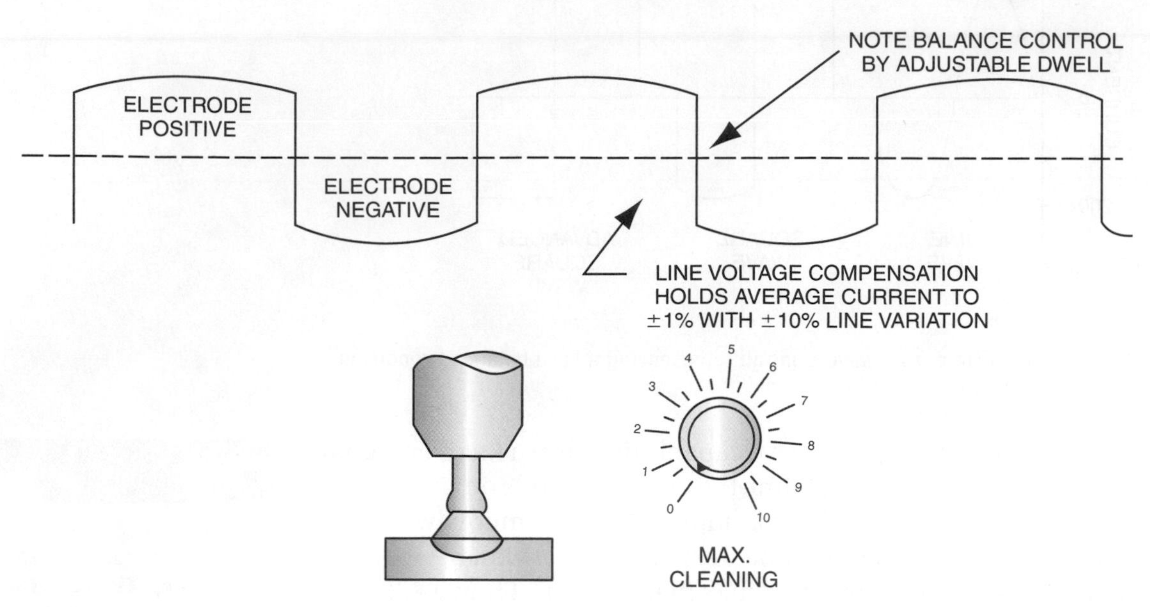

Figure 4.34
Maximum cleaning control setting

When the control knob is set in-between max cleaning and max penetration, it is called a balanced waveform. This is a good place to start for general AC applications, **Figure 4.35.** Balance control can be increased toward max EN if greater penetration or a pointed electrode is required; however, operators must be careful not to turn the balance control too far toward EN. When this happens, the oxides on top of the aluminum will not be completely removed and poor fusion or porosity may result. It is also important to note that transformer rectifiers, when set at max penetration, have 68% dwell time on EN, while advanced square wave inverter power supplies may produce as much as 99% dwell time on EN. In essence, this

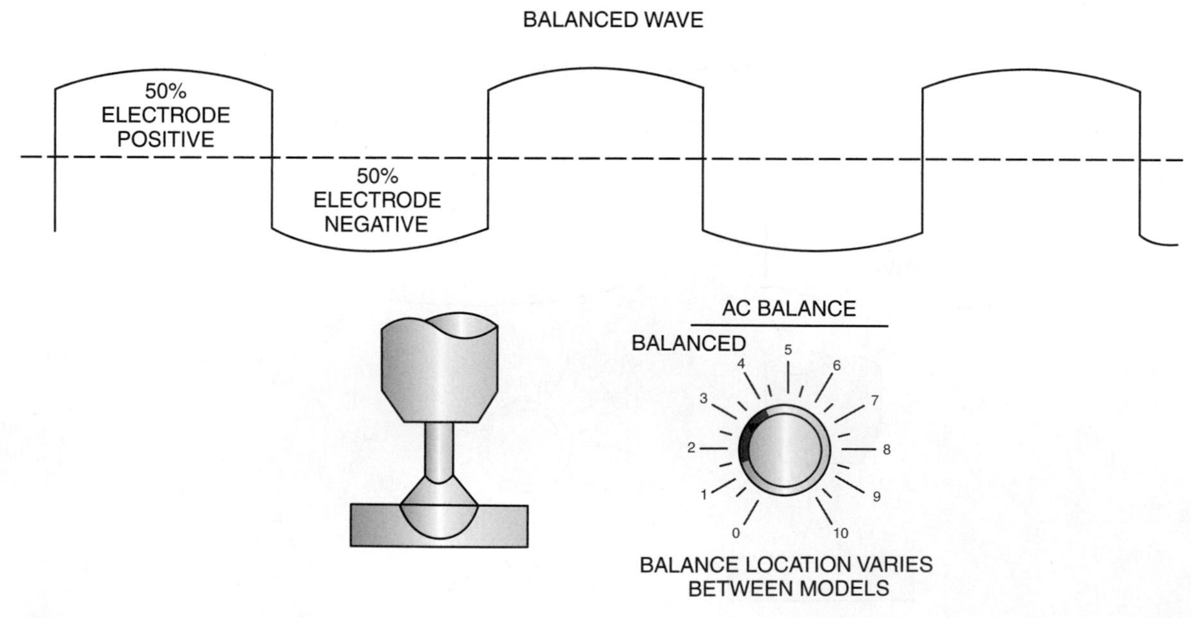

Figure 4.35
Balanced control setting

means that when a transformer square wave is set at 10 on its balance control dial, it delivers about the same amount of EN as when an advanced square wave inverter is set to 7.

If you have a square wave power supply, try adjusting the balance control toward max cleaning and max penetration when working on aluminum practices, and note the results in etching at the weld toes and the overall bead profile.

CAUTION

Never allow noninert gases, such as O_2, CO_2, or N, to come in contact with your inert gas system. Very small amounts can contaminate the inert gas, which may result in the weld failing.

SHIELDING GASES

The shielding gases used for the GTA welding process are argon (Ar), helium (He), hydrogen (H), nitrogen (N), or a mixture of two or more of these gases. The purpose of the shielding gas is to protect the molten weld pool and the tungsten electrode from the harmful effects of air. The shielding gas also affects the amount of heat produced by the arc and the resulting weld bead appearance.

Argon and helium are **inert gases.** This means that they will not combine chemically with any other material. Argon and helium may be found in mixtures but never as compounds. Because they are inert, they will not affect the molten weld pool in any way.

Argon

Argon is a by-product in air separation plants. Air is cooled to temperatures that cause it to liquefy; then its constituents are fractionally distilled. The primary products are oxygen and nitrogen. Before these gases were produced on a tonnage scale, argon was a rare gas. Now it is distributed in cylinders as gas or in bulk as a liquid.

Because argon is denser than air, it effectively shields welds in deep grooves in the flat position. However, this higher density can be a hindrance when welding overhead because higher flow rates are necessary. The argon is relatively easy to ionize and thus suitable for alternating current applications and easier starts. This property also permits fairly long arcs at lower voltages, making it virtually insensitive to changes in arc length. Argon is also the only commercial gas that produces the cleaning discussed earlier. These characteristics are most useful for manual welding, especially with filler metals added, as shown in **Figure 4.36.**

Figure 4.36
Highly concentrated ionized argon gas column
Courtesy of Larry Jeffus

Helium

Helium is a by-product of the natural gas industry. It is removed from natural gas as the gas undergoes separation (fractionation) for purification or refinement.

Helium offers the advantage of deeper penetration. The arc force with helium is sufficient to displace the molten weld pool with very short arcs. In some mechanized applications, the tip of the tungsten electrode is positioned below the workpiece surface to obtain very deep and narrow penetration. This technique is especially effective for welding aged aluminum alloys prone to overaging. It is also very effective at high welding speeds, as for tube mills. However, helium is less forgiving for manual welding. With helium, penetration and bead profile are sensitive to arc length, and the long arcs that are needed for feeding filler wires are more difficult to control.

Helium has been mixed with argon to gain the combined benefits of cathode cleaning and deeper penetration, particularly for manual welding. The most common of these mixtures is 75% helium and 25% argon.

Although the GTA process was developed with helium as the shielding gas, argon is now used whenever possible because it is much cheaper. Helium also has some disadvantages. Because it is lighter than air, helium doesn't allow for good shielding. Its flow rates must be about twice as high as argon's for acceptable stiffness in the gas stream, and proper protection is difficult in drafts unless high flow rates are used. It is difficult to ionize, necessitating higher voltages to support the arc and making the arc more difficult to ignite. Alternating current arcs are very unstable. However, helium is not used with alternating current because the cleaning action does not occur.

Hydrogen

Hydrogen is not an inert gas and is not used as a primary shielding gas. However, it can be added to argon in small amounts (1–3%) when deep penetration and high welding speeds are needed. It also improves the weld surface cleanliness and bead profile on some grades of stainless steel that are very sensitive to oxygen. Hydrogen additions are restricted to stainless steel because hydrogen is the primary cause of porosity in aluminum welds. It can cause porosity in carbon steel and, in highly restrained welds, underbead cracking in carbon and low alloy steel.

Nitrogen

Nitrogen is not an inert gas. Like hydrogen, nitrogen has been used as an additive to argon. But it cannot be used with some materials, such as ferritic steel, because it produces porosity. In other cases, such as with austenitic stainless steel, nitrogen is useful as an austenite stabilizer in the alloy. It is used to increase penetration when welding copper. Nitrogen is also sometimes used as a backing gas on some austenitic stainless steel pipe and tubing because of its low relative cost. When using nitrogen to protect the back side of a weld, care must be taken to be sure that all fitups are tight and that no nitrogen mixes with the shield gas at the top of the weld. Unfortunately, because of the general success with inert gas mix-

Table 4.3 Postwelding Gas Flow Times

Electrode Diameter		Postwelding Gas Flow Time*
in.	(mm)	
.01	(0.25)	5 sec
0.02	(0.5)	5 sec
0.04	(1)	5 sec
1/16	(2)	8 sec
3/32	(2.4)	10 sec
1/8	(3)	15 sec
5/32	(4)	20 sec
3/16	(5)	25 sec
1/4	(6)	30 sec

*The time may be longer if either the base metal or the tungsten electrode does not cool below the rapid oxidation temperatures within the postflow times shown.

The **postflow time** is the time during which the gas continues flowing after the welding current has stopped. This period serves to protect the molten weld pool, the filler rod, and the tungsten electrode as they cool to a temperature at which they will not oxidize rapidly. The time of the flow is determined by the welding current and the tungsten size, **Table 4.3.**

Gas Flow Rate

The shielding gas flow rate is measured in cubic feet per hour (cfh) or in metric measure as liters per minute (L/min). The rate of flow should be as low as possible and still give adequate coverage. High gas flow rates waste shielding gases and may lead to contamination. The contamination comes from turbulence in the gas at high flow rates. Air is drawn into the gas envelope by a venturi effect around the edge of the nozzle. Also, the air can be drawn in under the nozzle if the torch is held at too sharp an angle to the metal, **Figure 4.39.**

The larger the nozzle size, the higher is the flow rate permissible without causing turbulence. **Table 4.4** shows the average and maximum flow rates for most nozzle sizes. A gas lens can be used in combination with the nozzle to stabilize the gas flow, thus eliminating some turbulence. A gas lens will add to the turbulence problem if there is any spatter or contamination on its surface.

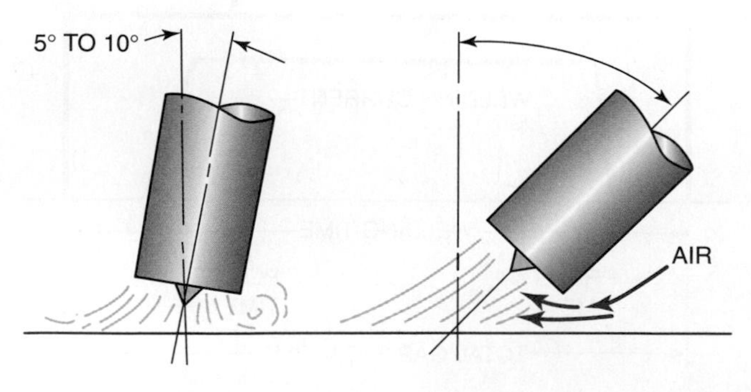

Figure 4.39
Too steep an angle between the torch and work may draw in air

Table 4.4 Suggested Argon Gas Flow Rate for Given Cup Sizes

Nozzle Inside Diameter		Gas Flow*	
in.	(mm)	cfh	(L/min)
1/4	(6)	10–14	(4.7–6.6)
5/16	(8)	11–15	(5.2–7.0)
3/8	(10)	12–16	(5.6–7.5)
7/16	(11)	13–17	(6.1–8.0)
1/2	(13)	17–20	(8.0–9.4)
5/8	(16)	17–20	(8.0–9.4)

*The flow rates may need to be increased or decreased depending upon the conditions under which the weld is to be performed.

REMOTE CONTROLS

A remote control can be used to start the weld, increase or decrease the current, and stop the weld. The remote control can be either a foot-operated or hand-operated device. The foot control works adequately if the welder can be seated. Welds that must be performed away from a welding station may use a hand or thumb control, or may not have any remote welding controls.

Most remote controls have an on-off switch that is activated at the first or last part of the control movement. A variable resistor increases the current as the control is pressed more. A variable resistor works in a manner similar to the accelerator pedal on a car to increase the power (current), **Figure 4.40.** The operating amperage range is determined by the value that has been set on the main controls of the machine.

Figure 4.40
A foot-operated device can be used to increase the current
Courtesy of Larry Jeffus

EXPERIMENT 4-4

Setting Up a GTA Welder

Using a GTA welding machine; remote control welding torch; gas flowmeter; gas source (cylinder or manifold); tungsten; nozzle; collet; collet body; cap; and any other hoses, special tools, and required equipment, you will set up the machine for GTA welding, **Figure 4.41.**

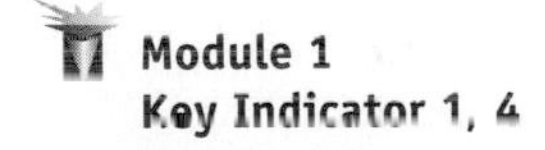

Module 1
Key Indicator 1, 4

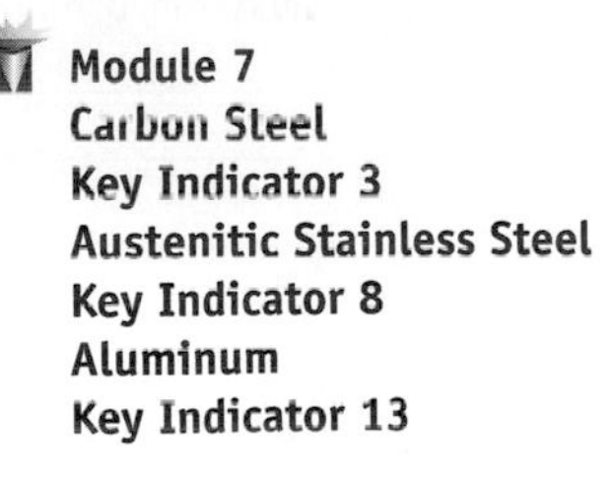

Module 7
Carbon Steel
Key Indicator 3
Austenitic Stainless Steel
Key Indicator 8
Aluminum
Key Indicator 13

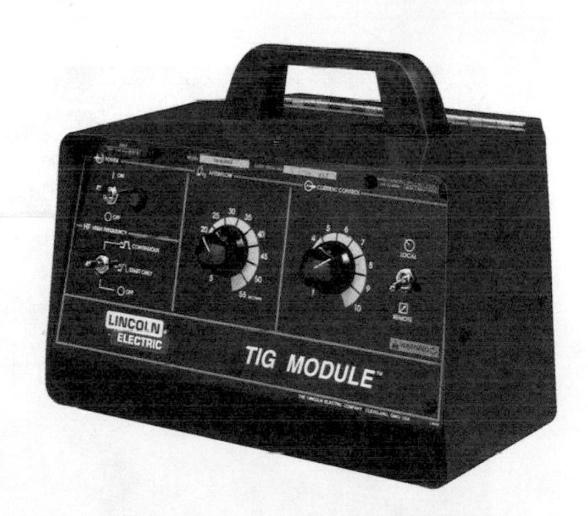

Figure 4.41
GTA welding unit that can be added to a standard power supply so that it can be used for GTA welding
Courtesy of Lincoln Electric Company

Figure 4.42
Always be sure the power is off when making machine connections
Courtesy of Larry Jeffus

Figure 4.43
Tighten each fitting as it is connected to avoid missing a connection
Courtesy of Larry Jeffus

1. Start with the power switch off, **Figure 4.42.** Use a wrench to attach the torch hose to the machine. The water hoses should have left-hand threads to prevent incorrectly connecting them. Tighten the fittings only as tightly as needed to prevent leaks, **Figure 4.43.** Attach the cooling water "in" to the machine solenoid and the water "out" to the power block.
2. The flowmeter or flowmeter regulator should be attached next. If a gas cylinder is used, secure it in place with a safety chain. Then remove the valve protection cap and crack the valve to blow out any dirt, **Figure 4.44.** Attach the flowmeter so that the tube is vertical.
3. Connect the gas hose from the meter to the gas "in" connection on the machine.
4. With both the machine and main power switched off, turn on the water and gas so that the connection to the machine can be checked for leaks. Tighten any leaking fittings to stop the leakages.

Figure 4.44
During transportation or storage, dirt may collect in the valve. Cracking the valve is the best way to remove any dirt.

Figure 4.45
Setting the current
Courtesy of Larry Jeffus

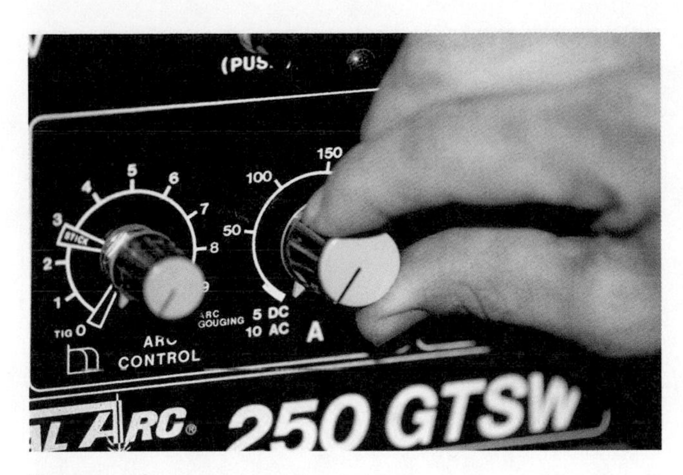

Figure 4.46
Setting the amperage range
Courtesy of Larry Jeffus

Figure 4.47
The high-frequency switch should be placed in the appropriate position
Courtesy of Larry Jeffus

5. Turn on both the machine and main power switches and watch for leaks in the torch hoses and fittings.
6. With the power off, switch the machine to the GTA welding mode.
7. Select the desired type of current and amperage range, **Figure 4.45** and **Figure 4.46.**
8. Set the fine current adjustment to the proper range, depending upon the size of tungsten used, **Table 4.5.**
9. Place the high-frequency switch in the appropriate position, auto (HF start) for DC or continuous for AC, **Figure 4.47.**
10. The remote control can be plugged in and the selector switch set, **Figure 4.48.**
11. The collet and collet body should be installed on the torch first, **Figure 4.49.**
12. On the Linde brand or copies of Linde torches, installing the back cap first will stop the collet body from being screwed into the torch fully. A poor connection will result in excessive electrical and thermal resistance, causing a heat buildup in the head.
13. The tungsten can be installed and the end cap tightened to hold the tungsten in place. Select and install the desired nozzle size. Adjust the tungsten length so that it does not stick out more than the diameter of the nozzle, **Figure 4.50.**
14. Check the manufacturer's operating manual for the machine to ensure that all connections and settings are correct.

Table 4.5 Amperage Range of Tungsten Electrodes

Electrode Diameter		DCEN	DCEP	AC
in.	(mm)			
0.04	(1)	15–60	Not recommended	10–50
1/16	(2)	70–100	10–20	50–90
3/32	(2.4)	90–200	15–30	80–130
1/8	(3)	150–350	25–40	100–200
5/32	(4)	300–450	40–55	160–300

Figure 4.48
Setting the remote control switch
Courtesy of Larry Jeffus

Figure 4.49
Inserting collet and collet body
Courtesy of Larry Jeffus

Figure 4.50
Install the nozzle (cup) to the torch body
Courtesy of Larry Jeffus

CAUTION

Turn off all power before attempting to stop any leaks in the water system.

The GTA welding system is now ready to be used.

15. Turn on the power, depress the remote control, and again check for leaks.
16. While the postflow is still engaged, set the gas flow by adjusting the valve on the flowmeter.

Complete a copy of the "Student Welding Report" listed in Appendix I or provided by your instructor.

EXPERIMENT 4-5

Striking an Arc

Module 1
Key Indicator 1, 4

Module 7
Key Indicator 1
Carbon Steel
Key Indicator 3, 4
Austenitic Stainless Steel
Key Indicator 8, 9
Aluminum
Key Indicator 13, 14

Using a properly set up GTA welding machine, proper safety gear, and clean scrap metal, you will strike a GTA welding arc.

1. Position yourself so that you are comfortable and can see the torch, tungsten, and plate while the tungsten tip is held about 1/4 in. (6 mm) above the metal. Try to hold the torch at a vertical angle ranging from 0° to 15°. Too steep an angle will not give adequate gas coverage, **Figure 4.51.**

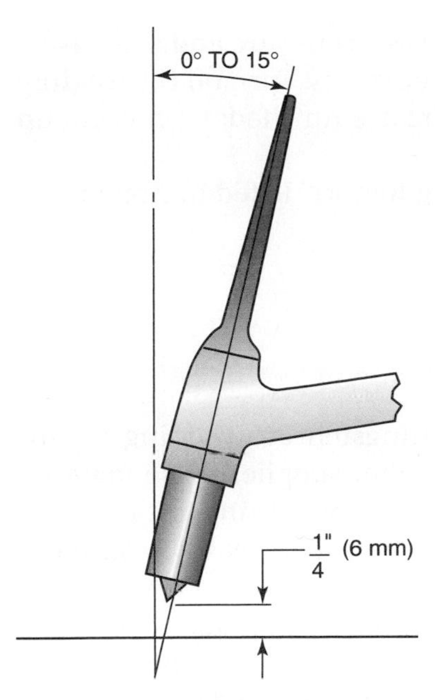

Figure 4.51
GTA torch position

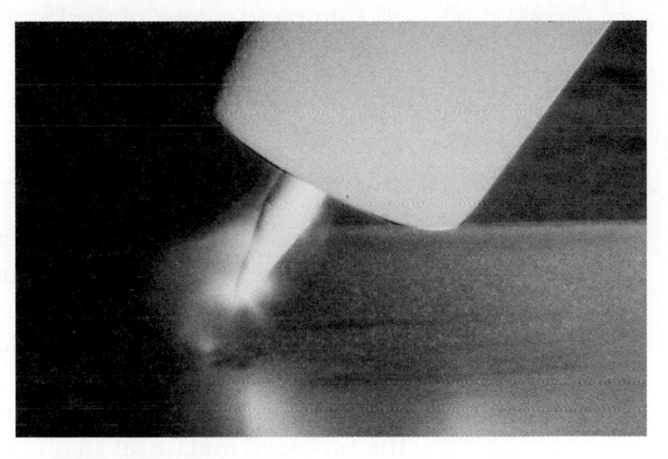

Figure 4.52
High frequency starting before arc starts
Courtesy of Larry Jeffus

Figure 4.53
Stable gas tungsten arc
Courtesy of Larry Jeffus

2. Lower your arc welding helmet and depress the remote control. A high-pitched, erratic arc should be immediately jumping across the gap between the tungsten and the plate. If the high-frequency arc is not established, lower the torch until it appears, **Figure 4.52.**
3. Slowly increase the current until the main welding arc appears, **Figure 4.53.**
4. Observe the color change of the tungsten as the arc appears.
5. Move the tungsten around in a small circle until a molten weld pool appears on the metal.
6. Slowly decrease the current and observe the change in the molten weld pool.
7. Reduce the current until the arc is extinguished.
8. Hold the torch in place over the weld until the postflow stops.
9. Raise your hood and inspect the weld.

CAUTION

Avoid touching the metal table with any unprotected skin or jewelry. The high frequency can cause an uncomfortable shock.

Repeat this procedure until you can easily start the arc and establish a molten weld pool using both AC and DCEN currents. Turn off the welding machine, water, and shielding gas when you are finished; then clean up your work area.

Complete a copy of the "Student Welding Report" listed in Appendix I or provided by your instructor.

SUMMARY

One of the prime considerations for gas tungsten arc welding equipment setup is the cleanliness of the equipment, supplies, base material or material, and the welders themselves. When everything is clean, you will find that the welding process will proceed more easily and more successfully.

Another major factor affecting your ability to produce quality welds is the tungsten end or tip shape. As you practice making the various welds, you will find that keeping the tungsten electrode tip shaped appropriately will assist you in producing uniform welds.

Often, new welders feel that there is some sort of attraction between the tungsten electrode, filler metal, and base metal during the welding process because it seems to continually become contaminated. This almost continuous contamination can be very frustrating. At times it may seem overwhelming; however, with continued practice and diligence, you will be able to control this problem. Even experienced welders in the field can be plagued from time to time with tungsten contamination. At other times they can weld an entire day without contaminating the tungsten. It is often beneficial for students to realize that tungsten contamination is just part of the process, and they must therefore try to ignore the possibility of it happening and concentrate on producing the welds.

REVIEW

1. What early advancements made the GTA welding process more effective and reduced its cost?
2. What metals were weldable only by the GTAW process before GMAW was developed?
3. Which two of tungsten's properties make it the most versatile choice for GTA welding?
4. Why must the tip of the tungsten be hot?
5. Why does some tungsten erosion occur?
6. What function regarding tungsten heat do the collet and torch play?
7. What problem can an excessively large tungsten cause?
8. What holds the molten ball of tungsten in place at the tip of the electrode during DCEP welding?
9. Using **Table 4.1,** answer the following:
   a. What color identifies EWTh-2?
   b. What is the composition of EWCe-2?
   c. What color identifies EWLa-1.5
10. What does adding thorium oxide do for the tungsten electrode?

11. How can the end of a tungsten electrode be shaped?
12. Why should a grinding stone that is used for sharpening tungsten not be used for other metals?
13. Why should the grinding marks run lengthwise on the tungsten electrode end?
14. What are three ways of breaking off the contaminated end of a tungsten electrode?
15. Why should the torch be as cool as possible?
16. What will happen to a water-cooled torch cable if the flow of cooling water stops?
17. Why must shielding gas hoses not be made from rubber?
18. Why should the water solenoid be on the supply side of the water system?
19. What problem can a long nozzle cause to the tungsten?
20. Why must the tube of a flowmeter be vertical?
21. What is the heat distribution with DCEN welding current?
22. What is the heat distribution with DCEP welding current?
23. What is the heat distribution with AC welding current?
24. Why must AC welding power provided by transformer power supplies use high frequencies in order to work?
25. Why are argon and helium known as inert gases?
26. Why is argon's ease of ionization a benefit?
27. What makes helium difficult to use for manual welding?
28. What are the benefits of adding hydrogen to argon for welding?
29. What is the purpose of a hot start?
30. Using **Table 4.3,** determine the gas postflow time for a 3/32-in. (2.4-mm) tungsten.
31. How can air be drawn into the shielding gas?
32. Using **Table 4.4,** determine the minimum gas flow rate for a 1/2-in. (13-mm) nozzle.
33. What functions can a remote control provide the welder?

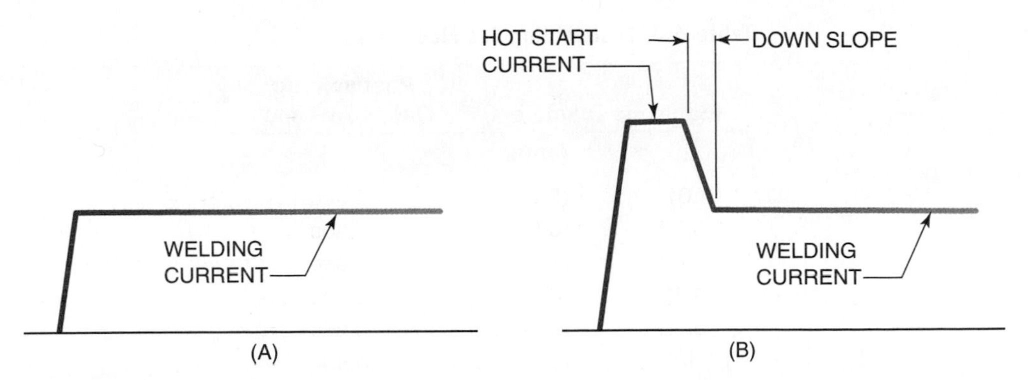

Figure 4.37
Standard method of starting welding current (A); hot start method of starting welding current (B)

tures and because of potential metallurgical problems, nitrogen has not received much attention as an additive for GTA welding.

Hot Start

The hot start allows a controlled surge of welding current as the arc is started to quickly establish a molten weld pool. Rapidly establishing a molten weld pool on metals with a high thermal conductivity is often hard without this higher than normal current. Adjustments can be made in the length of time and the percentage above the normal current, **Figure 4.37.**

Preflow Time and Postflow Time

Preflow time is the time during which gas flows to clear out any air in the nozzle or surrounding the weld zone. The operator sets the length of time that the gas flows before the welding current is started, **Figure 4.38.** Because some machines do not have preflow, many welders find it hard to hold a position while waiting for the current to start. One solution to this problem is to use the postflow for preflow. Switch on the current to engage the postflow. Now, with the current off, the gas is flowing, and the GTA torch can be lowered to the welding position. The welder's helmet should be lowered and the current restarted before the postflow stops. This allows welders to have preflow of shield gas and to start the arc when they are ready.

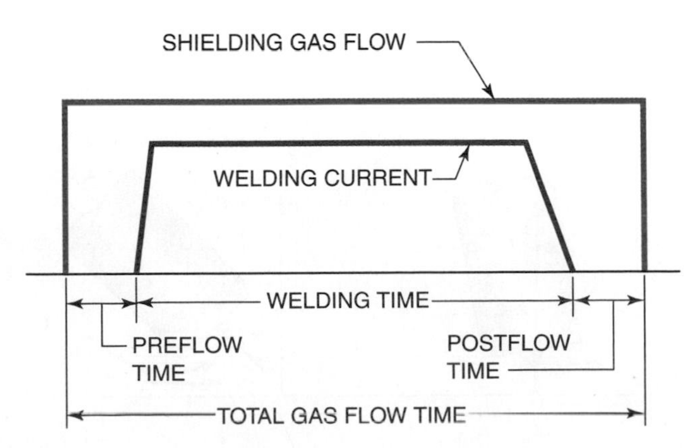

Figure 4.38
Welding time compared to shielding gas flow time

CHAPTER

5 Gas Tungsten Arc Welding of Plate

OBJECTIVES

After completing this chapter, the student should be able to

- list four applications for which the gas tungsten arc welding process is more commonly used
- describe four effects on the weld of varying torch angles
- explain why the filler rod end must be kept inside the protective zone of the shielding gas and how to accomplish this
- list four disadvantages to welding with a contaminated tungsten electrode
- list four techniques to reduce electrode contamination
- determine the correct machine settings for the minimum and maximum welding current for the machine used, the types and sizes of tungsten, and the metal types and thicknesses
- list four factors that affect the gas preflow and postflow times required to protect the electrode and the weld
- determine the minimum and maximum gas flow settings for each nozzle size, tungsten size, and amperage setting
- compare the characteristics of low carbon and mild steels, stainless steel, and aluminum with respect to GTA welding
- prepare carbon steel, stainless steel, and aluminum for GTA welding

KEY TERMS

chill plate	gas coverage	protective zone
contamination	oxide layer	surface tension

AWS SENSE EG2.0

Key Indicators Addressed in this Chapter

Module 1: Occupational Orientation

Key Indicator 1: Prepares time or job cards, reports or records

Key Indicator 3: Follows verbal instructions to complete work assignments

Key Indicator 4: Follows written instructions to complete work assignments

Module 2: Safety and Health of Welders

Key Indicator 1: Demonstrates proper use and inspection of personal protection equipment (PPE)
Key Indicator 2: Demonstrates safe operation practices in the work area
Key Indicator 3: Demonstrates proper use and inspection of ventilation equipment
Key Indicator 4: Demonstrates proper hot zone operation
Key Indicator 6: Understands proper use of precautionary labeling and MSDS information.
Key Indicator 7: Demonstrates proper inspection and operation of equipment for each welding or thermal cutting process used

Module 7: Gas Tungsten Arc Welding (GTAW)

Key Indicator 1: Performs Safety inspections of GTAW equipment and accessories
Key Indicator 2: Makes Minor external repairs to GTAW equipment and accessories

Carbon Steel

Key Indicator 3: Sets up for GTAW operations on carbon steel
Key Indicator 4: Operates GTAW equipment on carbon steel
Key Indicator 5: Makes GTAW fillet welds, in all positions, on carbon steel
Key Indicator 6: Makes GTAW groove welds, in all positions, on carbon steel
Key Indicator 7: Passes GTAW welder performance qualification test (workmanship sample) on carbon steel

Austenitic Stainless Steel

Key Indicator 8: Sets up for GTAW operations on austenitic stainless steel
Key Indicator 9: Operates GTAW equipment on austenitic stainless steel
Key Indicator 10: Makes GTAW fillet welds, in 1F, 2F, and 3F positions, on austenitic stainless steel
Key Indicator 11: Makes GTAW groove welds, in 1G and 2G positions, on austenitic stainless steel
Key Indicator 12: Passes GTAW welder performance qualification test (workmanship sample) on austenitic stainless steel

Aluminum

Key Indicator 13: Sets up for GTAW operations on aluminum
Key Indicator 14: Operates GTAW equipment on aluminum
Key Indicator 15: Makes GTAW fillet welds, in the 1F and 2F positions, on aluminum
Key Indicator 16: Makes GTAW groove welds, in the 1G position, on aluminum
Key Indicator 17: Passes GTAW welder performance qualification test (workmanship sample) on aluminum

Module 9: Welding Inspection and Testing

Key Indicator 1: Examines cut surfaces and edges of prepared base metal parts
Key Indicator 2: Examines tacks, root passes, intermediate layers, and completed welds.

INTRODUCTION

The gas tungsten arc welding process can be used to join nearly all types and thicknesses of metal. Welders can have a clear, unobstructed view of the molten weld pool because GTA welding is fluxless, slagless, and smokeless. The clear view of the weld allows welders to make changes in their welding technique, current, travel speed, and rate at which the filler metal is added to the weld as the weld progresses, ensuring that a quality weld is being made. This gives the welder very fine control of the welding process.

The fine control of the weld that is possible with GTA welding makes it an ideal process for very close-tolerance, high-quality welds. GTA welding is used to make critical welds, such as those on aircraft structures. If these welds fail, serious injury, death, and/or significant loss of property can result. Sometimes GTA welding is used to make the critical root pass of a weld that will be completed using another faster process. It is also used when weld appearance is important to the look of the finished part, as in some furniture, decorations, and/or sculptures.

The proper setup of GTA equipment can often affect the quality of the weld performed. Charts and graphs are available that give the correct amperage, gas flow rate, and time for various types of welds and metals. These charts are designed for optimum laboratory or classroom conditions. Actual conditions in the field will have an effect on these values. The experiments in this chapter are designed to help the welder understand the harmful effects on welding of less than ideal conditions. This will allow the welder to evaluate the appearance of a weld and make the necessary changes in technique or setup to improve the weld.

After a person has learned to weld in the lab, troubleshooting field welding problems will become much easier. The weld should be watched carefully to pick up any changes that could indicate a needed adjustment. When welders can do this, they have mastered the GTA process and have made themselves better potential employees. To make a weld is good; to solve a welding problem is better.

TORCH ANGLE

The torch should be held as close to perpendicular as possible in relation to the plate surface. The torch may be angled from 0° to 15° from perpendicular for better visibility and still have the proper shielding **gas coverage.** As the gas flows out it must form a **protective zone** around the weld. Tilting the torch changes the shape of this protective zone, **Figure 5.1.** Too much tilting of the torch will cause the protective shielding gas zone to become so distorted that the weld may not be protected from contamination from the air. The closer the torch is held to perpendicular, the better the weld is shielded.

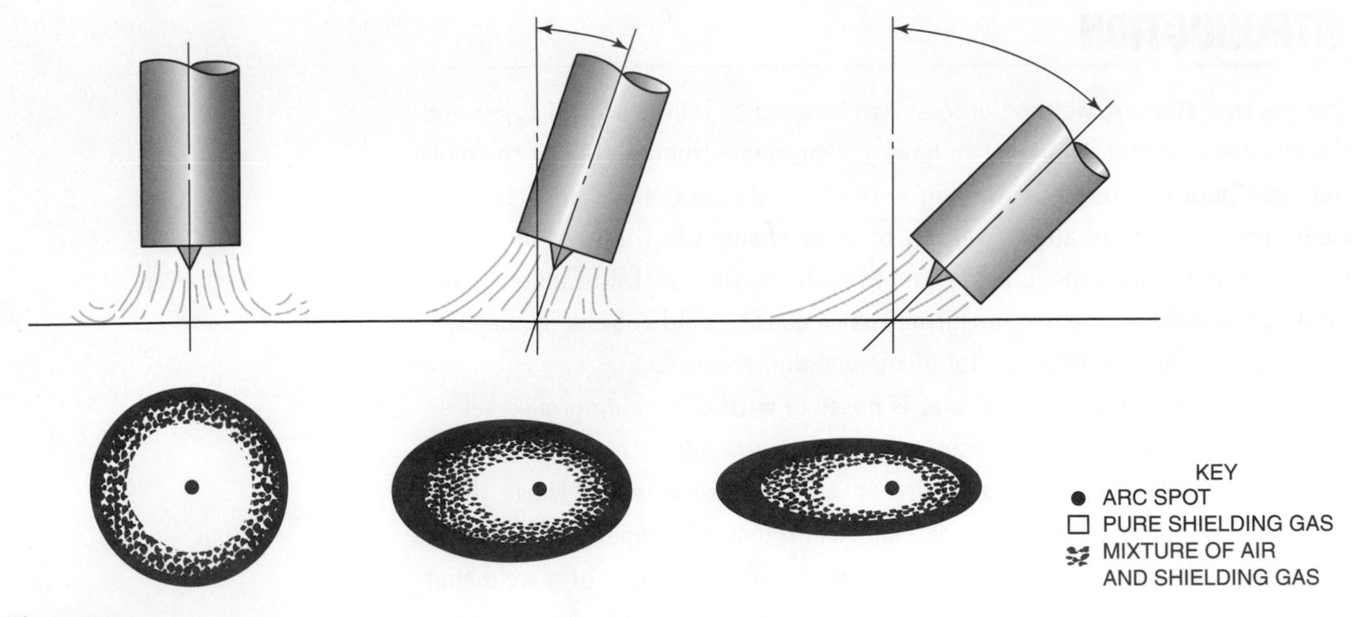

Figure 5.1
Gas coverage patterns for different GTA torch angles. Note how the area covered by the shielding gas becomes narrower and elongates as the angle of the torch increases from the perpendicular.

The velocity of the shielding gas also affects the protective zone as the torch angle changes. As the velocity increases, a low-pressure area develops behind the cup. When the low-pressure area becomes strong enough, air is pulled into the shielding gas. The sharper the angle and the higher the flow rate, the greater the possibility contamination will occur from the onset of turbulence in the gas stream. This causes air to become mixed with the shielding gas. Turbulence caused by the shielding gas striking the work will also cause air to mix with the shielding gas at high velocities.

FILLER ROD MANIPULATION

The filler rod end must be kept inside the protective zone of the shielding gas, **Figure 5.2.** The end of the filler rod is hot, and if it is removed from the gas protection, it will oxidize rapidly. The oxide will then be added to the

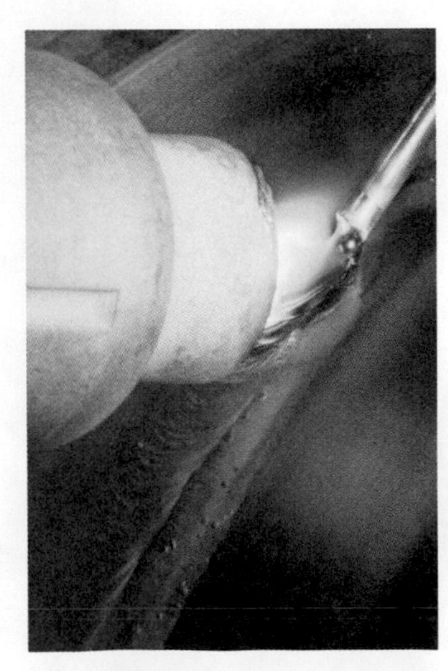

Figure 5.2
The hot filler rod end is well within the protective gas envelope
Courtesy of Larry Jeffus

Figure 5.3
(A) Filler properly protected, (B) some oxides on filler, and (C) excessive oxides caused by improper filler rod manipulation
Courtesy of Larry Jeffus

Figure 5.4
Filler being left in the molten weld pool as the arc is extinguished
Courtesy of Larry Jeffus

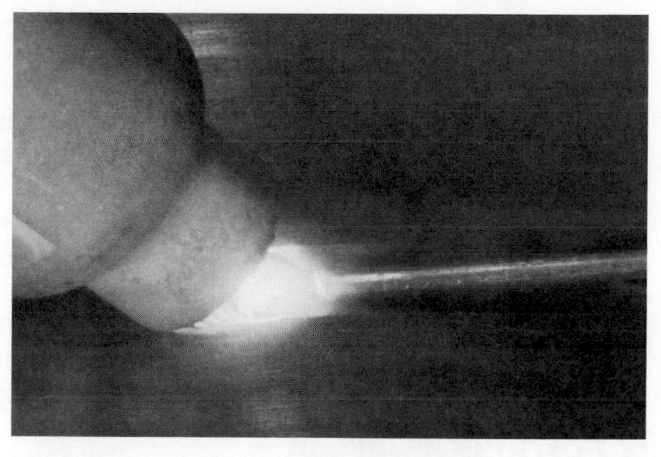

Figure 5.5
Filler being remelted as the weld is continued
Courtesy of Larry Jeffus

molten weld pool, **Figure 5.3.** When a weld is stopped so that the welder can change position, the shielding gas must be kept flowing around the rod end to protect it until it is cool. If the end of the rod becomes oxidized, it should be cut off before restarting. The following method can be used both to protect the rod end and reduce the possibility of crater cracking—that is, breaking the arc but keeping the torch over the crater while, at the same time, sticking the rod in the molten weld pool before it cools, **Figure 5.4.** When the weld is restarted, the rod is simply melted loose again, **Figure 5.5.**

The rod should enter the shielding gas as close to the base metal as possible, **Figure 5.6.** A 15° angle or less to the plate surface prevents air from being pulled into the welding zone behind the rod, **Figure 5.7.** As an example, if a rod is held in a stream of running water, air can be pulled in. The faster the water flows or the steeper the angle at which the rod is held, the more air is pulled in. The same action occurs with the shielding gas as its flow increases or as the rod angle increases.

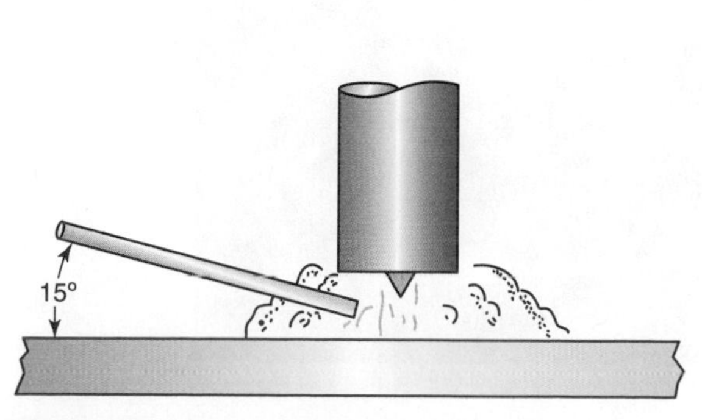

Figure 5.6
Keep the filled metal at approximately a 15° angle

Figure 5.7
Too much filler rod angle has caused oxides to be formed on the filler rod end
Courtesy of Larry Jeffus

TUNGSTEN CONTAMINATION

For new welding students, the most frequently occurring and most time-consuming problem is tungsten **contamination.** The tungsten becomes contaminated when it touches the molten weld pool or when it is touched by the filler metal. When this happens, especially with aluminum, **surface tension** pulls the contamination up onto the hot tungsten, **Figure 5.8.** The extreme heat causes some of the metal to vaporize and form a large, widely scattered **oxide layer.** On aluminum, this layer is black. On iron (steel and stainless steel), this layer is a reddish color.

The contamination caused by the tungsten touching the molten weld pool or filler metal forms a weak weld. On a welding job, both the weld and the tungsten must be cleaned before any more welding can be done. The weld crater must be ground or chiseled to remove the tungsten contamination, and the tungsten end must be reshaped. Extremely tiny tungsten particles will show up if the weld is X-rayed. Failure to remove the contamination properly will result in the failure of the weld.

When starting to weld, the beginning student may save weld practice time by burning off the contamination. On a scrap, usually copper plate, strike an arc using a higher than normal amperage setting. The arc will be erratic and discolored at first, but, as the contamination vaporizes, the arc will stabilize. Contamination can also be knocked off by quickly flipping the torch head.

CAUTION

This procedure should never be used with heavy contaminations or when a welder is on the job in the field. It is designed only to help the new student in the first few days of training to save time and increase weld production.

CURRENT SETTING

The amperage set on a machine and the actual welding current are often not the same. The amperage indicated on the machine's control is the same as that at the arc only for the following conditions:

- The power to the machine is exactly correct.
- The lead length is very short.
- All cable connections are perfect with zero resistance.
- The arc length is exactly the right length.
- The remote current control is in the full on position.

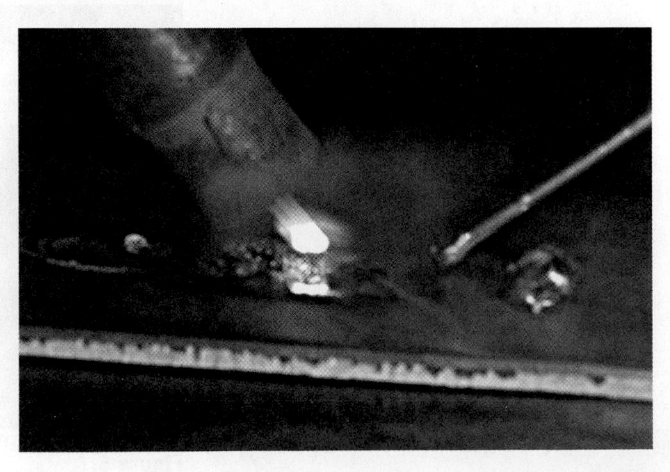

Figure 5.8
Contaminated tungsten
Courtesy of Larry Jeffus

Table 5.1 Sample Chart Used to Record GTA Welding Machine Settings

Current and Tungsten Electrode Size	Amperage/Machine Setting				
	Too Low	Low	Good	High	Too High

If any one of these factors changes, the actual welding amperage will change.

In addition to the difference between indicated and actual welding amperage, there is a more significant difference between amperage and welding power. The welding power, in watts, is based on the formula $W = E \times I$, or volts (E) multiplied by amperes (I) equals watts (W). Thus, the indicated power to a weld from two different types of welding machines set at 100 amperes will vary depending upon the voltage of the machine.

The welding machine setting will vary within a range from low to high (cool to hot). The range for one machine may be different from that of another machine. The setting will also be different for various types and sizes of tungstens, polarities, types and thicknesses of metal, joint position or design, and shielding gas used.

A chart, such as the one in **Table 5.1,** and a series of tests can be used to set the lower and upper limits for the amperage settings. As students' welding skills improve with practice, they will become familiar with the machine settings so that a table for these settings is no longer needed. In the welding industry, some welders will mark a line on the dial of the machine to help in resetting the machine. If a welder is required to make a number of different machine setups, a list or chart can be made and taped to the machine. This practice is more professional than marking the machine dials.

EXPERIMENTS

Experiments are designed to help new welders learn some basic skills that will help them troubleshoot welding problems. If you do the experiments listed in this chapter, you will be better able to determine what is causing a problem with your weld. As you learn more about welding, subtle changes will become more noticeable. Even experienced welders make changes in the setup, current, or welding technique as they try to resolve a problem.

Experiment 5-1 will help the welder determine the correct machine settings for the minimum and maximum welding current for the machine used, the types and sizes of tungstens, and the metal types and thicknesses. Most welding will be performed with a medium-range or mid-range machine setting. The exact setting is more important for machines without remote controls. The remote control allows changes in welding current to be made during the welding without having to stop.

EXPERIMENT 5-1

Setting the Welding Current

Module 1
Key Indicator 1, 4

Module 7
Key Indicator 1, 3, 4

Module 9
Key Indicator 1, 2

Using a properly set up GTA welding machine and torch, proper safety protection, one of each available tungsten size and type, and 16-gauge mild steel, 1/8 in. (3 mm) and 1/4 in. (6 mm) thick, you will work with a small group of students to develop a chart of the correct machine current setting for each type and size of tungsten.

Set the machine welding power switch for DCEN (DCSP) and the amperage control to its lowest setting, **Figure 5.9.** Sharpen a point on each tungsten and install one of the smaller diameter tungstens in the GTA torch. Select a nozzle with a 1/2-in. (13-mm)-diameter hole and attach it to the torch head. Set the preflow time to 0 and postflow to 10 to 15 seconds. Connect the remote control if it is available. Turn on the main power and hold the torch so that it cannot short out. Depress the remote controls to start the shielding gas so the flow rate can be set at 20 cfh (8 L/min). Switch the high frequency to start. All other functions, such as pulse, hot start, slope, and so on, should be in the off position.

Place the piece of 16-gauge sheet metal flat on the welding table. Hold the torch vertically with the tungsten about 1/4 in. (6 mm) above the metal. Lower your welding hood and fully depress the remote control. Watch the arc to see if it stabilizes and melts the metal. After a short period of time (15 to 30 seconds), stop, raise your hood, and check the plate for a melted spot. If melting occurred, note the size of the spot and depth of penetration, **Figure 5.10.** Increase the amperage setting by 5 or 10 A, note the setting on the chart, and repeat the process.

After each test, observe and record the results. The important settings to note are:

1. when the tungsten first heats up and the arc stabilizes
2. when the metal first melts
3. when 100% penetration of the metal first occurs
4. when burn-through first occurs
5. when the tungsten starts glowing white hot and/or melts

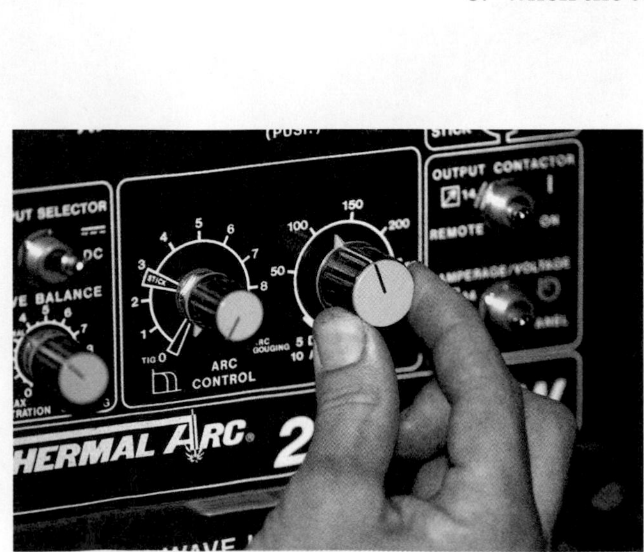

Figure 5.9
Lower the welding current to zero or as low as possible
Courtesy of Larry Jeffus

Figure 5.10
Melting first occurring
Courtesy of Larry Jeffus

The lowest (minimum) acceptable amperage setting is when the molten weld pool first appears on the base metal and the arc is stable. The highest (maximum) amperage setting is when the base metal burn-through or melting of the tungsten occurs. Any current setting between the high and low points is within the amperage range for that specific setup.

To establish the range for the next tungsten type or size, repeat the test. After each test, the metal should be cooled to prevent overheating. After each type and size of tungsten has been tested and an operating range established, repeat the procedure using the next thicker metal. Repeat this procedure until you have set up the operating ranges for all of the metals and tungstens you will be using. Turn off the welding machine, shielding gas, and cooling water, and clean up your work area when you are finished welding.

Complete a copy of the "Student Welding Report" listed in Appendix I or provided by your instructor.

GAS FLOW

The gas preflow and postflow times required to protect both the tungsten and the weld depend upon the following factors:

- wind or draft speed
- nozzle size used
- tungsten size used
- amperage
- joint design
- welding position
- type of metal welded

The weld quality can be adversely affected by improper gas flow settings. The lowest possible gas flow rates and the shortest preflow or postflow time can help reduce the cost of welding by saving the expensive shielding gas.

In Experiment 5-2, the minimum and maximum gas flow settings for each nozzle size, tungsten size, and amperage setting will be determined. The chart a welder prepares based on experiments is to improve that welder's skill and welding technique. Charts may differ slightly from one welder to another. As a welder's skill improves, the chart may change. As experience is gained, a welder will learn how to set the gas flow effectively without the need for this chart.

The minimum flow rates and times must be increased when welding in drafty areas or for out-of-position welds. The rates and times can be somewhat lower for tee joints or welds made in tight areas. The maximum flow rates must never be exceeded. Exceeding these flow rates causes weld contamination and increases the rejection rate.

Module 1
Key Indicator 1, 4

Module 7
Key Indicator 1
Carbon Steel
Key Indicator 3, 4

Module 9
Key Indicator 1, 2

EXPERIMENT 5-2

Setting Gas Flow

Using a properly set up GTA welding machine and torch, proper safety protection, one of each available tungsten size, metal that is 16 gauge to 1/4 in. (6 mm) thick, and the welding current chart developed in

Figure 5.11
Setting the postflow timer
Courtesy of Larry Jeffus

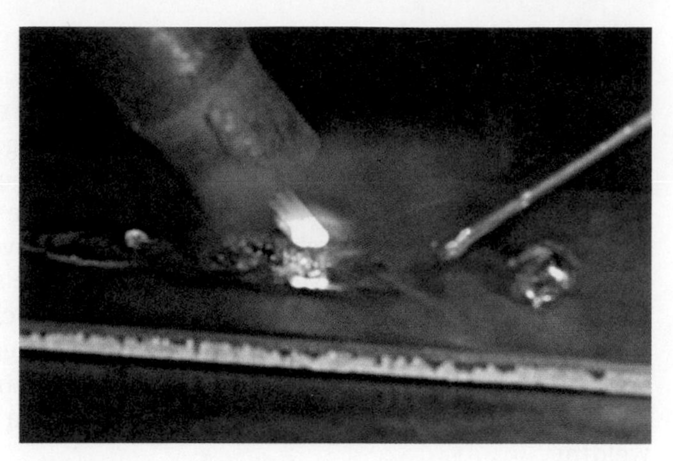

Figure 5.12
Oxides forming due to inadequate gas shielding
Courtesy of Larry Jeffus

Experiment 5-1, you will work with a small group of students to make a chart of the minimum and maximum flow rates and times for each nozzle size, tungsten size, and amperage setting. An assistant will also be needed to change and record the flow rate while you work.

Set the machine welding power switch for DCEN (DCSP). Set the amperage to the lowest setting for the size of tungsten used. Set the preflow time to 0 and postflow to 20 seconds, **Figure 5.11.** Turn on the main power. With the torch held so that it cannot short out, depress the remote control to start the shielding gas flow and set the flow at 20 cfh (9 L/min). Switch the high frequency to start. All other functions, such as pulse, hot start, slope, and so on, should be in the off position.

Starting with the smallest nozzle and tungsten size, strike an arc and establish a molten pool on a piece of metal in the flat position. Watch the molten weld pool and tungsten for signs of oxide formation as another person slowly lowers the gas flow rate. Have that person note this setting (where oxide formation begins), **Figure 5.12,** as the minimum flow rate on the chart next to the nozzle size and current setting. Now slowly increase the flow rate until the molten pool starts to be blown back or oxides start forming. This setting should be noted on the chart as the maximum flow rate for this current and nozzle size, **Table 5.2.** Lower the flow to a rate of 2 cfh or 3 cfh (1 L/min or 2 L/min) above the minimum value noted on the chart, and then stop the arc. Record the length of time from the point when the arc stops and the tungsten stops glowing as the postflow time. Repeat this test at a medium and then high current setting for this nozzle

Table 5.2 Sample Chart for Setting Shielding Gas Flow Rate and Time

Electrode and Nozzle Size	Flow Rate					Postflow Time		
	Too Low	Low	Good	High	Too High	Too Short	OK	Too Long

and tungsten size. When using high current settings, it may be necessary to move the torch or use thicker plate to prevent burn-through.

Repeat this test procedure with each available nozzle and tungsten size. Stainless steel or aluminum is preferred for this experiment because the oxides are more quickly noticeable than when mild steel is used. If aluminum is used, the welding current must be AC, and the high-frequency switch should be set on continuous.

To establish the minimum preflow time for each nozzle and tungsten size, set the amperage to a medium-high setting. Hold the torch above the metal so that an arc will be instantly started. Set the preflow timer to 0 and the gas flow to just above the minimum value noted on the chart. Quickly strike an arc on metal thin enough to cause a weld pool to form instantly at that power setting. Stop the arc and examine the weld pool and tungsten for oxides. Repeat this procedure, increasing the preflow time until no oxides are formed on either the plate or tungsten. Record this time on the chart as the minimum preflow time. Repeat this test with each available nozzle and tungsten size. Turn off the welding machine, shielding gas, and cooling water, and clean up your work area when you are finished welding.

Complete a copy of the "Student Welding Report" listed in Appendix I or provided by your instructor.

PRACTICE WELDS

The practice welds are grouped according to the weld position and type of joint, and not by the type of metal. The order in which a person decides to do the welds is that person's choice. It is suggested that the stringer beads be done in each metal and position before the different joints are tried. Each metal has its own characteristics that may make one metal easier for a person to work on than another metal.

Mild steel is inexpensive and requires the least amount of cleaning. Slight changes in the metal have little effect on the welding skill required. Stainless steel is somewhat affected by cleanliness, requiring little preweld cleaning. However, the weld pool shows overheating or poor gas coverage. With aluminum, cleanliness is a critical factor. Oxides on aluminum may prevent the molten weld pool from flowing together. The surface tension helps hold the metal in place, giving excellent bead contour and appearance.

The degree of difficulty a welder encounters with each of these metals depends upon the individual's experience. Try each weld with each metal to determine which metal will be easiest to master first. The type of welding machine and materials used will also affect a welder's progress. Practice will help welders overcome any obstacle to their progress.

Low-Carbon and Mild Steels

Low-carbon and mild steel are two basic steel classifications. These steels are the most common type of steels a new GTA welding student will experience welding. Carbon is the primary alloy in these classifications of steel, and it ranges from 0.15% or less for low carbon and 0.15% to 0.30% for mild steel. The GTA welding techniques required for welding steels in both

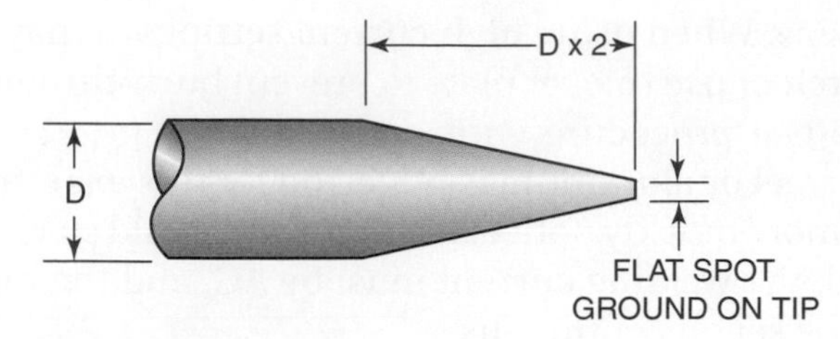

Figure 5.13
Tungsten tip shape for mild steel or stainless steel

classifications are the same. You start with EWTh-1, EWTh-2, EWCe-2, or one of the EWLa pointed tungsten electrodes, **Figure 5.13,** with the welding machine set for DCEN (DCSP) welding current. **Table 5.3** lists the types of filler metal used for both low-carbon and mild steels.

During the manufacturing process, small pockets of primarily carbon dioxide gas sometimes become trapped inside low-carbon and mild steels. There are only a few molecules of gas trapped inside the microscopic pockets within the steel, so they do not affect the steel strength. During most other types of welding, fluxes on the filler metal capture these gas pockets and they are removed. However, because GTA filler metals do not have fluxes when these gas pockets become hot during welding, they expand and can sometimes cause weld porosity. You are most likely to see porosity caused by these gases when you are not using a filler metal. Most GTA filler metals have some alloys, called deoxidizers, that can help prevent porosity caused by gases trapped in the base metal. RG45 gas welding rods do not have these deoxidizers and are not recommended for GTA welding.

Stainless Steel

The setup and manipulation techniques required for stainless steel are nearly the same as those for low-carbon and mild steels, and skills transfer is easy. The major difference is that most welds on steels do not show the effects of contamination as easily as do welds on stainless steels. To make a weld on stainless steel you must do a better job of precleaning the base metal and filler metal; make sure you have adequate shielding gas coverage and do not overheat the weld.

Table 5.3 Filler Metals for Low-Carbon and Mild Steels

SAE No.	Carbon %	AWS Filler Metal No.
	Low Carbon	
1006	0.08 max	RG60 or ER70S-3
1008	0.10 max	RG60 or ER70S-3
1010	0.08 to 0.15	RG60 or ER70S-3
	Mild Steel	
1015	0.11 to 0.16	RG60 or ER70S-3
1016	0.13 to 0.18	RG60 or ER70S-3
1018	0.15 to 0.20	RG60 or ER70S-3
1020	0.18 to 0.23	RG60 or ER70S-3
1025	0.22 to 0.29	RG60 or ER70S-3

Table 5.4 Temperatures at Which Various Colored Oxide Layers Form on Steel

Surface Color	Approximate Temperature at Which Color is Formed	
	°F	(°C)
Light Straw	400	(200)
Tan	450	(230)
Brown	525	(275)
Purple	575	(300)
Dark Blue	600	(315)
Black	800	(425)

Table 5.5 Filler Metals for Stainless Steels

AISI No.	AWS Filler No.	AISI No.	AWS Filler No.
303	ER308	310	ER310
304	ER308	316	ER316L
304L	ER308L	316L	ER316L
309	ER309	410	ER410

The most common sign that there is a problem with a stainless steel weld is the bead color after the weld. The greater the contamination, the darker the color. The exposure of the weld bead to the atmosphere before it has cooled will also change the bead color. It is impossible, however, to determine the extent of contamination of a weld with only visual inspection. Both light-colored and dark-colored welds may not be free from oxides. Thus, it is desirable to take the time and necessary precautions to make welds that are no darker than dark blue, **Table 5.4.** Welds with only slight oxide layers are better for multiple passes.

Using a low arc current setting with faster travel speeds is important when welding stainless steel, because some stainless steels are subject to carbide precipitation. Carbide precipitation, the combining of carbon with chromium, occurs in some stainless steels when they are kept at a temperature between 800°F and 1500°F (625°C and 815°C) for a long time. There are a number of ways of controlling carbide precipitation, including the adding of alloys to the stainless steel and the lowering of the percentage of carbon. Also, during welding, special alloy filler metals can be used to control the problem, but the most important thing a welder can do is travel fast and use as little welding heat as possible.

Black crusty spots may appear on weld beads. These spots are often caused by improper cleaning of the filler rod or failure to keep the end of the rod inside the shielding gas.

Table 5.5 lists some common types of stainless steels and the recommended filler metals.

Aluminum

Aluminum is GTA welded using EWP, EWZr, EWCe-2, or EWLa rounded tip tungsten electrodes on transformer rectifier welding power supplies, **Figure 5.14,** with the welding machine set for ACHF welding current. EWZr, EWCe-2, or EWLa electrodes with a pointed tip may be used with some inverter welding power supplies due to the heat control afforded by

Figure 5.14
Tungsten tip shape for aluminum

Table 5.6 **Filler Metals for Aluminum Alloys**

AISI No.	AWS Filler No.	AISI No.	AWS Filler No.
1100	ER1100	3004	ER4043
3003	ER4043	6061	ER4043

their advanced circuitry. The alternating current provides good arc cleaning, and the continuous high frequency restarts the arc as the current changes direction.

The molten aluminum weld pool has high surface tension, which allows large weld beads to be controlled easily. **Table 5.6** lists some basic types of filler metals used for aluminum welding.

The high thermal conductivity of the metal may make starting a weld on thick sections difficult without first preheating the base metal. In most cases the preheat temperature is around 300°F (150°C) but will vary depending on metal thickness and alloy type. Specific preheat temperatures are available from the metal supplier.

The processes of cleaning and keeping the metal clean take a lot of time. Removal of the oxide layer is easy using a chemical or mechanical method. Ten minutes after cleaning, however, the oxide layer may again be thick enough to require recleaning. The oxide that forms reduces the ability of the weld pool to flow together. Keep your hands and gloves clean and oil free so the base metal or filler rods do not become recontaminated.

Although aluminum resists oxidation at room temperature, it rapidly oxidizes at welding temperatures. If the filler rod is not kept inside the shielding gas, it will quickly oxidize, but, because of the low melting temperature of the filler rod, the end will melt before it is added to the weld pool if it is held too closely to the arc, **Figure 5.15** and **Figure 5.16.**

CAUTION

The manufacturer's recommendations for using these products must be followed. Failure to do so may result in chemical burns, fires, fumes, or other safety hazards that could lead to serious injury. If anyone should come in contact with any chemicals, immediately refer to the material safety data sheet (MSDS) for the proper corrective action.

Metal Preparation

Both the base metal and the filler metal used in the GTAW process must be thoroughly cleaned before welding. Contamination left on the metal will be deposited in the weld because there is no flux to remove it. Oxides, oils, and dirt are the most common types of contaminants. They can be

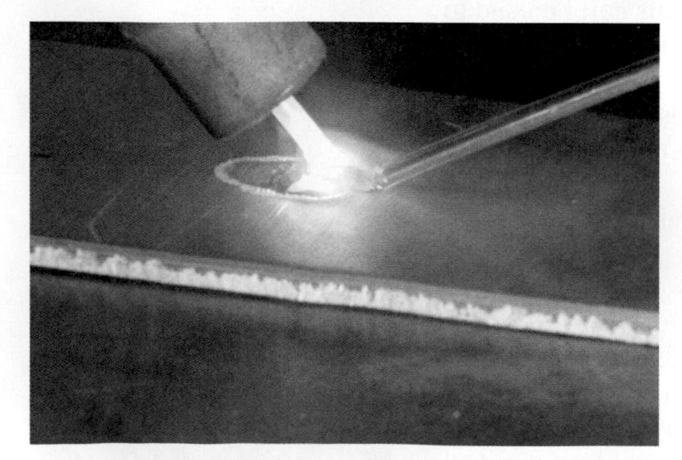

Figure 5.15
Aluminum filler being correctly added to the molten weld pool
Courtesy of Larry Jeffus

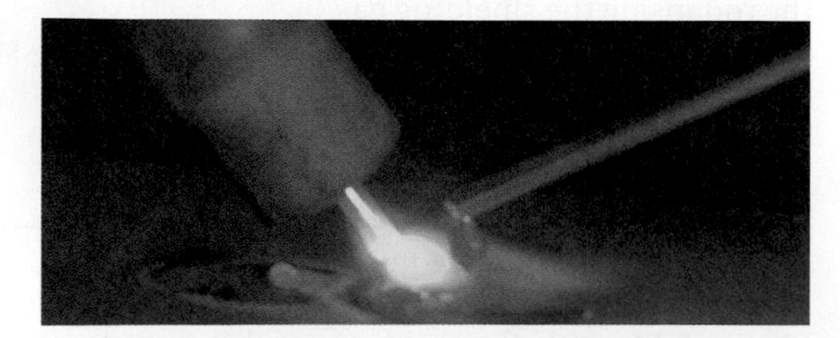

Figure 5.16
Filler rod being melted before it is added to the molten pool
Courtesy of Larry Jeffus

removed mechanically or chemically. Mechanical metal cleaning may be done by grinding, wire brushing, scraping, machining, or filing. Chemical cleaning may be done by using acids, alkalies, solvents, or detergents.

PRACTICE 5-1

Stringer Beads, Flat Position, on Mild Steel

Using a properly set up and adjusted GTA welding machine on DCEN, proper safety protection, and one or more pieces of mild steel, 6 in. (152 mm) long and 16 gauge and 1/8 in. (3 mm) thick, you will push a weld pool in a straight line down the plate, **Figure 5.17.** Maintain uniform weld pool size and penetration.

- Starting at one end of the piece of metal that is 1/8 in. (3 mm) thick, hold the torch as close as possible to a 90° angle.
- Lower your hood, strike an arc, and establish a weld pool.
- Move the torch in a stepping or circular oscillation pattern down the plate toward the other end, **Figure 5.18.**
- If the size of the weld pool changes, speed up or slow down the travel rate to keep the weld pool the same size for the entire length of the plate.

Module 1
Key Indicator 1, 3, 4

Module 2
Key Indicator 1, 2, 3, 4, 6, 7

Module 7
Carbon Steel
Key Indicator 3, 4

Module 9
Key Indicator 1,2

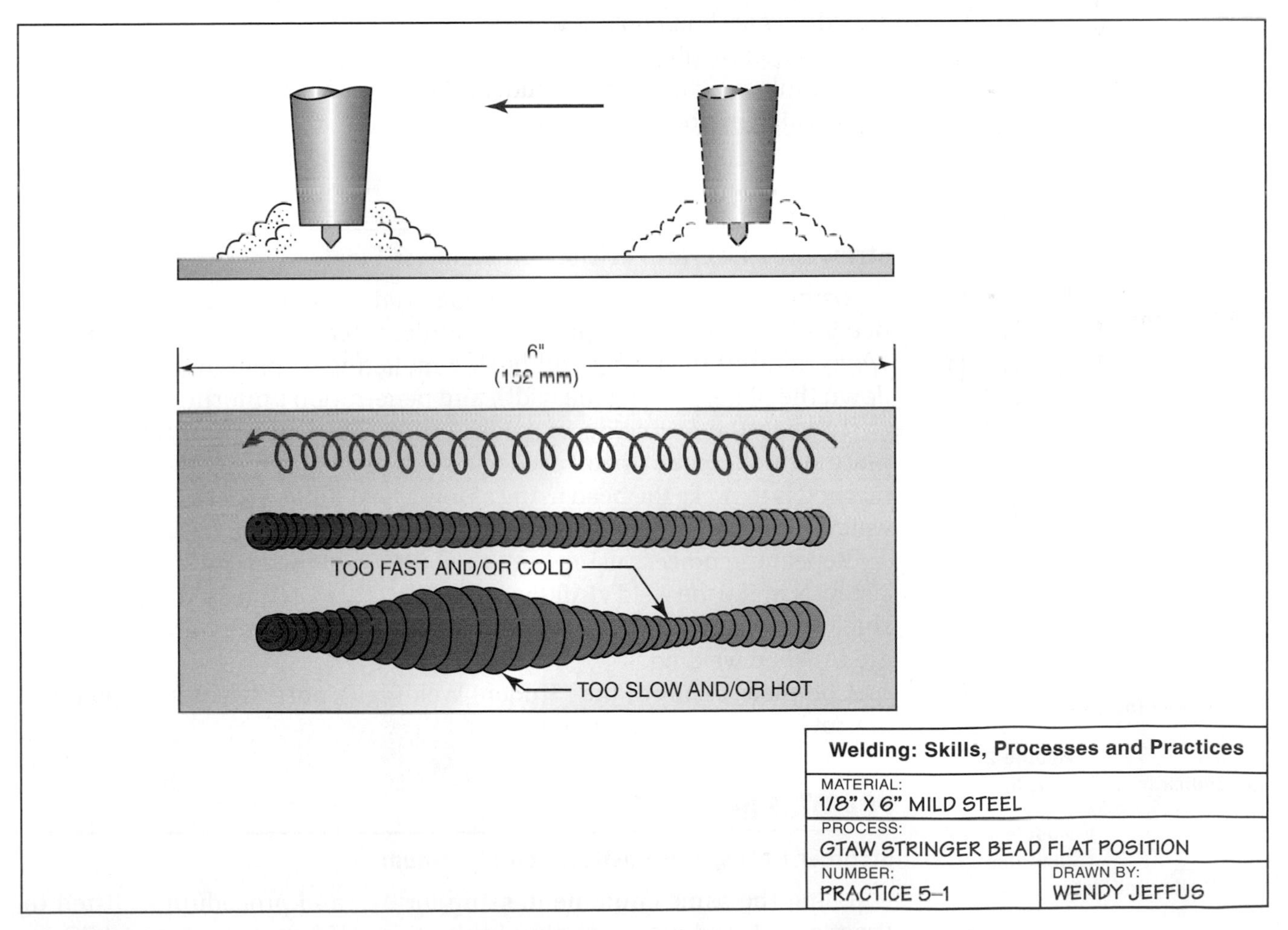

Figure 5.17
Surfacing weld in the flat position

Figure 5.18
Surfacing weld
Courtesy of Larry Jeffus

The ability to maintain uniformity in width and keep a straight line increases as you are able to see more than just the weld pool. As your skill improves, you will relax, and your field of vision will increase.

Repeat the process using both thicknesses of metal until you can consistently make the weld visually defect free. Turn off the welding machine, shielding gas, and cooling water, and clean up your work area when you are finished welding.

Complete a copy of the "Student Welding Report" listed in Appendix I or provided by your instructor.

Module 1
Key Indicator 1, 3, 4

Module 2
Key Indicator 1, 2, 3, 4, 6, 7

Module 7
Austenitic Stainless Steel
Key Indicator 8, 9

Module 9
Key Indicator 1, 2

PRACTICE 5-2

Stringer Beads, Flat Position, on Stainless Steel

Using the same equipment and material thicknesses as listed in Practice 5-1 and one or more pieces of stainless steel, 6 in. (152 mm) long and 1/4 in. (6 mm) thick, you will push a molten weld pool in a straight line down the plate, keeping the width and penetration uniform.

To keep the formation of oxides on the bead to a minimum, a **chill plate** (a thick piece of metal used to absorb heat) may be required. Another method is to make the bead using as low a heat input as possible. When the weld is finished, the weld bead should be no darker than dark blue.

Repeat the process using both thicknesses of metal until you can consistently make the weld visually defect free. Turn off the welding machine, shielding gas, and cooling water, and clean up your work area when you are finished welding.

Complete a copy of the "Student Welding Report" listed in Appendix I or provided by your instructor.

Module 1
Key Indicator 1, 3, 4

Module 2
Key Indicator 1, 2, 3, 4, 6, 7

Module 7
Aluminum
Key Indicator 13, 14

Module 9
Key Indicator 1, 2

PRACTICE 5-3

Stringer Beads, Flat Position, on Aluminum

Using the same equipment, setup for AC, and procedure as listed in Practice 5-1 and pieces of aluminum, 6 in. (152 mm) long and 1/16 in. (2 mm), 1/8 in. (3 mm), and 1/4 in. (6 mm) thick, you will push a weld pool in a straight line, maintaining uniform width and penetration for the length of the plate.

A high current setting will allow faster travel speeds. The faster speed helps control excessive penetration. Hot cracking may occur on some types of aluminum after a surfacing weld. This is not normally a problem when filler metal is added. If hot cracking should occur during this practice, do not be concerned.

Repeat the process using all thicknesses of metal until you can consistently make the weld visually defect free. Turn off the welding machine, shielding gas, and cooling water, and clean up your work area when you are finished welding.

Complete a copy of the "Student Welding Report" listed in Appendix I or provided by your instructor.

Module 1
Key Indicator 1, 3, 4

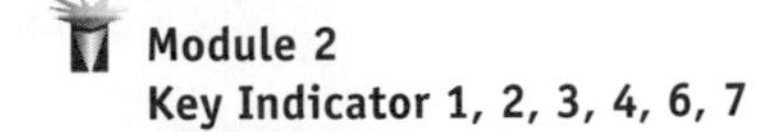

Module 2
Key Indicator 1, 2, 3, 4, 6, 7

Module 7
Carbon Steel
Key Indicator 3, 4
Austenitic Stainless Steel
Key Indicator 8, 9
Aluminum
Key Indicator 13, 14

Module 9
Key Indicator 1, 2

PRACTICE 5-4

Flat Position, Using Mild Steel, Stainless Steel, Aluminum

For this practice, you will need a properly set up and adjusted GTA welding machine, proper safety protection, and filler rods, 36 in. (0.9 m) long × 1/16 in. (2 mm), 3/32 in. (2.4 mm), and 1/8 in. (3 mm) in diameter, one or more pieces of mild steel, stainless steel, and aluminum, 6 in. (152 mm) long × 1/16 in. (2 mm) and 1/8 in. (3 mm) thick, and aluminum plate 1/4 in. (6 mm) thick, **Table 5.7, Table 5.8,** and **Table 5.9.** In this practice,

Table 5.7 Suggested Setting for GTA Welding of Mild Steel

Tungsten			Welding			Shielding			Filler	
Type	Size	Tip	Amp	Current	HF	Type	Flow	Size	Type	Size
EWTh-1 or EWTh-2 or EWCe-2 or EWLa	1/16 in. (2 mm)	Point	50 to 100	DCEN DCSP	Start or auto	Argon	16 cfh 7 L/min auto	3/8 in. (10 mm)	RG60 or ER70S-3	1/16–3/32 in. (2–2.4 mm)
	3/32 in. (2.4 mm)	Point	70 to 150	DCEN DCSP	Start or auto	Argon	16 cfh 7 L/min auto	3/8 in. (10 mm)	RG60 or ER70S-3	1/16–3/32 in. (2–2.4 mm)
	1/8 in. (3 mm)	Point	90 to 250	DCEN DCSP	Start or auto	Argon	20 cfh 9 L/min	1/2 in. (13 mm)	RG60 or ER70S-3	3/32–1/8 in. (2.4–3 mm)

Table 5.8 Suggested Setting for GTA Welding of Stainless Steel

Tungsten			Welding			Shielding			Filler	
Type	Size	Tip	Amp	Current	HF	Type	Flow	Size	Type	Size
EWTh-1 or EWTh-2	1/16 in. (2 mm)	Point	70 to 100	DCEN DCSP	Start or auto	Argon	16 cfh 7 L/min	3/8 in. (10 mm)	ER308 or ER316	1/16–3/32 in. (2–2.4 mm)
EWTh-1 or EWCe-2 or EWLa	3/32 in. (2.4 mm)	Point	70 to 150	DCEN DCSP	Start or auto	Argon	16 cfh 7 L/min	3/8 in. (10 mm)	ER308 or ER316	1/16–3/32 in. (2–2.4 mm)
	1/8 in. (3 mm)	Point	90 to 250	DCEN DCSP	Start or auto	Argon	20 cfh 9 L/min	1/2 in. (13 mm)	ER308 or ER316	3/32–1/8 in. (2.4–3 mm)

Table 5.9 Suggested Setting for GTA Welding of Aluminum

Tungsten		Welding				Shielding			Filler	
Type	Size	Tip	Amp	Current	HF	Type	Flow	Size	Type	Size
EWZr or EWLa or EWCe-2 or EWP	1/16 in.	Round 2 mm	50 to 90	AC	Continues or on	Argon	17 cfh 8 L/min	7/16 in. (11 mm)	ER1100 or ER4043	1/16–3/32 in. (2–2.4 mm)
	3/32 in.	Round 2.4 mm	80 to 130	AC	Continues or on	Argon	20 cfh 9 L/min	1/2 in. (13 mm)	ER1100 or ER4043	1/16–3/32 in. (2–2.4 mm)
	1/8 in.	Round 3 mm	100 to 200	AC	Continues or on	Argon	20 cfh 9 L/min	5/8 in. (16 mm)	ER1100 or ER4043	3/32–1/8 in. (2.4–3 mm)

Note: A pointed electrode may be preferred with AC and inverter GTAW power supplies with enhanced balance control capabilities.

you will make a straight stringer bead, 6 in. (152 mm) long, that is uniform in width, reinforcement, and penetration, **Figure 5.19.** Use DCEN current on the steel and AC current on the aluminum.

Starting with the metal that is 1/8 in. (3 mm) thick and the filler rod having a 3/32-in. (2.4-mm) diameter, strike an arc and establish a weld pool, **Figure 5.20.** Move the torch in a circle as in the practice beading. When the torch is on one side, add filler rod to the other side of the molten weld pool, **Figure 5.21.** The end of the rod can be held lightly in the leading edge of the molten weld pool, or it can be dipped into the molten weld pool. If you are using the dipping method, be sure not to allow the tip to melt and drip into the weld pool, **Figure 5.22.** Change to another size filler rod and determine its effect on the weld pool.

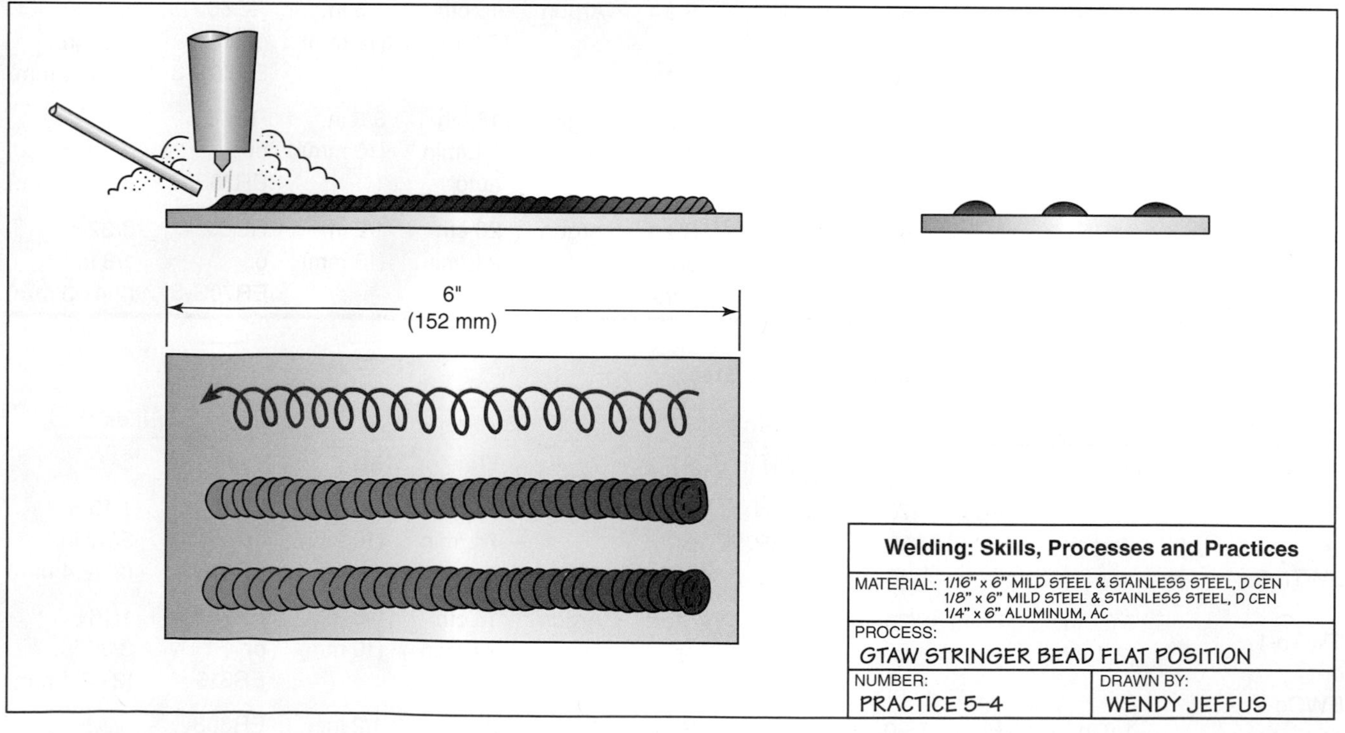

Figure 5.19
Stringer beads in the flat position

Figure 5.20
Establish a molten weld pool and dip the filler rod into it
Courtesy of Larry Jeffu

Figure 5.21
Note the difference in the weld produced when different size filler rods are used
Courtesy of Larry Jeffus

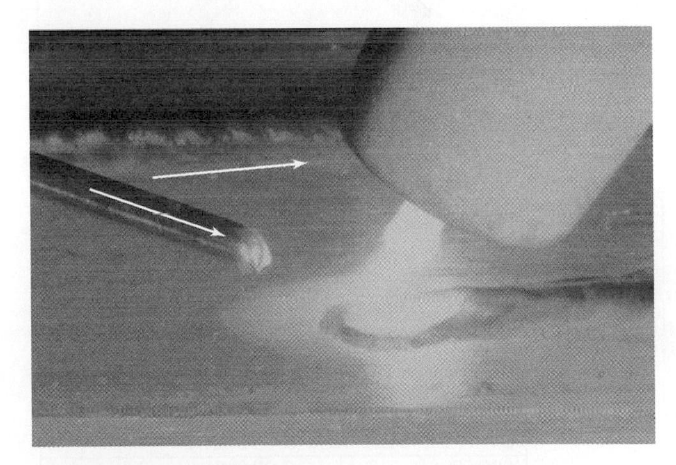

Figure 5.22
Move the electrode back as the filler rod is added
Courtesy of Larry Jeffus

Maintain a smooth and uniform rhythm as filler metal is added. This will help to keep the bead uniform. Vary the rhythms to determine which one is easiest for you. If the rod sticks, move the torch toward the rod until it melts free.

When the full 6-in. (152-mm)-long weld bead is completed, cool and inspect it for uniformity and defects. Repeat the process using all thicknesses of metal until you can consistently make the weld visually defect free. Turn off the welding machine, shielding gas, and cooling water, and clean up your work area when you are finished welding.

Complete a copy of the "Student Welding Report" listed in Appendix I or provided by your instructor.

Module 1
Key Indicator 1, 3, 4

Module 2
Key Indicator 1, 2, 3, 4, 6, 7

Module 7
Key Indicator 1, 2
Carbon Steel
Key Indicator 3, 4
Austenitic Stainless Steel
Key Indicator 8, 9
Aluminum
Key Indicator 13, 14

Module 9
Key Indicator 1, 2

PRACTICE 5-5

Outside Corner Joint, 1G Position, Using Mild Steel, Stainless Steel, Aluminum

Using the same equipment and materials listed in Practice 5-4, weld an outside corner joint in the flat position, **Figure 5.23.**

- Place one of the pieces of metal flat on the table and hold or brace the other piece of metal horizontally on it.
- Tack weld both ends of the plates together, **Figure 5.24.**
- Set the plates up and add two or three more tack welds on the joint as required, **Figure 5.25.**

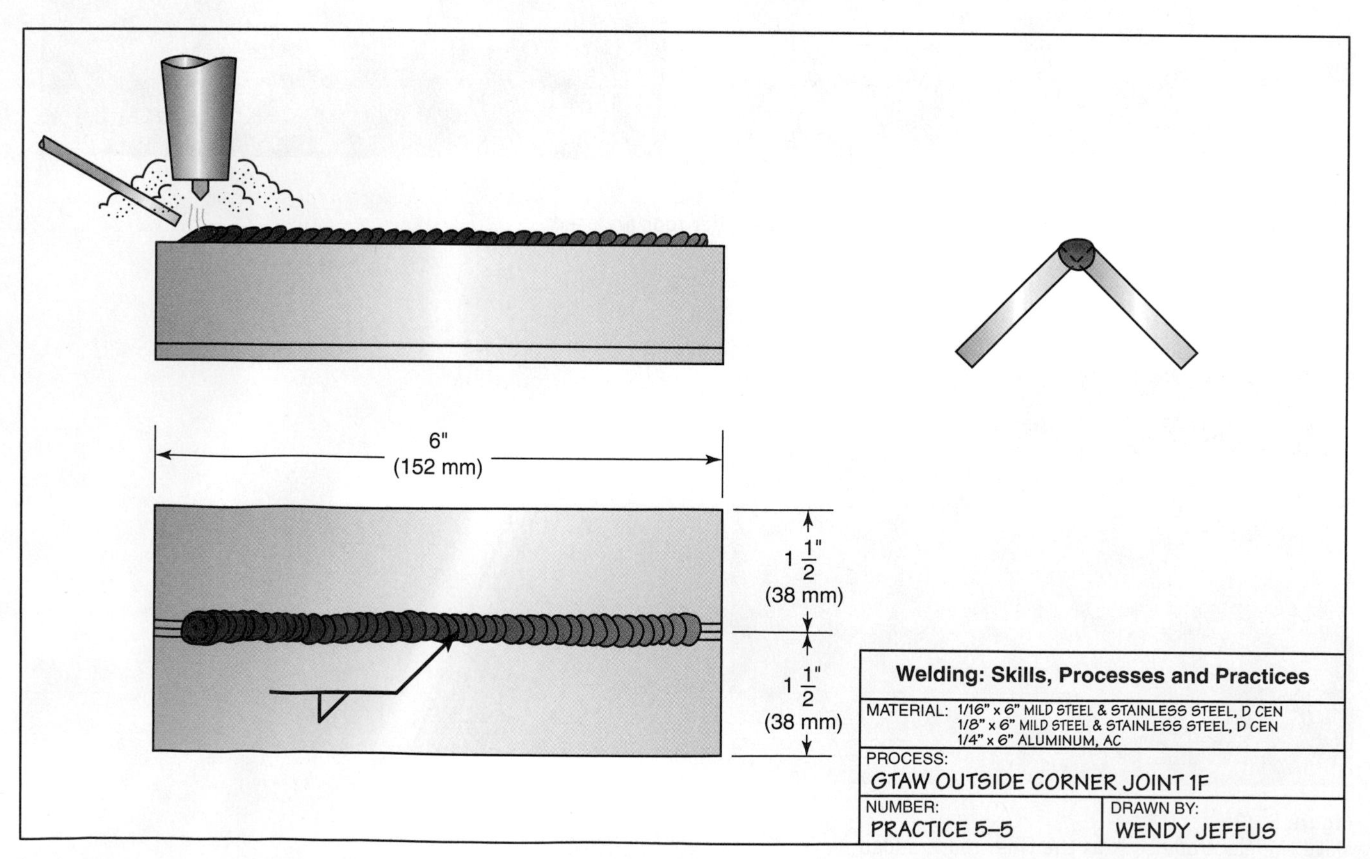

Figure 5.23
Outside corner joint in the flat position

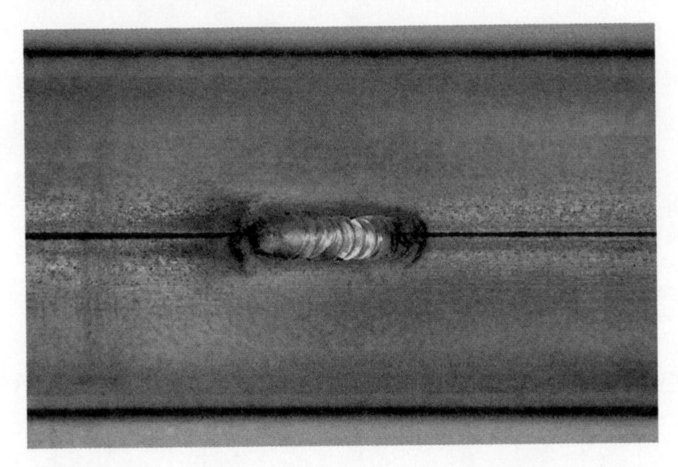

Figure 5.24
Tack weld. Note the good fusion at the start and crater fill at the end.
Courtesy of Larry Jeffus

Figure 5.25
Outside corner tack welded together
Courtesy of Larry Jeffus

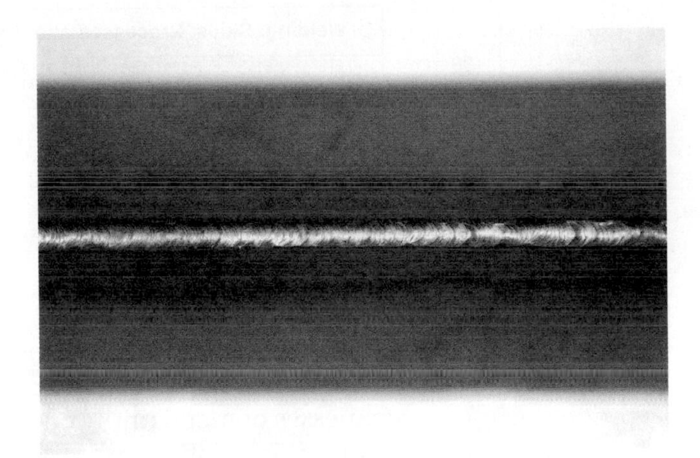

Figure 5.26
Outside corner joint. Note precleaning along weld.
Courtesy of Larry Jeffus

- Starting at one end, make a uniform weld, adding filler metal as needed. In **Figure 5.26,** note the metal areas that arc precleaned before the weld is made.

Repeat each weld as needed until all are mastered. Turn off the welding machine, shielding gas, and cooling water, and clean up your work area when you are finished welding.

Complete a copy of the "Student Welding Report" listed in Appendix I or provided by your instructor.

Module 1
Key Indicator 1, 3, 4

Module 2
Key Indicator 1, 2, 3, 4, 6, 7

Module 7
Carbon Steel
Key Indicator 6
Austenitic Stainless Steel
Key Indicator 11
Aluminum
Key Indicator 16

Module 9
Key Indicator 1, 2

PRACTICE 5-6

Butt Joint, 1G Position, Using Mild Steel, Stainless Steel, Aluminum

Using the same equipment and materials as listed in Practice 5-4, you will weld a butt joint in the flat position, **Figure 5.27.**

Place the metal flat on the table and tack weld both ends together, **Figure 5.28.** Two or three additional tack welds can be made along the

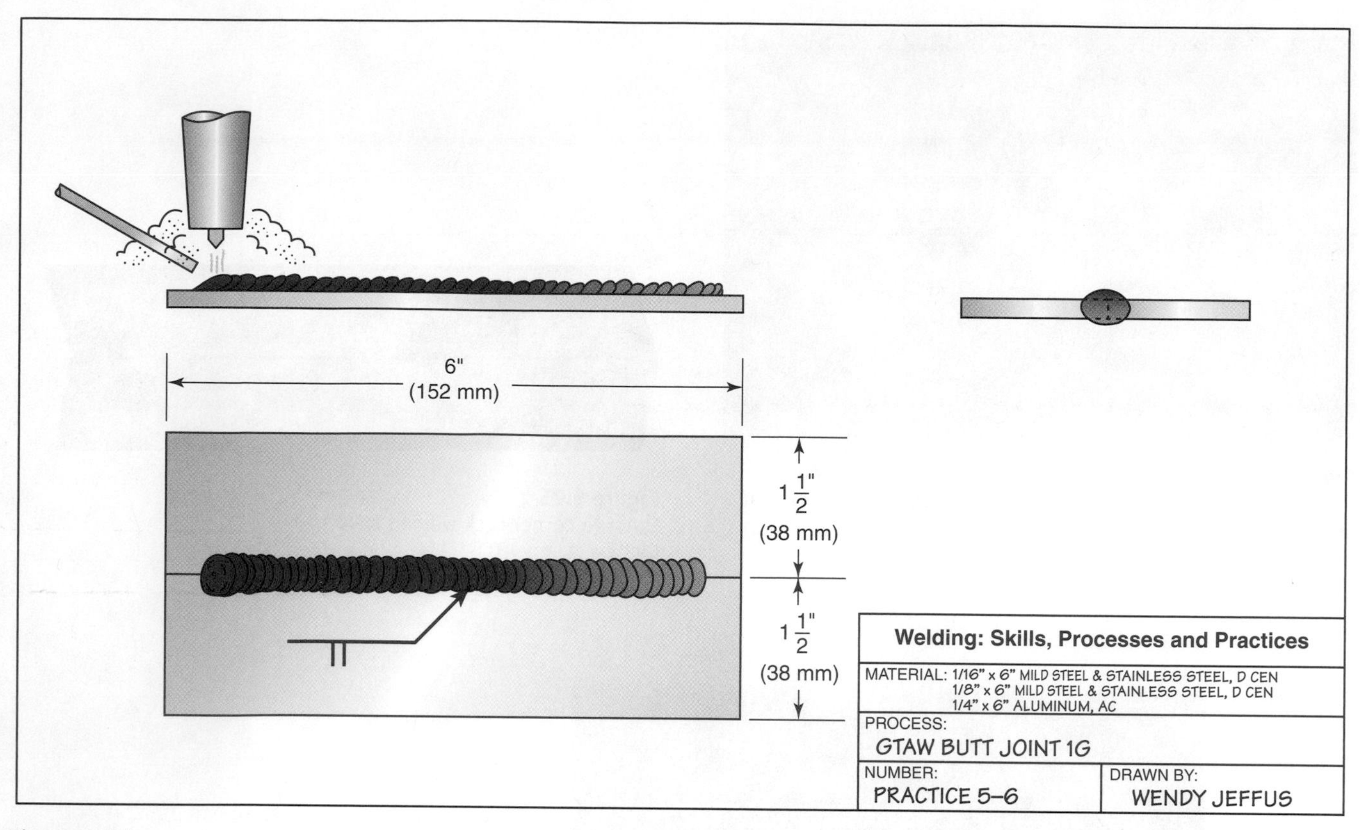

Figure 5.27
Square butt joint in the flat position

joint as needed. Starting at one end, make a uniform weld along the joint. Add filler metal as required to make a uniform weld.

Repeat the process using all thicknesses of metal until you can consistently make the weld visually defect free. Turn off the welding machine, shielding gas, and cooling water, and clean up your work area when you are finished welding.

Figure 5.28
Tack weld on butt joint
Courtesy of Larry Jeffus

Complete a copy of the "Student Welding Report" listed in Appendix I or provided by your instructor.

PRACTICE 5-7

Butt Joint, 1G Position, with 100% Penetration, to Be Tested, Using Mild Steel, Stainless Steel, Aluminum

Using the same equipment and materials as listed in Practice 5-4, you will weld a butt joint with 100% penetration, **Figure 5.29,** along the entire 6-in. (152-mm) length of the joint. After the weld is completed, shear out strips 1 in. (25 mm) wide and bend-test them as shown in **Figure 5.30.**

Repeat each weld until all have 100% root penetration. Turn off the welding machine, shielding gas, and cooling water, and clean up your work area when you are finished welding.

Complete a copy of the "Student Welding Report" listed in Appendix I or provided by your instructor.

Module 1
Key Indicator 1, 3, 4

Module 2
Key Indicator 1, 2, 3, 4, 6, 7

Module 7
Carbon Steel
Key Indicator 6
Austenitic Stainless Steel
Key Indicator 11
Aluminum
Key Indicator 16

Module 9
Key Indicator 1, 2

PRACTICE 5-8

Butt Joint, 1G Position, with Minimum Distortion, Using Mild Steel, Stainless Steel, Aluminum

Using the same equipment and materials as listed in Practice 5-4, you will weld a flat butt joint, while controlling both distortion and penetration, **Figure 5.31.**

Tack weld the plates together as shown in **Figure 5.32.** Using a backstepping weld sequence, make a series of welds approximately 1 in. (25 mm) long along the joint. Be sure to fill each weld crater adequately to reduce crater cracking.

Repeat the process using all thicknesses of metal until you can consistently make the weld visually defect free. Turn off the welding machine, shielding gas, and cooling water, and clean up your work area when you are finished welding.

Complete a copy of the "Student Welding Report" listed in Appendix I or provided by your instructor.

Module 1
Key Indicator 1, 3, 4

Module 2
Key Indicator 1, 2, 3, 4, 6, 7

Module 7
Carbon Steel
Key Indicator 6
Austenitic Stainless Steel
Key Indicator 11
Aluminum
Key Indicator 16

Module 9
Key Indicator 1, 2

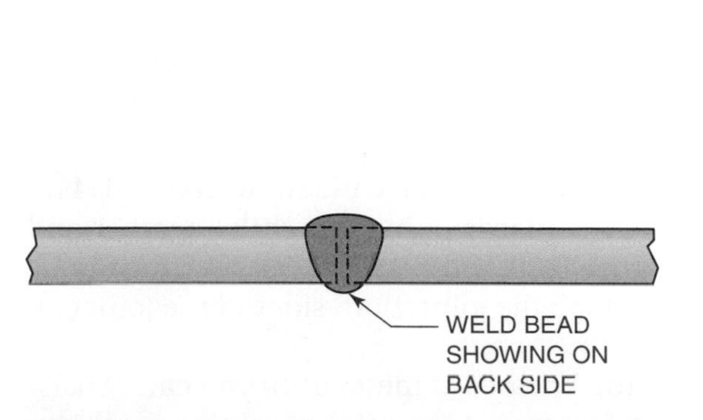

Figure 5.29
100% weld penetration

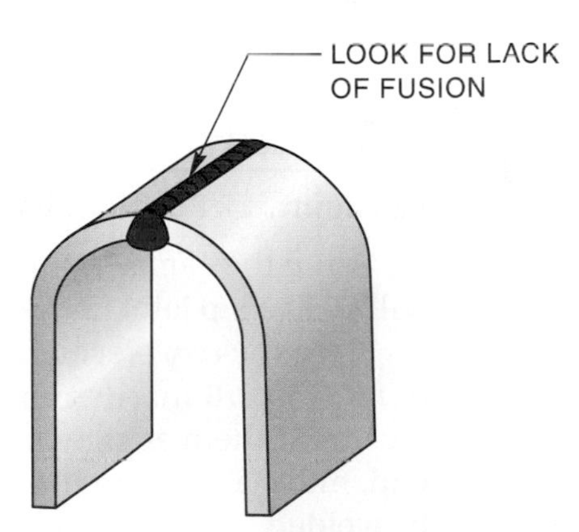

Figure 5.30
Bend the 1-in. (25-mm) strip of butt joint backward and look at the root for 100% penetration

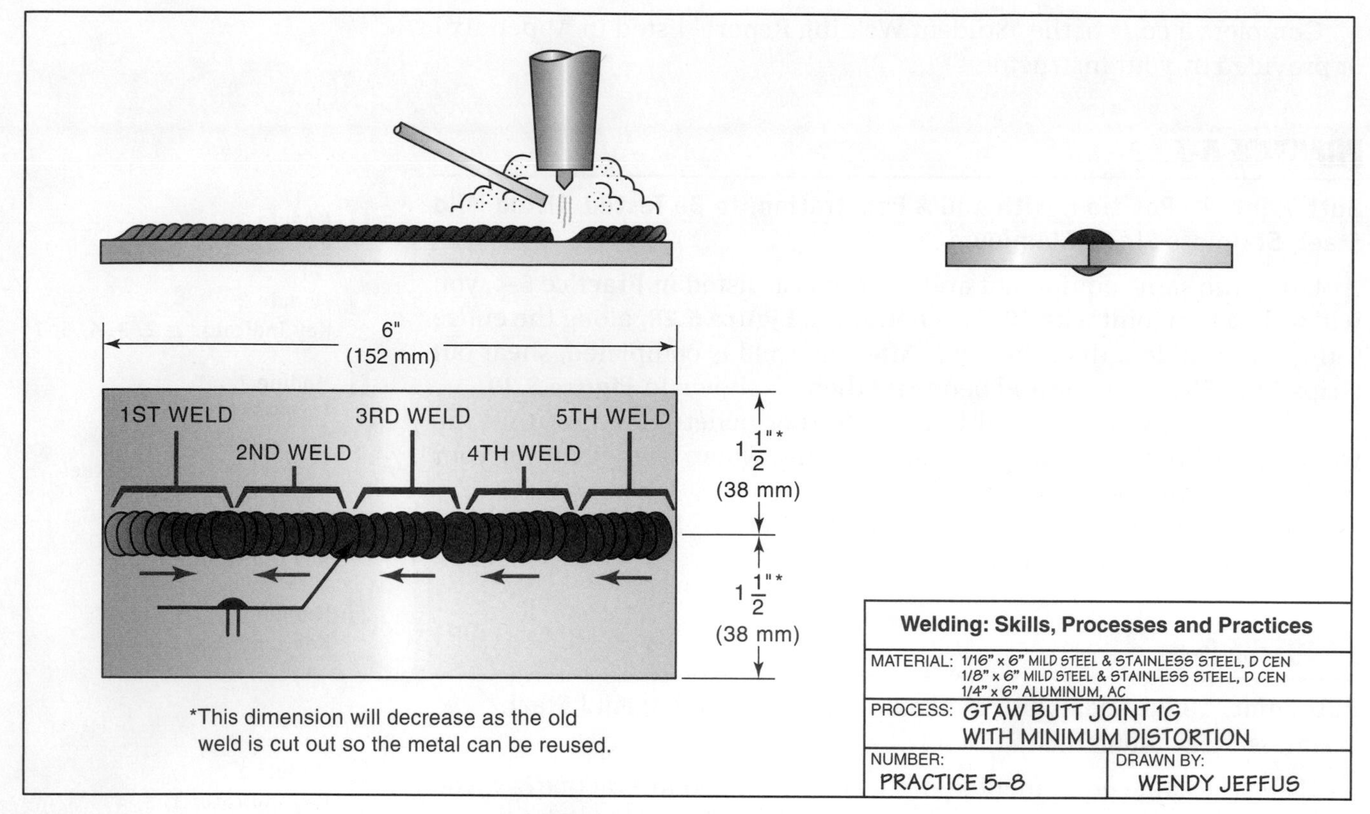

Figure 5.31
Square butt joint in the flat position with minimum distortion

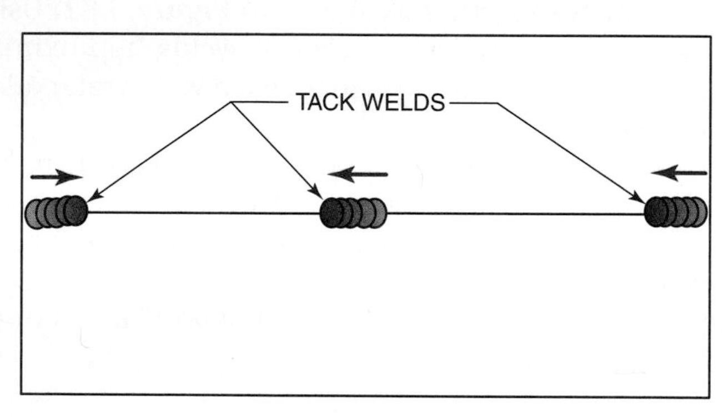

Figure 5.32
Tack welds on a butt joint

PRACTICE 5-9

Lap Joint, 1F Position, Using Mild Steel, Stainless Steel, Aluminum

Module 1
Key Indicator 1, 3, 4

Module 2
Key Indicator 1, 2, 3, 4, 6, 7

Module 7
Carbon Steel
Key Indicator 6
Austenitic Stainless Steel
Key Indicator 10
Aluminum
Key Indicator 15

Module 9
Key Indicator 1, 2

Using the same equipment and materials as listed in Practice 5-4, you will weld a lap joint in the flat position, **Figure 5.33.**

Place the two pieces of metal flat on the table with an overlap of 1/4 in. (6 mm) to 3/8 in. (10 mm). Hold the pieces of metal tightly together and tack weld them as shown in **Figure 5.34** and **Figure 5.35.** Starting at one end, make a uniform fillet weld along the joint. Both sides of the joint can be welded.

Repeat the process using all thicknesses of metal until you can consistently make the weld visually defect free. Turn off the welding machine, shielding gas, and cooling water, and clean up your work area when you are finished welding.

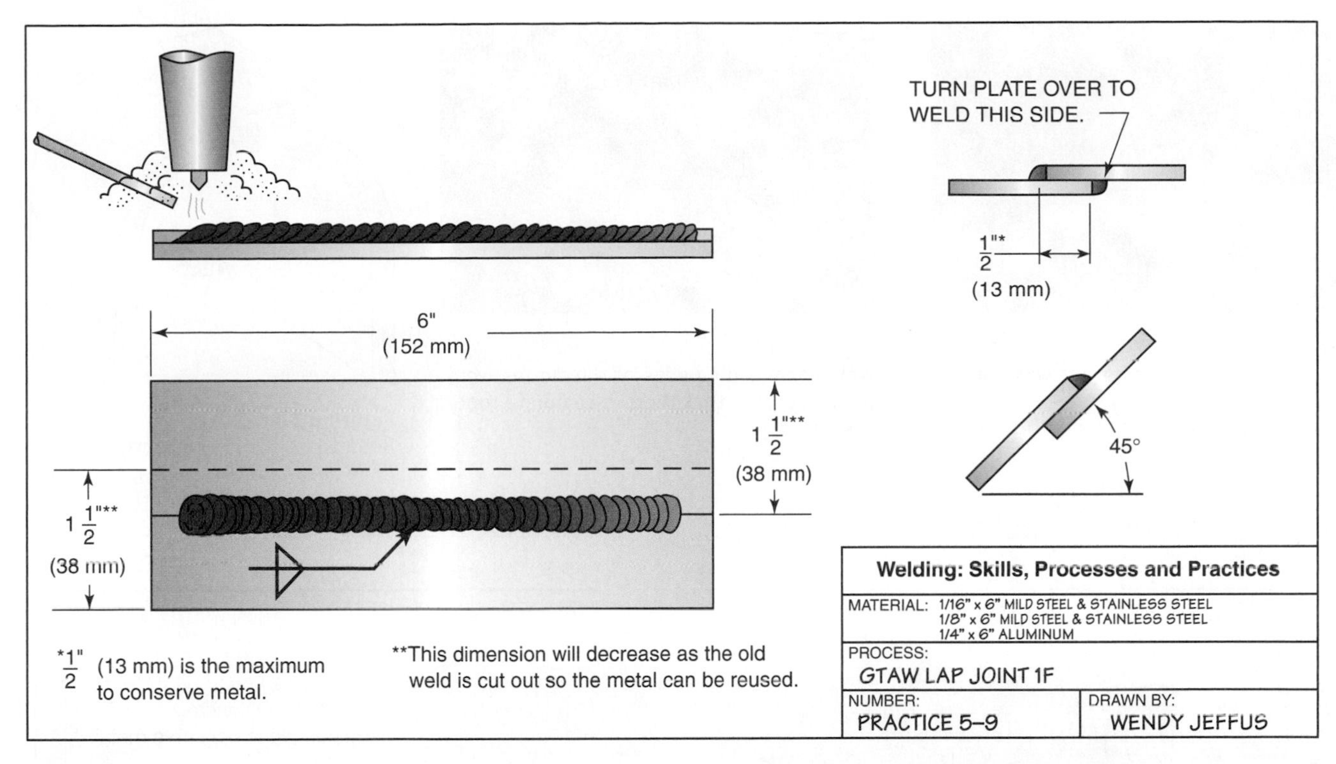

Figure 5.33
Lap joint in the flat position

Figure 5.34
Be sure both the top and bottom pieces are melted before adding filler metal
Courtesy of Larry Jeffus

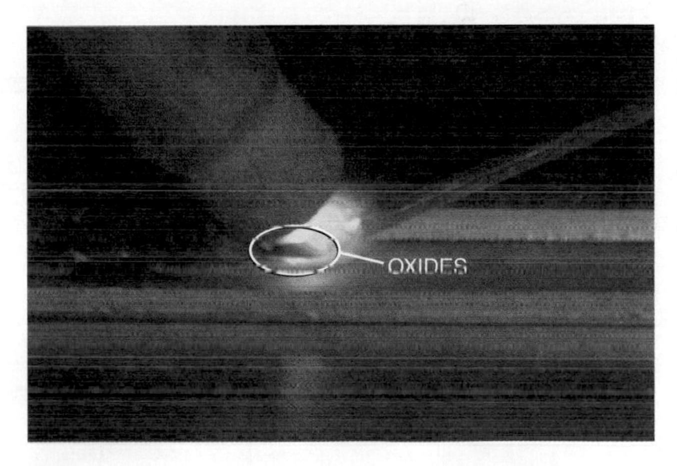

Figure 5.35
Oxides form during tack welding. Do not complete the tack welds. These oxides will become part of the finished weld if the tack is completed.
Courtesy of Larry Jeffus

Complete a copy of the "Student Welding Report" listed in Appendix I or provided by your instructor.

Module 1
Key Indicator 1, 3, 4

Module 2
Key Indicator 1, 2, 3, 4, 6, 7

Module 7
Carbon Steel
Key Indicator 5
Austenitic Stainless Steel
Key Indicator 10
Aluminum
Key Indicator 15

Module 9
Key Indicator 1, 2

PRACTICE 5-10

Lap Joint, 1F Position, to Be Tested, Using Mild Steel, Stainless Steel, Aluminum

Using the same equipment and materials as listed in Practice 5-4, you will make a fillet weld on one side of a lap joint and test it for 100% root penetration, **Figure 5.36, Figure 5.37,** and **Figure 5.38.** After the weld is completed, shear out strips 1 in. (25 mm) wide on one side of the joint and bend-test them as shown in **Figure 5.39.**

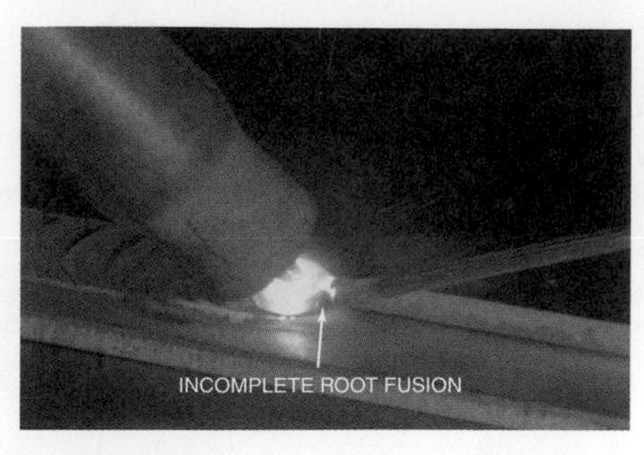

Figure 5.36
A notch indicates that the root was not properly melted and fused
Courtesy of Larry Jeffus

Figure 5.37
Watch the leading edge of the molten weld pool to ensure that there is complete root fusion
Courtesy of Larry Jeffus

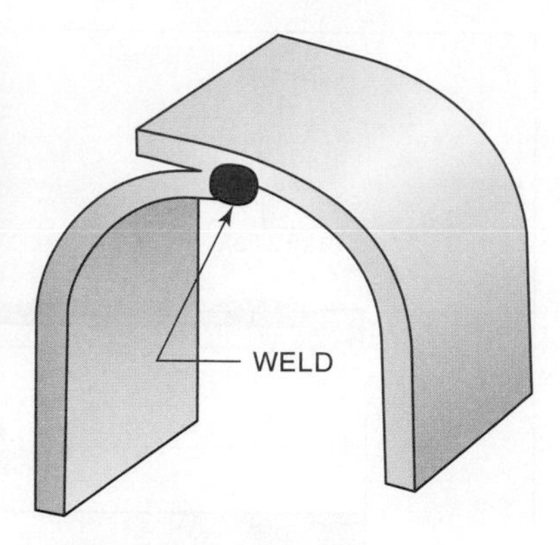

Figure 5.38
Bend the 1-in (25-mm) strip of lap joint backward and look at the root for 100% penetration

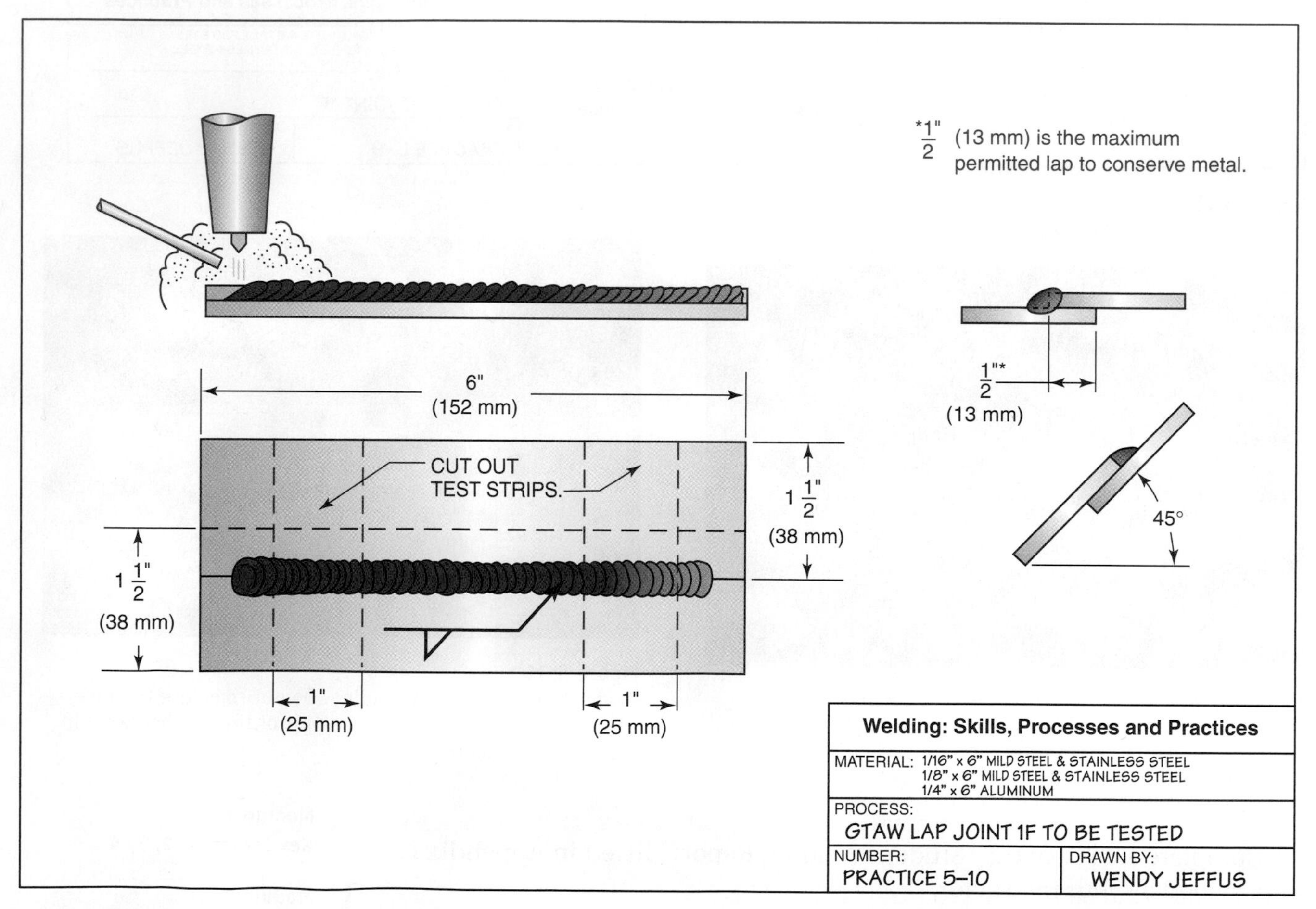

Figure 5.39
Remove strips and test for root fusion

Repeat each weld until all have 100% root penetration. Turn off the welding machine, shielding gas, and cooling water, and clean up your work area when you are finished welding.

Complete a copy of the "Student Welding Report" listed in Appendix I or provided by your instructor.

PRACTICE 5-11

Tee Joint, 1F Position, Using Mild Steel, Stainless Steel, Aluminum

Using the same equipment and materials as listed in Practice 5-4, you will weld a tee joint in the flat position, **Figure 5.40.**

Place one of the pieces of metal flat on the table and hold or brace the other piece of metal horizontally on it. Tack weld both ends of the plates together, **Figure 5.41.** Set up the plates in the flat position and add two or three more tack welds to the joint as required, **Figure 5.42.**

On the metal that is 1/16 in. (1.5 mm) thick, it may not be possible to weld both sides, but on thicker material a fillet weld can usually be made on both sides. The exception to this is if carbide precipitation occurs on the stainless steel during welding.

Starting at one end, make a uniform weld, adding filler metal as needed.

Repeat the process using all thicknesses of metal until you can consistently make the weld visually defect free. Turn off the welding machine, shielding gas, and cooling water, and clean up your work area when you are finished welding.

Complete a copy of the "Student Welding Report" listed in Appendix I or provided by your instructor.

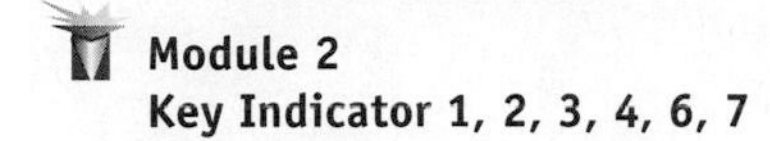

Module 1
Key Indicator 1, 3, 4

Module 2
Key Indicator 1, 2, 3, 4, 6, 7

Module 7
Carbon Steel
Key Indicator 5
Austenitic Stainless Steel
Key Indicator 10
Aluminum
Key Indicator 15

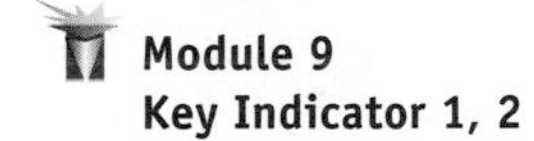

Module 9
Key Indicator 1, 2

PRACTICE 5-12

Tee Joint, 1F Position, to Be Tested, Using Mild Steel, Stainless Steel, Aluminum

Using the same equipment and materials as listed in Practice 5-4, you will weld a tee joint and test it for 100% root penetration, **Figure 5.43.** After

Module 1
Key Indicator 1, 3, 4

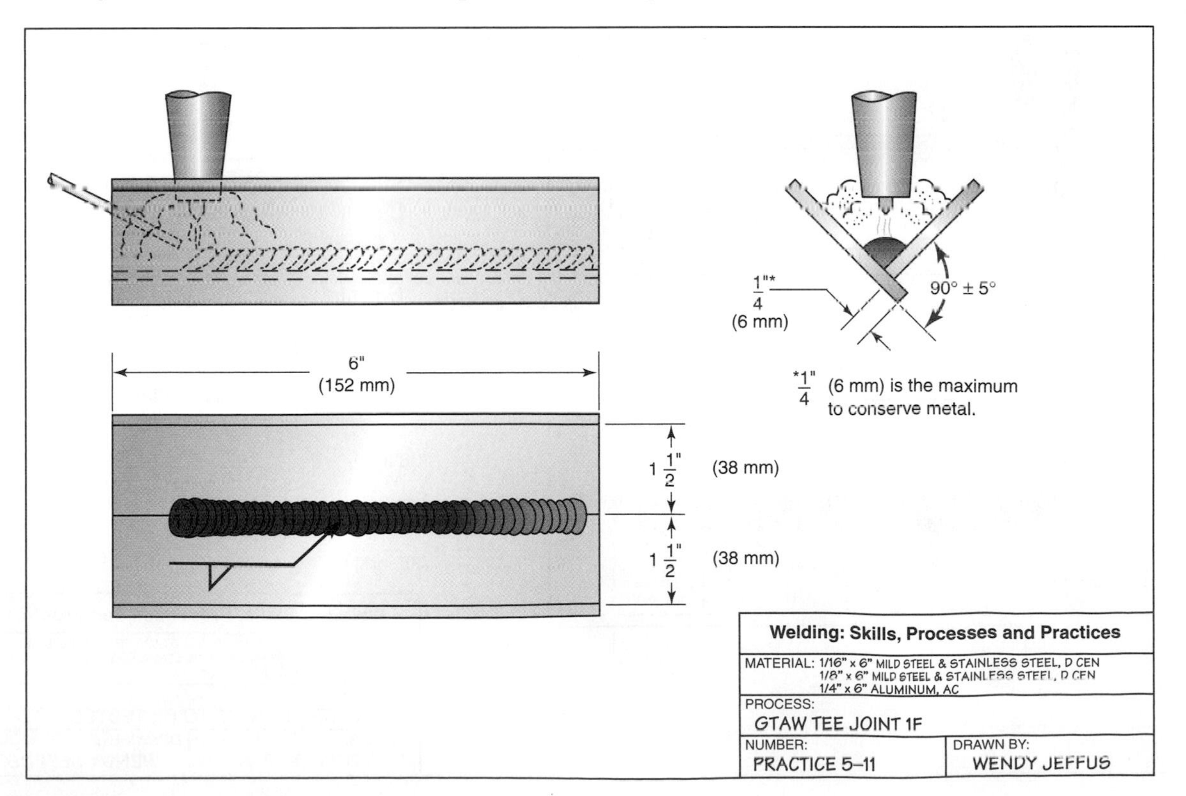

Figure 5.40
Tee joint in the flat position

Figure 5.41
Tack weld on a tee joint
Courtesy of Larry Jeffus

Figure 5.42
Keep the tack welds small so that they will not affect the weld
Courtesy of Larry Jeffus

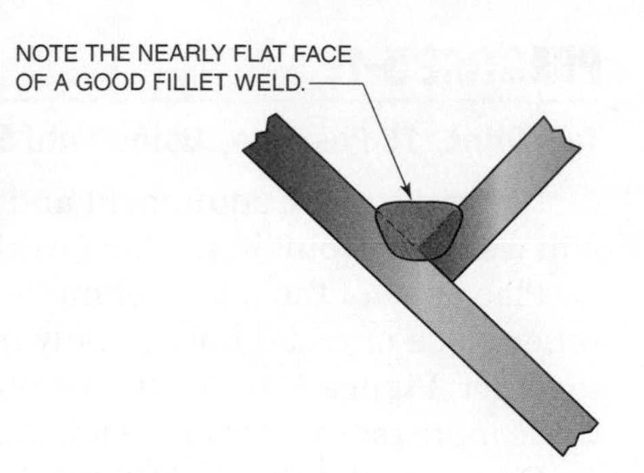

Figure 5.43
100% weld penetration

Module 2
Key Indicator 1, 2, 3, 4, 6, 7

Module 7
Carbon Steel
Key Indicator 5
Austenitic Stainless Steel
Key Indicator 10
Aluminum
Key Indicator 15

Module 9
Key Indicator 1, 2

the weld is completed, cut or shear out strips 1 in. (25 mm) wide and bend-test them as shown in **Figure 5.44** and **Figure 5.45.**

Repeat each weld until all have 100% root penetration. Turn off the welding machine, shielding gas, and cooling water, and clean up your work area when you are finished welding.

Complete a copy of the "Student Welding Report" listed in Appendix I or provided by your instructor.

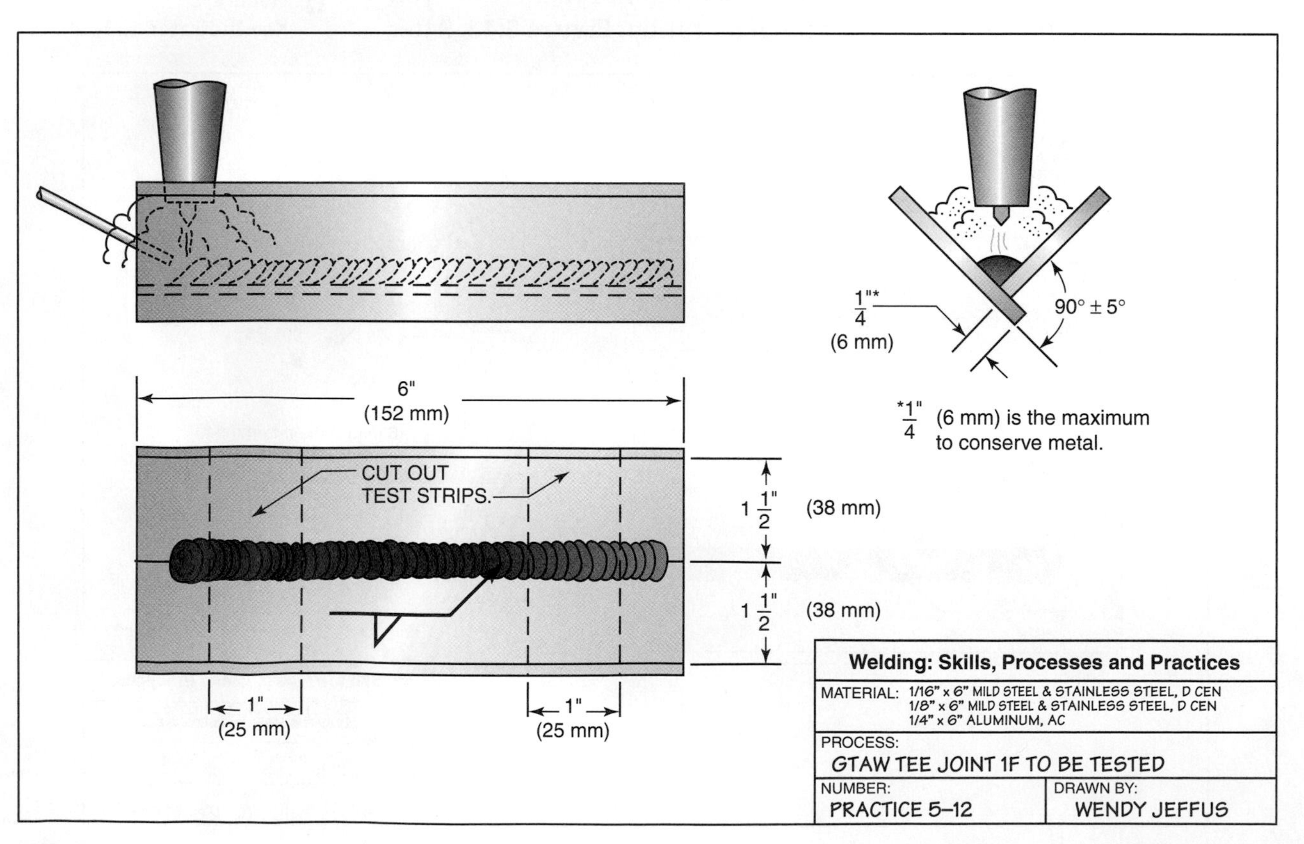

Figure 5.44
Tee joint in the flat position to be tested

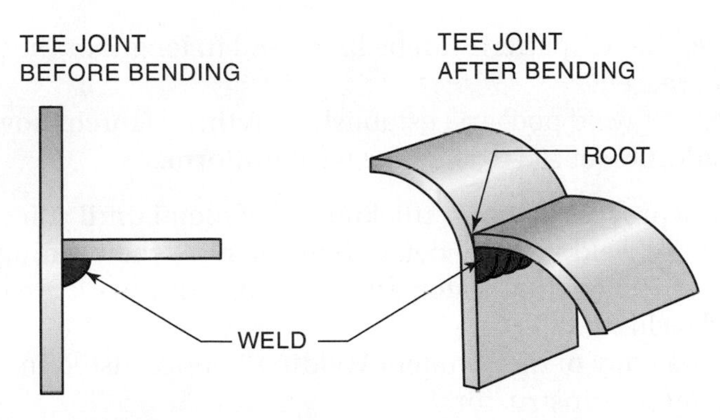

Figure 5.45
Bend the 1-in. (25-mm) strip of the tee joint backward and look at the root for 100% penetration

PRACTICE 5-13

Stringer Bead at a 45° Vertical Angle, Using Mild Steel, Stainless Steel, Aluminum

Using the same equipment and materials listed in Practice 5-4, you will make a stringer bead in the vertical up position.

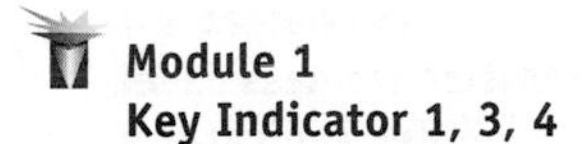

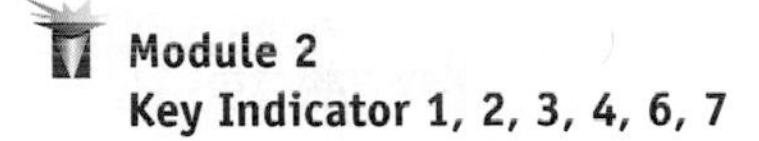

- Starting at the bottom and welding in an upward direction, add the filler metal to the top edge of the weld pool and move the torch in a circle or "C" pattern, **Figure 5.46.** If the weld pool size starts to in-

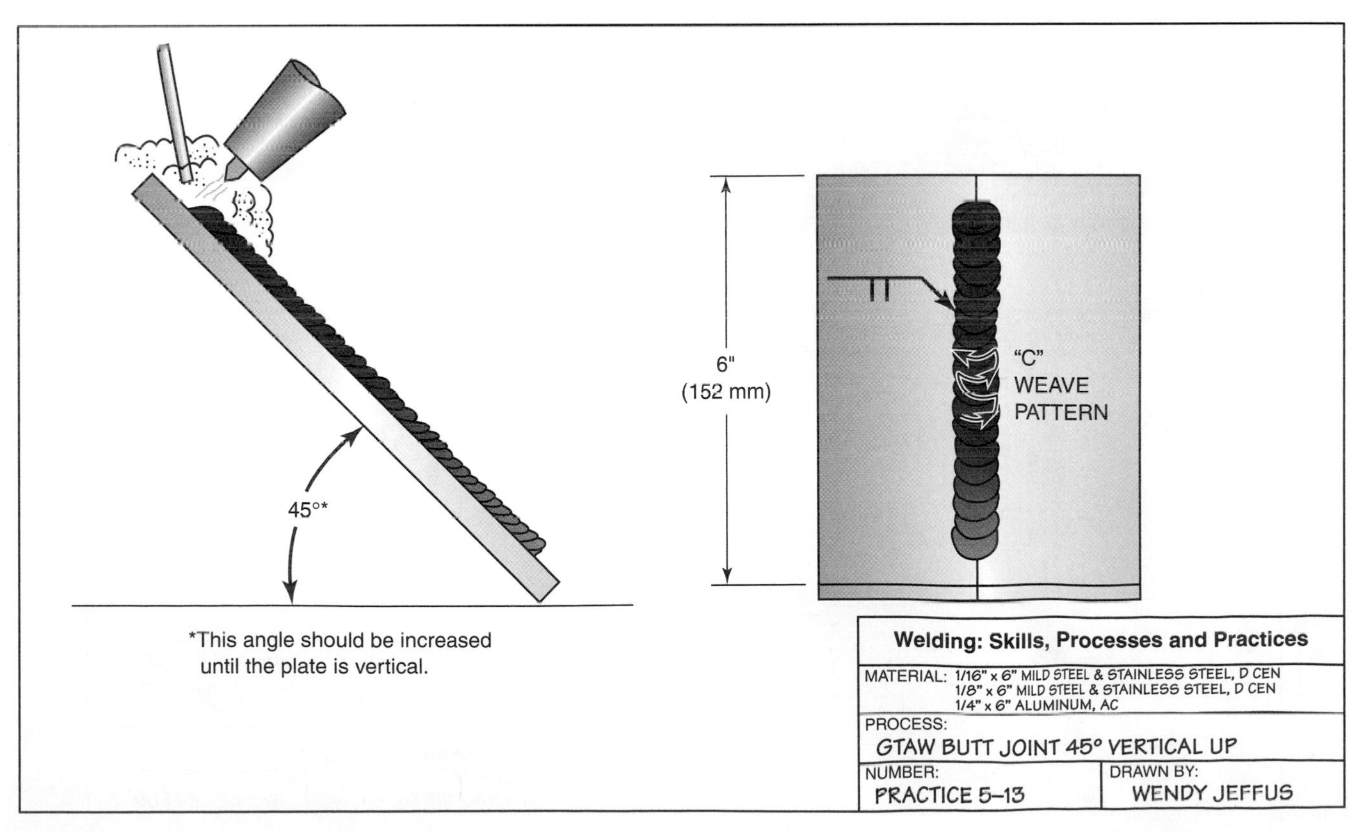

Figure 5.46
45° vertical up

Module 7
Carbon Steel
Key Indicator 4
Austenitic Stainless Steel
Key Indicator 9
Aluminum
Key Indicator 14

Module 9
Key Indicator 1, 2

crease, the "C" pattern can be increased in length or the power can be decreased.

- Watch the weld pool and establish a rhythm of torch movement and addition of rod to keep the weld uniform.

Repeat the process using all thicknesses of metal until you can consistently make the weld visually defect free. Turn off the welding machine, shielding gas, and cooling water, and clean up your work area when you are finished welding.

Complete a copy of the "Student Welding Report" listed in Appendix I or provided by your instructor.

Module 1
Key Indicator 1, 3, 4

Module 2
Key Indicator 1, 2, 3, 4, 6, 7

Module 7
Carbon Steel
Key Indicator 4
Austenitic Stainless Steel
Key Indicator 9
Aluminum
Key Indicator 14

Module 9
Key Indicator 1, 2

PRACTICE 5-14

Stringer Bead, 3G Position, Using Mild Steel, Stainless Steel, Aluminum

Repeat Practice 5-13. Gradually increase the angle as you develop skill until the weld is being made in the vertical up position, **Figure 5.47.** Repeat the process using all thicknesses of metal until you can consistently make the weld visually defect free. Turn off the welding machine, shielding gas, and cooling water, and clean up your work area when you are finished welding.

Complete a copy of the "Student Welding Report" listed in Appendix I or provided by your instructor.

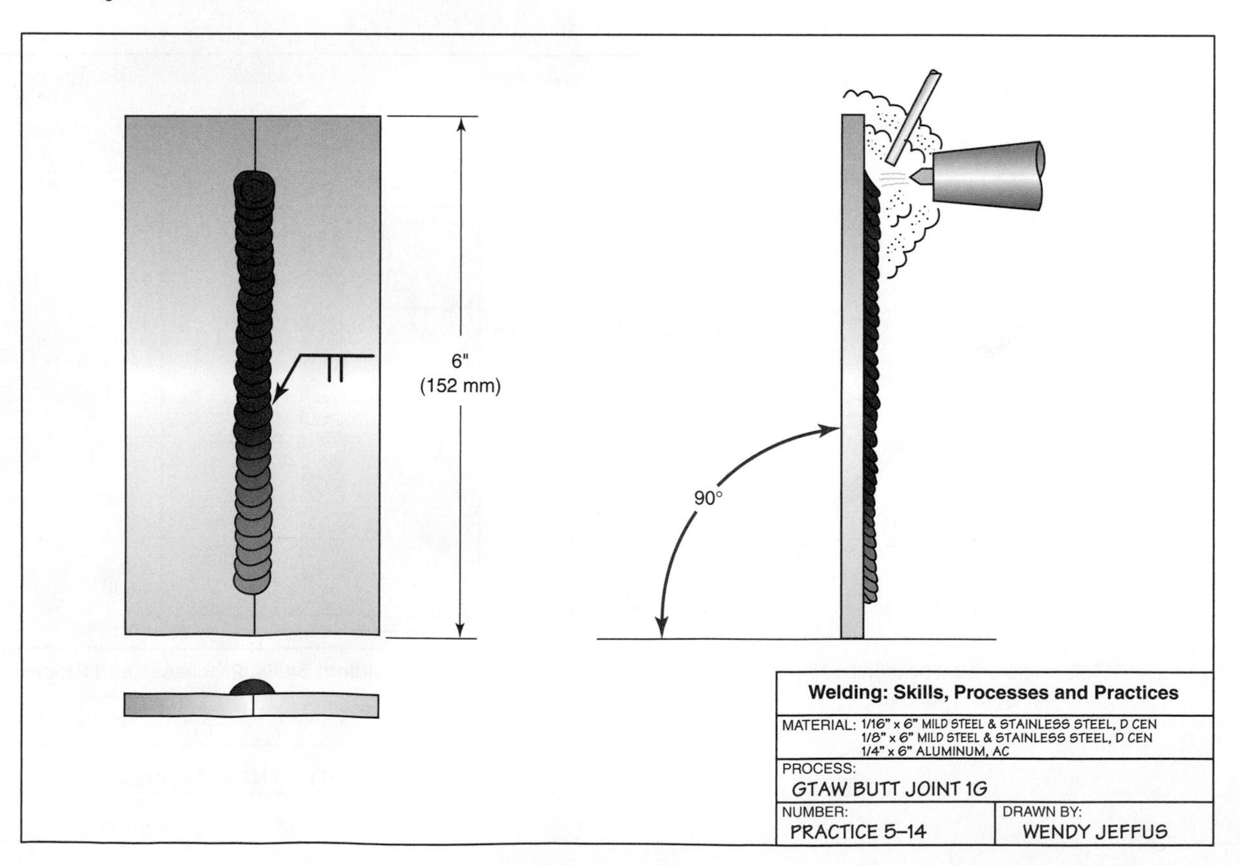

Figure 5.47
Vertical up position

PRACTICE 5-15

Butt Joint at a 45° Vertical Angle, Using Mild Steel, Stainless Steel, Aluminum

Using the same equipment and materials as listed in Practice 5-4, you will weld a butt joint in the vertical up position.

After tack welding the plates together, start the weld at the bottom and weld in an upward direction. The same rhythmic torch and rod movement practiced for the 45° stringer bead should be used to control the weld.

Repeat the process using all thicknesses of metal until you can consistently make the weld visually defect free. Turn off the welding machine, shielding gas, and cooling water, and clean up your work area when you are finished welding.

Complete a copy of the "Student Welding Report" listed in Appendix I or provided by your instructor.

Module 1
Key Indicator 1, 3, 4

Module 2
Key Indicator 1, 2, 3, 4, 6, 7

Module 7
Carbon Steel
Key Indicator 6
Austenitic Stainless Steel
Key Indicator 11
Aluminum
Key Indicator 16

Module 9
Key Indicator 1, 2

PRACTICE 5-16

Butt Joint, 3G Position, Using Mild Steel, Stainless Steel, Aluminum

Repeat Practice 5-15. Gradually increase the plate angle after each weld. As you develop skill, continue increasing the angle until the weld is being made in the vertical up position. Repeat the process using all thicknesses of metal until you can consistently make the weld visually defect free. Turn off the welding machine, shielding gas, and cooling water, and clean up your work area when you are finished welding.

Complete a copy of the "Student Welding Report" listed in Appendix I or provided by your instructor.

Module 1
Key Indicator 1, 3, 4

Module 2
Key Indicator 1, 2, 3, 4, 6, 7

Module 7
Carbon Steel
Key Indicator 6
Austenitic Stainless Steel
Key Indicator 11
Aluminum
Key Indicator 16

Module 9
Key Indicator 1, 2

PRACTICE 5-17

Butt Joint, 3G Position, with 100% Penetration, to Be Tested, Using Mild Steel, Stainless Steel, Aluminum

Repeat Practice 5-16. Make the needed changes in the root opening to allow 100% penetration, **Figure 5.48.** It may be necessary to provide a backing gas to protect the root from atmospheric contamination. After the weld is completed, visually inspect it for uniformity and defects. Then shear out strips 1 in. (25 mm) wide and bend-test them.

Repeat each weld until all have 100% root penetration. Turn off the welding machine, shielding gas, and cooling water, and clean up your work area when you are finished welding.

Complete a copy of the "Student Welding Report" listed in Appendix I or provided by your instructor.

Module 1
Key Indicator 1, 3, 4

Module 2
Key Indicator 1, 2, 3, 4, 6, 7

Module 7
Carbon Steel
Key Indicator 6
Austenitic Stainless Steel
Key Indicator 11
Aluminum
Key Indicator 16

Module 9
Key Indicator 1, 2

PRACTICE 5-18

Lap Joint at a 45° Vertical Angle, Using Mild Steel, Stainless Steel, Aluminum

Using the same equipment and materials as listed in Practice 5-4, you will make a vertical up fillet weld on a lap joint.

Module 1
Key Indicator 1, 3, 4

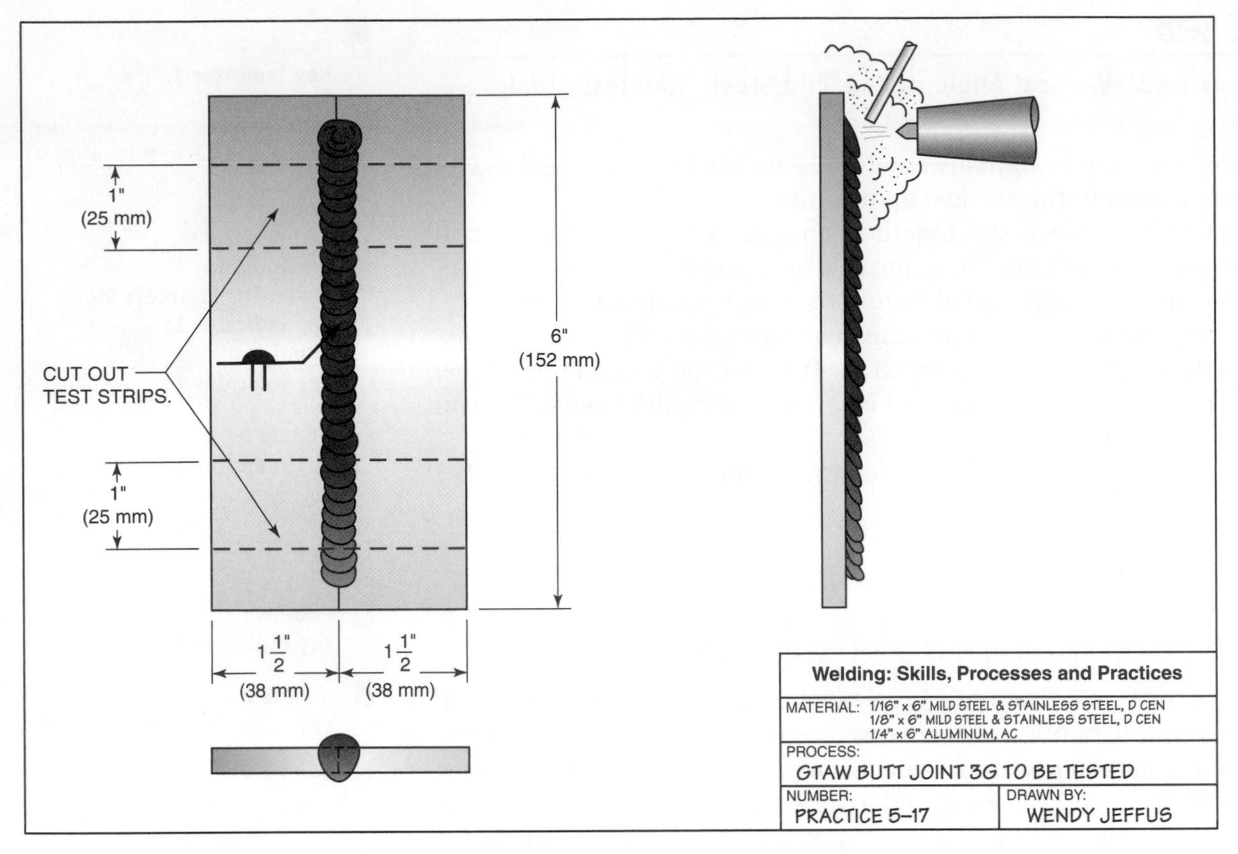

Figure 5.48
Vertical up butt joint with 100% penetration. Strips are to be cut for testing.

Module 2
Key Indicator 1, 2, 3, 4, 6, 7

Module 7
Carbon Steel
Key Indicator 5
Austenitic Stainless Steel
Key Indicator 10
Aluminum
Key Indicator 15

Module 9
Key Indicator 1, 2

After tack welding the plates together, start the weld at the bottom and weld in an upward direction. It is important to maintain a uniform weld rhythm so that a nice-looking weld bead is formed. It may be necessary to move the torch in and around the base of the weld pool to ensure adequate root fusion, **Figure 5.49.** The filler metal should be added along the top edge of the weld pool near the top plate.

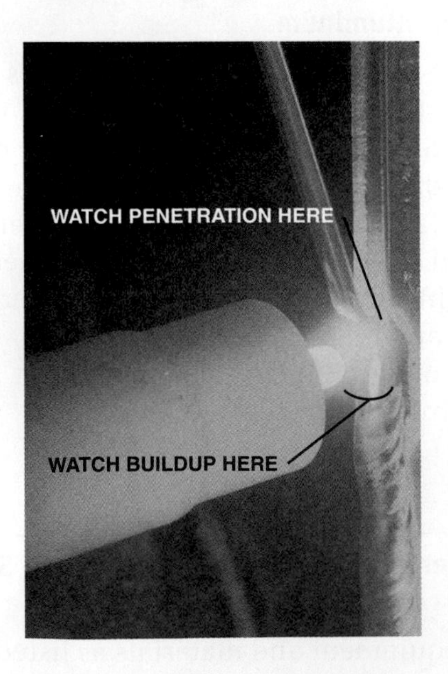

Figure 5.49
Lapjoint
Courtesy of Larry Jeffus

Repeat the process using all thicknesses of metal until you can consistently make the weld visually defect free. Turn off the welding machine, shielding gas, and cooling water, and clean up your work area when you are finished welding.

Complete a copy of the "Student Welding Report" listed in Appendix I or provided by your instructor.

PRACTICE 5-19

Lap Joint, 3F Position, Using Mild Steel, Stainless Steel, Aluminum

Repeat Practice 5-18. Gradually increase the plate angle after each weld as you develop your skill. Increase the angle until the weld is being made in the vertical up position, **Figure 5.50.**

Repeat the process using all thicknesses of metal until you can consistently make the weld visually defect free. Turn off the welding machine, shielding gas, and cooling water, and clean up your work area when you are finished welding.

Complete a copy of the "Student Welding Report" listed in Appendix I or provided by your instructor.

Module 1
Key Indicator 1, 3, 4

Module 2
Key Indicator 1, 2, 3, 4, 6, 7

Module 7
Carbon Steel
Key Indicator 5
Austenitic Stainless Steel
Key Indicator 10
Aluminum
Key Indicator 15

Module 9
Key Indicator 1, 2

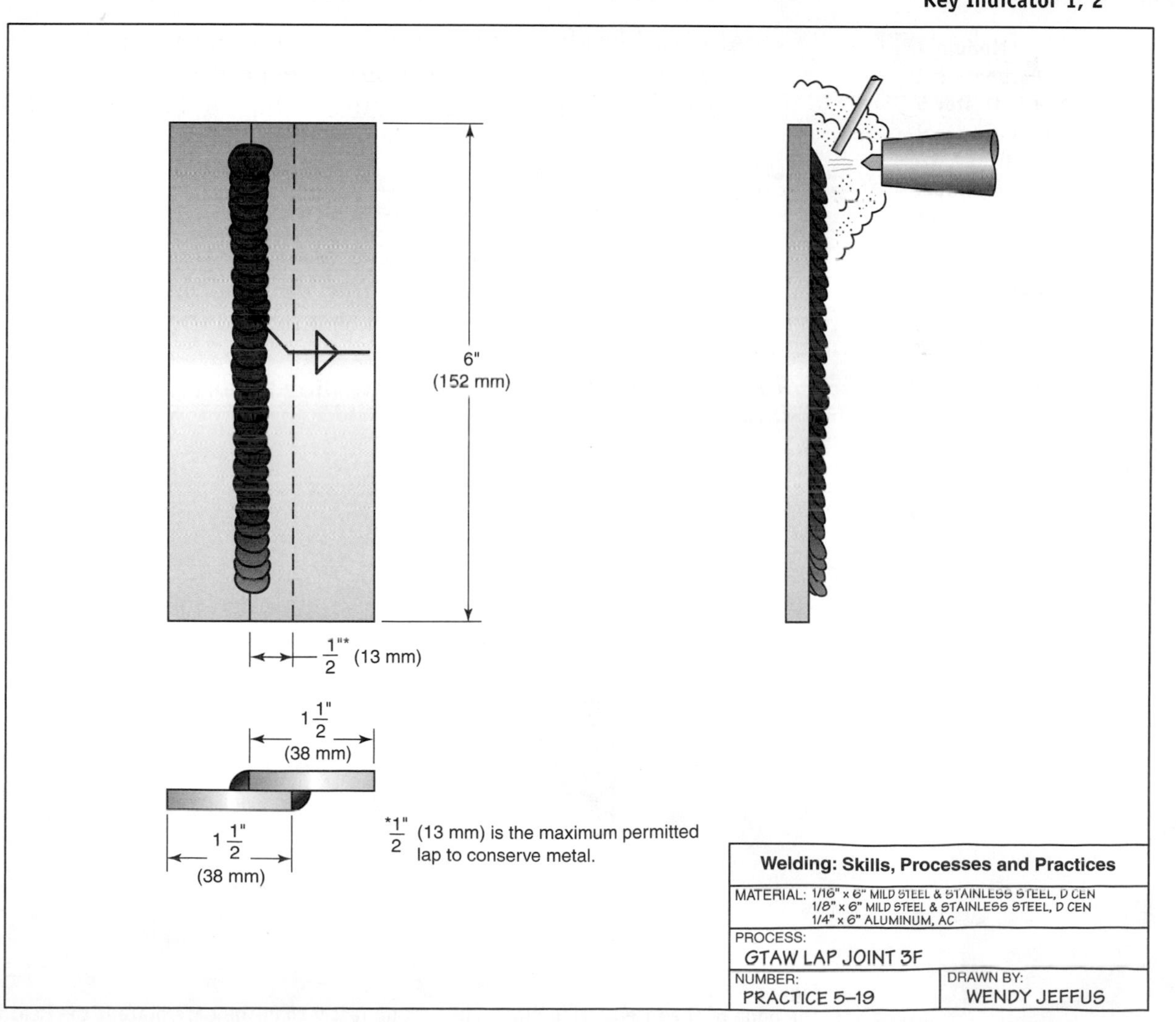

Figure 5.50
Vertical up lap joint

Module 1
Key Indicator 1, 3, 4

Module 2
Key Indicator 1, 2, 3, 4, 6, 7

Module 7
Carbon Steel
Key Indicator 5
Austenitic Stainless Steel
Key Indicator 10
Aluminum
Key Indicator 14

Module 9
Key Indicator 1, 2

PRACTICE 5-20

Lap Joint, 3F Position, with 100% Root Penetration, to Be Tested, Using Mild Steel, Stainless Steel, Aluminum

Repeat Practice 5-19 and make the fillet weld in a lap joint with 100% root penetration. After the weld is completed, visually inspect it for uniformity and defects before shearing out strips 1 in. (25 mm) wide and bend-testing them.

Repeat each weld until all have 100% root penetration. Turn off the welding machine, shielding gas, and cooling water, and clean up your work area when you are finished welding.

Complete a copy of the "Student Welding Report" listed in Appendix I or provided by your instructor.

Module 1
Key Indicator 1, 3, 4

Module 2
Key Indicator 1, 2, 3, 4, 6, 7

Module 7
Carbon Steel
Key Indicator 5
Austenitic Stainless Steel
Key Indicator 10
Aluminum
Key Indicator 15

Module 9
Key Indicator 1, 2

PRACTICE 5-21

Tee Joint at a 45° Vertical Angle, Using Mild Steel, Stainless Steel, Aluminum

Using the same equipment and materials listed in Practice 5-4, you will make a vertical up fillet weld on a tee joint.

- After tack welding the plates together, start the weld at the bottom and weld in an upward direction. The edge of the side plate, **Figure 5.51,** will heat up more quickly than the back plate. This rapid heating often leads to undercutting along this edge of the weld.
- To control undercutting, keep the arc on the back plate and add the filler metal to the weld pool near the side plate.

Repeat the process using all thicknesses of metal until you can consistently make the weld visually defect free. Turn off the welding machine, shielding gas, and cooling water, and clean up your work area when you are finished welding.

Complete a copy of the "Student Welding Report" listed in Appendix I or provided by your instructor.

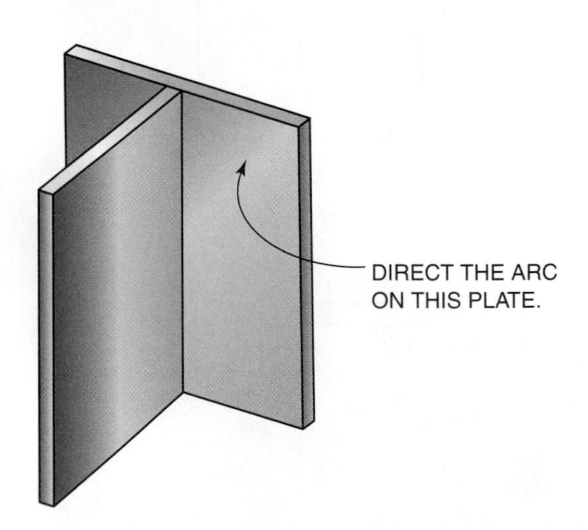

Figure 5.51
The edge of the intersecting plate will heat up faster than the base plate if the heat is not directed away from it

PRACTICE 5-22

Tee Joint, 3F Position, Using Mild Steel, Stainless Steel, Aluminum

Repeat Practice 5-21. Gradually increase the plate angle after each weld as you develop your skill. Increase the angle until the weld is being made in the vertical up position.

Repeat the process using all thicknesses of metal until you can consistently make the weld visually defect free. Turn off the welding machine, shielding gas, and cooling water, and clean up your work area when you are finished welding.

Complete a copy of the "Student Welding Report" listed in Appendix I or provided by your instructor.

Module 1
Key Indicator 1, 3, 4

Module 2
Key Indicator 1, 2, 3, 4, 6, 7

Module 7
Carbon Steel
Key Indicator 5
Austenitic Stainless Steel
Key Indicator 10
Aluminum
Key Indicator 14

Module 9
Key Indicator 1, 2

PRACTICE 5-23

Tee Joint, 3F Position, with 100% Root Penetration, to Be Tested, Using Mild Steel, Stainless Steel, Aluminum

Repeat Practice 5-21 and make the fillet weld in a lap joint with 100% root penetration. After the weld is completed, visually inspect it for uniformity and defects before shearing out strips 1 in. (25 mm) wide and bend-testing them.

Repeat each weld until all have 100% root penetration. Turn off the welding machine, shielding gas, and cooling water, and clean up your work area when you are finished welding.

Complete a copy of the "Student Welding Report" listed in Appendix I or provided by your instructor.

Module 1
Key Indicator 1, 3, 4

Module 2
Key Indicator 1, 2, 3, 4, 6, 7

Module 7
Carbon Steel
Key Indicator 5
Austenitic Stainless Steel
Key Indicator 10
Aluminum
Key Indicator 14

Module 9
Key Indicator 1, 2

PRACTICE 5-24

Stringer Bead at a 45° Reclining Angle, Using Mild Steel, Stainless Steel, Aluminum

Using the same equipment and materials as listed in Practice 5-4, you will make a weld bead on a plate at a 45° reclining angle, **Figure 5.52.** Add the filler metal along the top leading edge of the weld pool. Surface tension will help hold the weld pool on the top if the bead is not too large. The weld should be uniform in width and reinforcement.

Repeat the process using all thicknesses of metal until you can consistently make the weld visually defect free. Turn off the welding machine, shielding gas, and cooling water, and clean up your work area when you are finished welding.

Complete a copy of the "Student Welding Report" listed in Appendix I or provided by your instructor.

Module 1
Key Indicator 1, 3, 4

Module 2
Key Indicator 1, 2, 3, 4, 6, 7

Module 7
Carbon Steel
Key Indicator 6
Austenitic Stainless Steel
Key Indicator 11
Aluminum
Key Indicator 16

Module 9
Key Indicator 1, 2

PRACTICE 5-25

Stringer Bead, 2G Position, Using Mild Steel, Stainless Steel, Aluminum

Repeat Practice 5-24. Gradually increase the plate angle as you develop your skill until the weld is being made in the horizontal position on a vertical plate.

Module 1
Key Indicator 1, 3, 4

Module 2
Key Indicator 1, 2, 3, 4, 6, 7

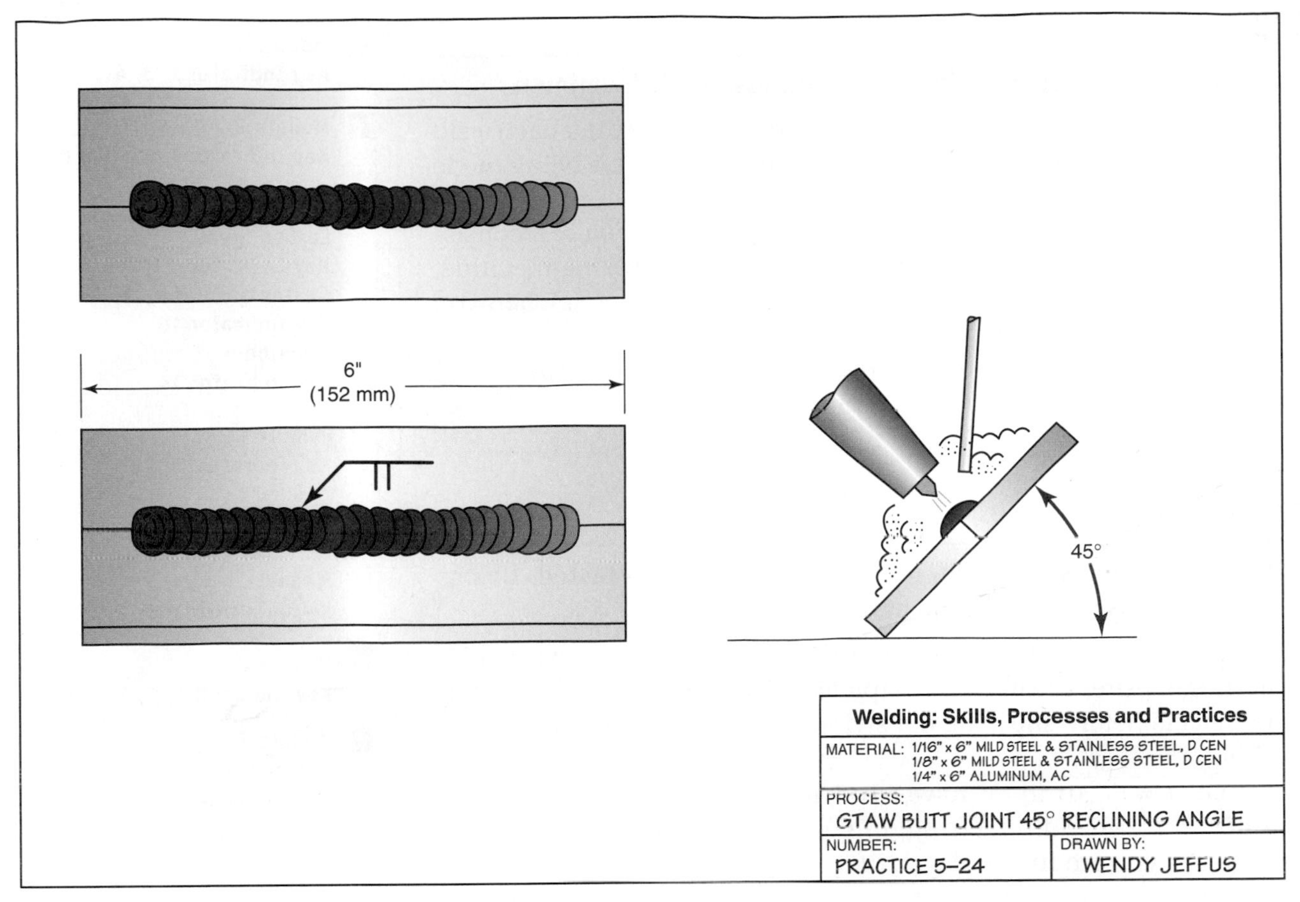

Figure 5.52
45° reclining angle

Module 7
Carbon Steel
Key Indicator 3, 4
Austenitic Stainless Steel
Key Indicator 8, 9
Aluminum
Key Indicator 13, 14

Module 9
Key Indicator 1, 2

Repeat the process using all thicknesses of metal until you can consistently make the weld visually defect free. Turn off the welding machine, shielding gas, and cooling water, and clean up your work area when you are finished welding.

Complete a copy of the "Student Welding Report" listed in Appendix I or provided by your instructor.

PRACTICE 5-26

Butt Joint, 2G Position, Using Mild Steel, Stainless Steel, Aluminum

Module 1
Key Indicator 1, 3, 4

Module 2
Key Indicator 1, 2, 3, 4, 6, 7

Module 7
Carbon Steel
Key Indicator 6
Austenitic Stainless Steel
Key Indicator 11
Aluminum
Key Indicator 16

Module 9
Key Indicator 1, 2

Using the same equipment and materials as listed in Practice 5-4, you will weld a butt joint in the horizontal position.

The welding techniques are the same as those used in Practice 5-25. Add the filler metal to the top plate, and keep the bead size small so it will be uniform.

Repeat the process using all thicknesses of metal until you can consistently make the weld visually defect free. Turn off the welding machine, shielding gas, and cooling water, and clean up your work area when you are finished welding.

Complete a copy of the "Student Welding Report" listed in Appendix I or provided by your instructor.

PRACTICE 5-27

Butt Joint, 2G Position, with 100% Penetration, to Be Tested, Using Mild Steel, Stainless Steel, Aluminum

Repeat Practice 5-26. It may be necessary to increase the root opening to ensure 100% penetration. A backing gas may be required to prevent atmospheric contamination. Using a "J" weave pattern will help to maintain a uniform weld bead. After the weld is completed, visually inspect it for uniformity and defects. Then shear out strips 1 in. (25 mm) wide and bend-test them.

Repeat each weld until all have 100% root penetration. Turn off the welding machine, shielding gas, and cooling water, and clean up your work area when you are finished welding.

Complete a copy of the "Student Welding Report" listed in Appendix I or provided by your instructor.

Module 1
Key Indicator 1, 3, 4

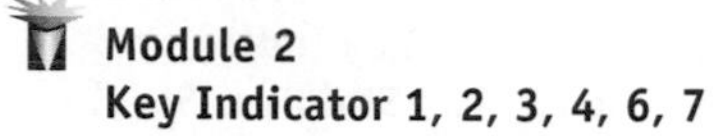

Module 2
Key Indicator 1, 2, 3, 4, 6, 7

Module 7
Carbon Steel
Key Indicator 6
Austenitic Stainless Steel
Key Indicator 11
Aluminum
Key Indicator 14

Module 9
Key Indicator 1, 2

PRACTICE 5-28

Lap Joint, 2F Position, Using Mild Steel, Stainless Steel, Aluminum

Using the same equipment and materials as listed in Practice 5-4, make a horizontal fillet weld on a lap joint.

- After tack welding the plates together, start the weld at one end. The bottom plate will act as a shelf to support the molten weld pool, **Figure 5.53.**
- Add the filler metal along the top edge of the weld pool to help control undercutting.

Repeat the process using all thicknesses of metal until you can consistently make the weld visually defect free. Turn off the welding machine, shielding gas, and cooling water, and clean up your work area when you are finished welding.

Complete a copy of the "Student Welding Report" listed in Appendix I or provided by your instructor.

Module 1
Key Indicator 1, 3, 4

Module 2
Key Indicator 1, 2, 3, 4, 6, 7

Module 7
Carbon Steel
Key Indicator 5
Austenitic Stainless Steel
Key Indicator 10
Aluminum
Key Indicator 14, 15

Module 9
Key Indicator 1, 2

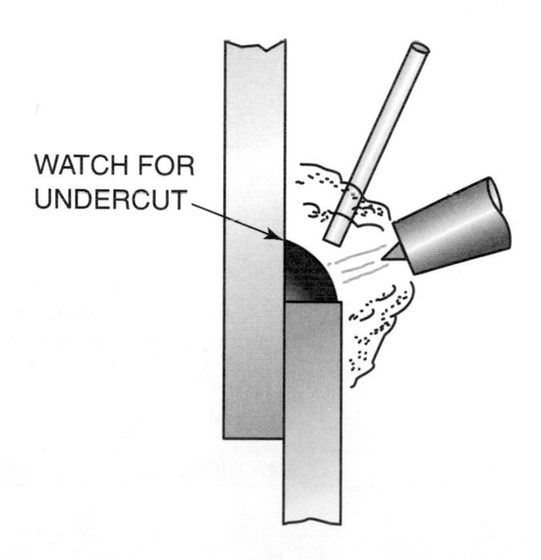

Figure 5.53
Horizontal lap joint

Module 1
Key Indicator 1, 3, 4

Module 2
Key Indicator 1, 2, 3, 4, 6, 7

Module 7
Carbon Steel
Key Indicator 5
Austenitic Stainless Steel
Key Indicator 10
Aluminum
Key Indicator 15

Module 9
Key Indicator 1, 2

PRACTICE 5-29

Lap Joint, 2F Position, with 100% Root Penetration, to Be Tested, Using Mild Steel, Stainless Steel, Aluminum

Repeat Practice 5-28. Be sure the weld is penetrating the root 100%. After the weld is completed, visually inspect it for uniformity and defects before shearing out strips 1 in. (25 mm) wide and bend-testing them.

Repeat each weld until all have 100% root penetration. Turn off the welding machine, shielding gas, and cooling water, and clean up your work area when you are finished welding.

Complete a copy of the "Student Welding Report" listed in Appendix I or provided by your instructor.

Module 1
Key Indicator 1, 3, 4

Module 2
Key Indicator 1, 2, 3, 4, 6, 7

Module 7
Carbon Steel
Key Indicator 5
Austenitic Stainless Steel
Key Indicator 10
Aluminum
Key Indicator 15

PRACTICE 5-30

Tee Joint, 2F Position, Using Mild Steel, Stainless Steel, Aluminum

Using the same equipment and materials as listed in Practice 5-4, you will make a horizontal fillet weld on a tee joint.

After tack welding the plates together, start the weld at one end. The bottom plate will act as a shelf to support the molten weld pool, **Figure 5.54.** As with the horizontal lap joint, add the filler metal along the top leading edge of the weld pool. This will help control undercut.

Repeat the process using all thicknesses of metal until you can consistently make the weld visually defect free. Turn off the welding machine, shielding gas, and cooling water, and clean up your work area when you are finished welding.

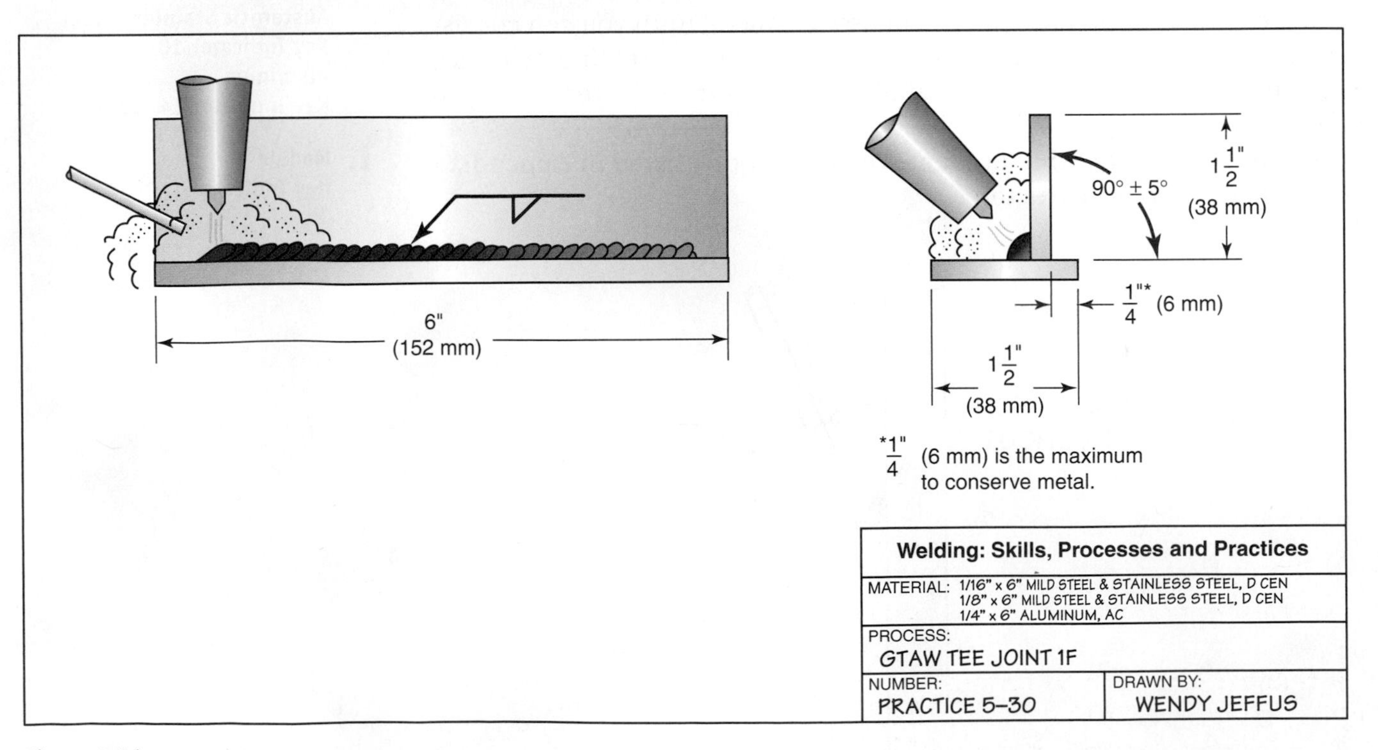

Figure 5.54
Tee joint in the horizontal position

Complete a copy of the "Student Welding Report" listed in Appendix I or provided by your instructor.

Module 9
Key Indicator 1, 2

PRACTICE 5-31

Tee Joint, 2F Position, with 100% Root Penetration, to Be Tested, Using Mild Steel, Stainless Steel, Aluminum

Repeat Practice 5-30. Be sure the weld is penetrating the root 100%. After the weld is completed, visually inspect it for uniformity and defects before shearing out strips 1 in. (25 mm) wide and bend-testing them.

Repeat each weld until all have 100% root penetration. Turn off the welding machine, shielding gas, and cooling water, and clean up your work area when you are finished welding.

Complete a copy of the "Student Welding Report" listed in Appendix I or provided by your instructor.

Module 1
Key Indicator 1, 3, 4

Module 2
Key Indicator 1, 2, 3, 4, 6, 7

Module 7
Carbon Steel
Key Indicator 4
Austenitic Stainless Steel
Key Indicator 9
Aluminum
Key Indicator 14

Module 9
Key Indicator 1, 2

PRACTICE 5-32

Stringer Bead, 4G Position, Using Mild Steel, Stainless Steel, Aluminum

Using the same equipment and materials as listed in Practice 5-4, you will make a weld bead on a plate in the overhead position.

The surface tension on the molten metal will hold the welding bead on the plate providing that it is not too large. A wide weld with little buildup will be easier to control and less likely to undercut along the edge.

Repeat the process using all thicknesses of metal until you can consistently make the weld visually defect free. Turn off the welding machine, shielding gas, and cooling water, and clean up your work area when you are finished welding.

Complete a copy of the "Student Welding Report" listed in Appendix I or provided by your instructor.

Module 1
Key Indicator 1, 3, 4

Module 2
Key Indicator 1, 2, 3, 4, 6, 7

Module 7
Carbon Steel
Key Indicator 6
Austenitic Stainless Steel
Key Indicator 9, 10
Aluminum
Key Indicator 14

Module 9
Key Indicator 1, 2

PRACTICE 5-33

Butt Joint, 4G Position, Using Mild Steel, Stainless Steel, Aluminum

Using the same equipment and materials as listed in Practice 5-4, you will weld a butt joint in the overhead position.

The same techniques used to make the stringer beads in Practice 5-32 are also used with the butt joint. The size of the bead should be kept small enough so that you can control the weld. Add the filler metal along the leading edge of the weld pool, **Figure 5.55.** The completed weld should be uniform and free from defects.

Repeat the process using all thicknesses of metal until you can consistently make the weld visually defect free. Turn off the welding machine, shielding gas, and cooling water, and clean up your work area when you are finished welding.

Complete a copy of the "Student Welding Report" listed in Appendix I or provided by your instructor.

Module 1
Key Indicator 1, 3, 4

Module 2
Key Indicator 1, 2, 3, 4, 6, 7

Module 7
Carbon Steel
Key Indicator 5
Austenitic Stainless Steel
Key Indicator 9
Aluminum
Key Indicator 14

Module 9
Key Indicator 1, 2

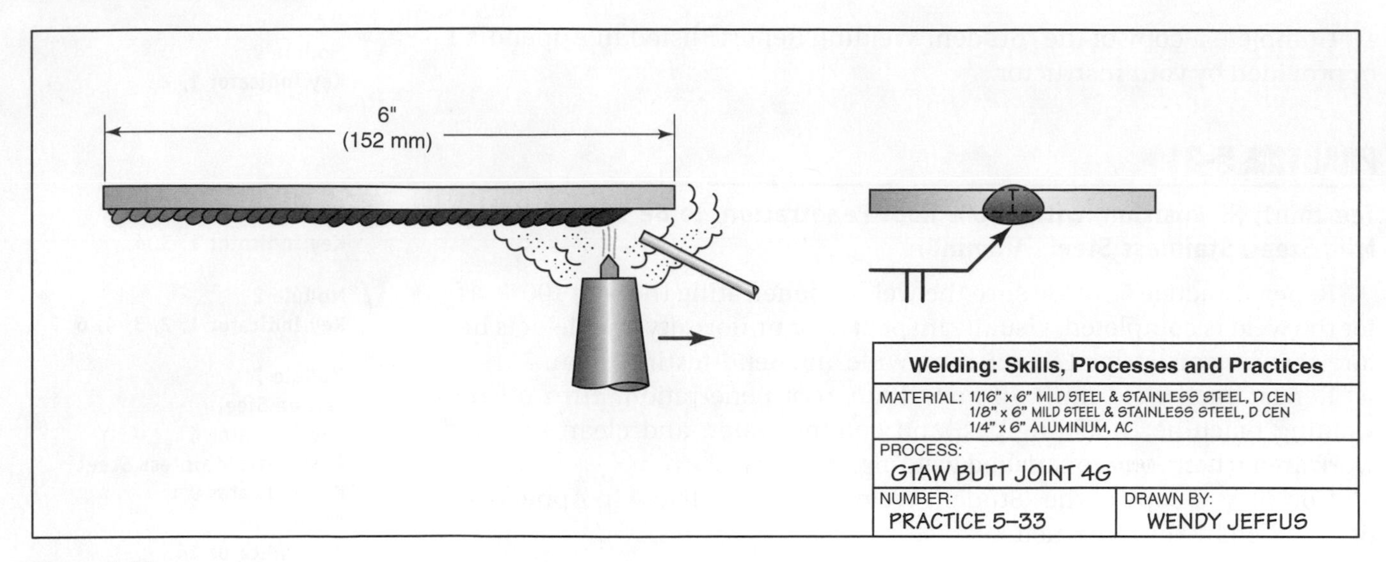

Figure 5.55
Overhead butt joint

PRACTICE 5-34

Lap Joint, 4F Position, Using Mild Steel, Stainless Steel, Aluminum

Module 1
Key Indicator 1, 3, 4
Module 2
Key Indicator 1, 2, 3, 4, 6, 7
Module 7
Carbon Steel
Key Indicator 5
Austenitic Stainless Steel
Key Indicator 9
Aluminum
Key Indicator 14
Module 9
Key Indicator 1, 2

Using the same equipment and materials as listed in Practice 5-4, you will make a fillet weld on a lap joint in the overhead position.

The major concentration of heat and filler metal should be on the top plate. Gravity and an occasional sweep of the torch along the bottom plate will pull the weld pool down. Undercutting along the top edge of the weld can be controlled by putting most of the filler metal along the top edge. The completed weld should be uniform and free from defects.

Repeat the process using all thicknesses of metal until you can consistently make the weld visually defect free. Turn off the welding machine, shielding gas, and cooling water, and clean up your work area when you are finished welding.

Complete a copy of the "Student Welding Report" listed in Appendix I or provided by your instructor.

PRACTICE 5-35

Tee Joint, 4F Position, Using Mild Steel, Stainless Steel, Aluminum

Module 1
Key Indicator 1, 3, 4
Module 2
Key Indicator 1, 2, 3, 4, 6, 7
Module 7
Key Indicator 1, 2
Carbon Steel
Key Indicator 3, 4, 5, 6, 7
Module 9
Key Indicator 1, 2

Using the same equipment and materials as listed in Practice 5-4, you will make a fillet weld on a tee joint in the overhead position.

The same techniques used to make the overhead lap weld in Practice 5-34 are used with the tee joint. As with the lap joint, most of the heat and filler metal should be concentrated on the top plate. A "J" weave pattern will help pull down any needed metal to the side plate. The completed weld should be uniform and free from defects.

Repeat the process using all thicknesses of metal until you can consistently make the weld visually defect free. Turn off the welding machine, shielding gas, and cooling water, and clean up your work area when you are finished welding.

Complete a copy of the "Student Welding Report" listed in Appendix I or provided by your instructor.

PRACTICE 5-36

Gas Tungsten Arc Welding (GTAW) on Plain Carbon Steel Workmanship Sample

Welding Procedure Specification (WPS)

Welding Procedure Specification No.: Practice 5-36. Date:

Title:

Welding GTAW of sheet to sheet.

Scope:

This procedure is applicable for square groove and fillet welds within the range of 18 gauge through 10 gauge.

Welding may be performed in the following positions: 1G and 2F.

Base Metal:

The base metal shall conform to carbon steel M-1, Group 1.

Backing material specification: none .

Filler Metal:

The filler metal shall conform to AWS specification no. 1/16 in. (2 mm) to 3/32 in. (2.4 mm) diameter. E70S-3 from AWS specification A5.18. This filler metal falls into F-number F-6 and A-number A-1.

Electrode:

The tungsten electrode shall conform to AWS specification no. EWTh-2, EWCe-2, or EWLa from AWS specification A5.12. The tungsten diameter shall be 1/8 in. (3.2 mm) maximum.

The tungsten end shape shall be tapered at two to three times its length to its diameter.

Shielding Gas:

The shielding gas, or gases, shall conform to the following compositions and purity: welding grade argon.

Joint Design and Tolerances:

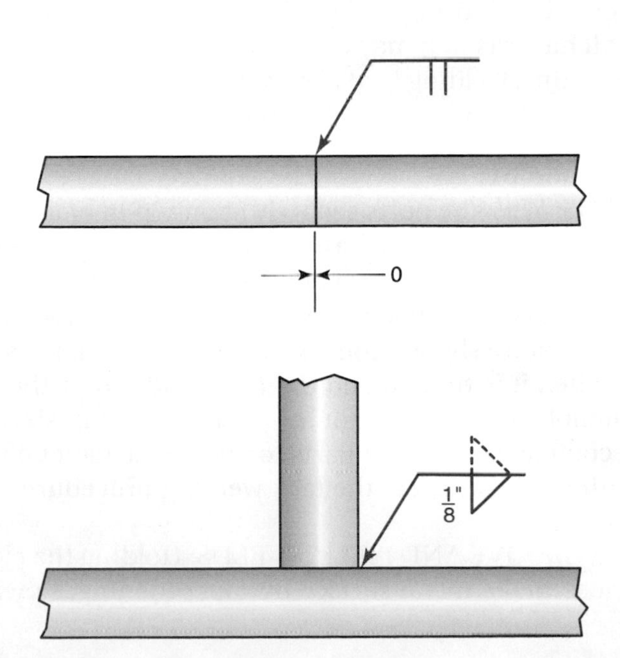

Preparation of Base Metal:

All hydrocarbons and other contaminations, such as cutting fluids, grease, oil, and primers, must be cleaned off all parts and filler metals before welding. This cleaning can be done with any suitable solvents or detergents. The joint face and inside and outside plate surface within 1 in. (25 mm) of the joint must be mechanically cleaned of slag, rust, and mill scale. Cleaning must be done with a wire brush or grinder down to bright metal.

Electrical Characteristics:

The current shall be direct current electrode negative (DCEN). The base metal shall be on the positive side of the line.

Metal Specifications		Gas Flow			Nozzle Size in. (mm)	Amperage Min. Max.
Thickness	Diameter of E70S-3*	Rates cfm) (L/min)	Flow Times Preflow	Postflow		
18 ga	1/16 in. (2 mm)	15 to 20 (7 to 9)	10 to 15 sec	10 to 25 sec	1/4 to 3/8 (6 to 10)	45 to 65
17 ga	1/16 in. (2 mm)	15 to 20 (7 to 9)	10 to 15 sec	10 to 25 sec	1/4 to 3/8 (6 to 10)	45 to 70
16 ga	1/16 in. (2 mm)	15 to 20 (7 to 9)	10 to 15 sec	10 to 25 sec	1/4 to 3/8 (6 to 10)	50 to 75
15 ga	1/16 in. (2 mm)	15 to 20 (7 to 9)	10 to 15 sec	10 to 25 sec	1/4 to 3/8 (6 to 10)	55 to 80
14 ga	3/32 in. (2.4 mm)	20 to 25 (9 to 12)	10 to 20 sec	10 to 30 sec	3/8 to 5/8 (10 to 16)	60 to 90
13 ga	3/32 in. (2.4 mm)	20 to 25 (9 to 12)	10 to 20 sec	10 to 30 sec	3/8 to 5/8 (10 to 16)	60 to 100
12 ga	3/32 in. (2.4 mm)	20 to 25 (9 to 12)	10 to 20 sec	10 to 30 sec	3/8 to 5/8 (10 to 16)	60 to 110
11 ga	3/32 in. (2.4 mm)	20 to 25 (9 to 12)	10 to 20 sec	10 to 30 sec	3/8 to 5/8 (10 to 16)	65 to 120
10 ga	3/32 in. (2.4 mm)	20 to 25 (9 to 12)	10 to 20 sec	10 to 30 sec	3/8 to 5/8 (10 to 16)	70 to 130

*Other E70S-X filler metal may be used.

Preheat:

The parts must be heated to a temperature higher than 50°F (10°C) before any welding is started.

Backing Gas:

None

Safety:

Proper protective clothing and equipment must be used. The area must be free of all hazards that may affect the welder or others in the area. The welding machine, welding leads, work clamp, electrode holder, and other equipment must be in safe working order.

Welding Technique:

TACK WELDS: With the parts securely clamped in place with the correct root gap, the tack welds are to be performed. Holding the electrode so that it is very close to the root face but not touching, slowly increase the current until the arc starts and a molten weld pool is formed. Add filler metal as required to maintain a slightly convex weld face and a flat or slightly concave root face. When it is time to end the tack weld, lower the current slowly so that the molten weld pool can be tapered down in size. When all tack welds are complete, allow the parts to cool as needed before assembling the remaining parts. Repeat the tack welding procedure until the entire part is assembled.

SQUARE GROOVE AND FILLET WELDS: Holding the electrode so that it is very close to the metal surface but not touching, slowly increase the

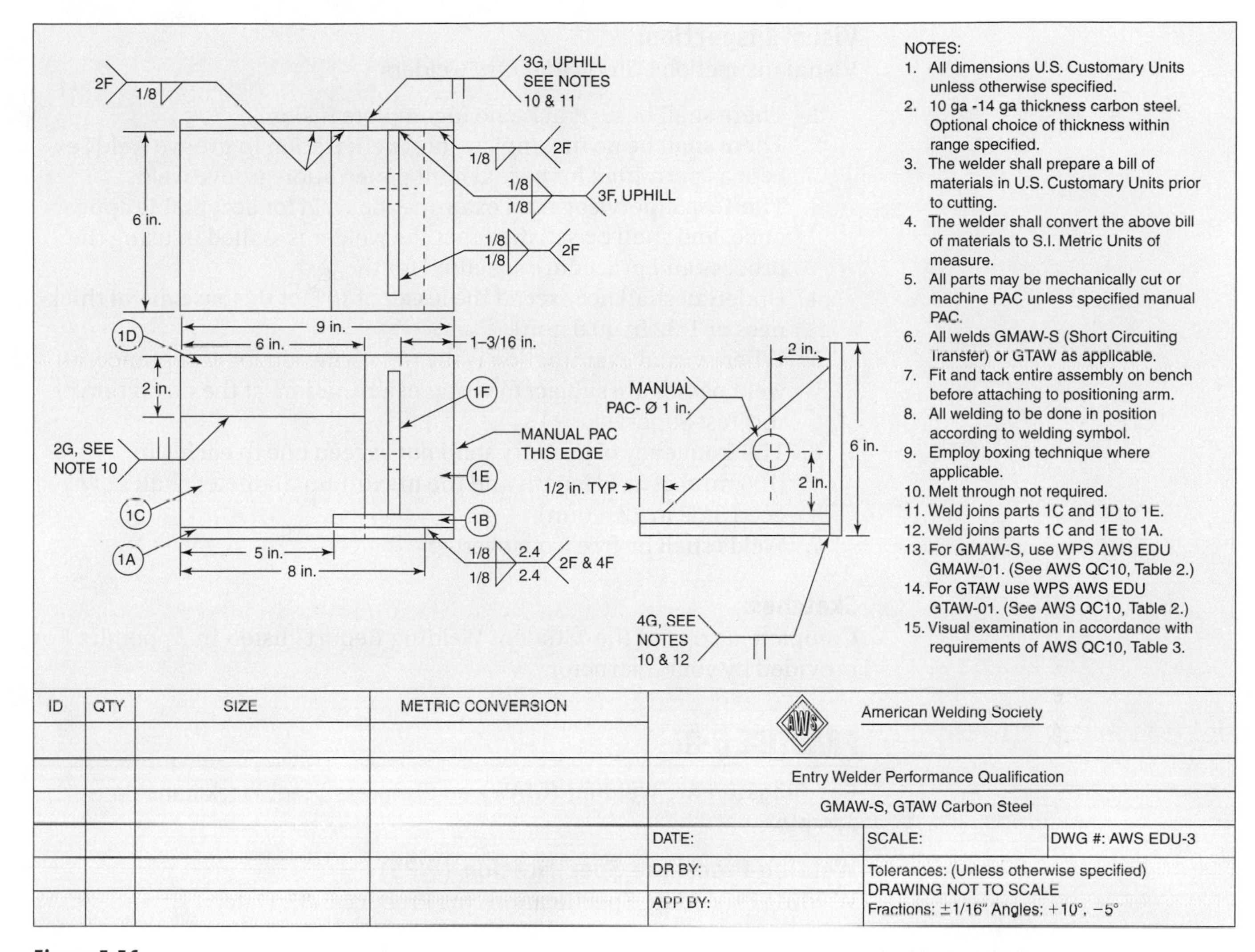

Figure 5.56
GTAW Plain Carbon Steel Workmanship Qualification Test
Courtesy of the American Welding Society

current until the arc starts and a molten weld pool is formed. As the weld progresses, add filler metal as required to maintain a flat or slightly convex weld face. If it is necessary to stop the weld or to reposition yourself or if the weld is completed, the current must be lowered slowly so that the molten weld pool can be tapered down in size.

Interpass Temperature:

The plate should not be heated to a temperature higher than 120°F (49°C) during the welding process. After each weld pass is completed, allow it to cool but never to a temperature below 50°F (10°C). The weldment must not be quenched in water.

Cleaning:

Recleaning may be required if the parts or filler metal becomes contaminated or reoxides to a degree that the weld quality will be affected. Reclean using the same procedure used for the original metal preparation. Any slag must be cleaned off between passes.

Visual Inspection:

Visual inspection criteria for entry welders*:

1. There shall be no cracks, no incomplete fusion.
2. There shall be no incomplete joint penetration in groove welds except as permitted for partial joint penetration groove welds.
3. The Test Supervisor shall examine the weld for acceptable appearance, and shall be satisfied that the welder is skilled in using the process and procedure specified for the text.
4. Undercut shall not exceed the lesser of 10% of the base metal thickness or 1/32 in. (0.8 mm).
5. Where visual examination is the only criterion for acceptance, all weld passes are subject to visual examination, at the discretion of the Test Supervisor.
6. The frequency of porosity shall not exceed one in each 4 in. (100 mm) of weld length and the maximum diameter shall not exceed 3/32 in. (2.4 mm).
7. Welds shall be free from overlap.

Sketches:

Complete a copy of the "Student Welding Report" listed in Appendix I or provided by your instructor.

PRACTICE 5-37

Gas Tungsten Arc Welding (GTAW) on Stainless Steel Workmanship Sample

Welding Procedure Specification (WPS)

Welding Procedure Specification No.: Practice 5-37. Date:

Module 1
Key Indicator 1, 3, 4

Module 2
Key Indicator 1, 2, 3, 4, 6, 7

Module 7
Key Indicator 1, 2
Austenitic Stainless Steel
Key Indicator 8, 9, 10, 11, 12

Module 9
Key Indicator 1, 2

Title:

Welding GTAW of sheet to sheet.

Scope:

This procedure is applicable for square groove and fillet welds within the range of 18 gauge through 10 gauge.

Welding may be performed in the following positions: 1G and 2F.

Base Metal:

The base metal shall conform to austenitic stainless steel M-8 or P-8. Backing material specification: none.

Filler Metal:

The filler metal shall conform to AWS specification no. ER3XX from AWS specification A5.9. This filler metal falls into F-number F-6 and A-number A-8.

Electrode:

The tungsten electrode shall conform to AWS specification no. EWTh-2, EWCe-2 or EWLa from AWS specification A5.12. The tungsten diameter shall be 1/8 in. (3.2 mm) maximum. The tungsten end shape shall be tapered at two to three times its length to its diameter.

Shielding Gas:

The shielding gas, or gases, shall conform to the following compositions and purity: welding grade argon.

*Courtesy of the American Welding Association.

Joint Design and Tolerances:

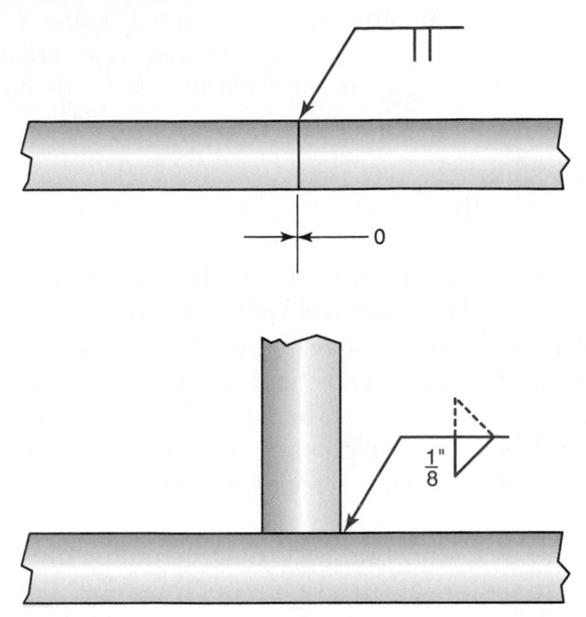

Preparation of Base Metal:

All hydrocarbons and other contaminations, such as cutting fluids, grease, oil, and primers, must be cleaned off all parts and filler metals before welding. This cleaning can be done with any suitable solvents or detergents. The joint face and inside and outside plate surface within 1 in. (25 mm) of the joint must be cleaned of slag, oxide, and scale. Cleaning can be mechanical or chemical. Mechanical metal cleaning can be done by grinding, stainless steel wire brushing, scraping, machining, or filing. Chemical cleaning can be done by using acids, alkalies, solvents, or detergents. Cleaning must be done down to bright metal.

Electrical Characteristics:

The current shall be direct current electrode negative (DCEN). The base metal shall be on the positive side of the line.

Metal Specifications		Gas Flow			Nozzle Size in. (mm)	Amperage Min. Max.
Thickness	Diameter of ER3XX*	Rates cfm (L/min	Flow Times Preflow	Postflow		
18 ga	1/16 in. (2 mm)	15 to 20 (7 to 9)	10 to 15 sec	10 to 25 sec	1/4 to 3/8 (6 to 10)	35 to 60
17 ga	1/16 in. (2 mm)	15 to 20 (7 to 9)	10 to 15 sec	10 to 25 sec	1/4 to 3/8 (6 to 10)	40 to 65
16 ga	1/16 in. (2 mm)	15 to 20 (7 to 9)	10 to 15 sec	10 to 25 sec	1/4 to 3/8 (6 to 10)	40 to 75
15 ga	1/16 in. (2 mm)	15 to 20 (7 to 9)	10 to 15 sec	10 to 25 sec	1/4 to 3/8 (6 to 10)	50 to 80
14 ga	3/32 in. (2.4 mm)	20 to 25 (9 to 12)	10 to 20 sec	10 to 30 sec	3/8 to 5/8 (10 to 16)	50 to 90
13 ga	3/32 in. (2.4 mm)	20 to 25 (9 to 12)	10 to 20 sec	10 to 30 sec	3/8 to 5/8 (10 to 16)	55 to 100
12 ga	3/32 in. (2.4 mm)	20 to 25 (9 to 12)	10 to 20 sec	10 to 30 sec	3/8 to 5/8 (10 to 16)	60 to 110
11 ga	3/32 in. (2.4 mm)	20 to 25 (9 to 12)	10 to 20 sec	10 to 30 sec	3/8 to 5/8 (10 to 16)	65 to 120
10 ga	3/32 in. (2.4 mm)	20 to 25 (9 to 12)	10 to 20 sec	10 to 30 sec	3/8 to 5/8 (10 to 16)	70 to 130

*Any ER3XX stainless steel A5.9 filler metal may be used.

Preheat:

The parts must be heated to a temperature higher than 50°F (10°C) before any welding is started.

Backing Gas:

None

Safety:

Proper protective clothing and equipment must be used. The area must be free of all hazards that may affect the welder or others in the area. The welding machine, welding leads, work clamp, electrode holder, and other equipment must be in safe working order.

Welding Technique:

TACK WELDS: With the parts securely clamped in place with the correct root gap, the tack welds are to be performed. Holding the electrode so that it is very close to the root face but not touching, slowly increase the current until the arc starts and a molten weld pool is formed. Add filler metal as required to maintain a slightly convex weld face and a flat or slightly concave root face. When it is time to end the tack weld, lower the current slowly so that the molten weld pool can be tapered down in size. When all tack welds are complete, allow the parts to cool as needed before assembling the remaining parts. Repeat the tack welding procedure until the entire part is assembled.

SQUARE GROOVE AND FILLET WELDS: Holding the electrode so that it is very close to the metal surface but not touching, slowly increase the current until the arc starts and a molten weld pool is formed. As the weld progresses, add filler metal as required to maintain a flat or slightly convex weld face. If it is necessary to stop the weld or to reposition yourself or if the weld is completed, the current must be lowered slowly so that the molten weld pool can be tapered down in size.

Interpass Temperature:

The plate should not be heated to a temperature higher than 350°F (180°C) during the welding process. After each weld pass is completed, allow it to cool but never to a temperature below 50°F (10°C). The weldment must not be quenched in water.

Cleaning:

Recleaning may be required if the parts or filler metal become contaminated or oxidized to a degree that the weld quality will be affected. Reclean using the same procedure used for the original metal preparation.

Visual Inspection:

Visual inspection criteria for entry welders*:

1. There shall be no cracks, no incomplete fusion.
2. There shall be no incomplete joint penetration in groove welds except as permitted for partial joint penetration groove welds.
3. The Test Supervisor shall examine the weld for acceptable appearance, and shall be satisfied that the welder is skilled in using the process and procedure specified for the text.
4. Undercut shall not exceed the lesser of 10% of the base metal thickness or 1/32 in. (0.8 mm).
5. Where visual examination is the only criterion for acceptance, all weld passes are subject to visual examination, at the discretion of the Test Supervisor.
6. The frequency of porosity shall not exceed one in each 4 in. (100 mm) of weld length and the maximum diameter shall not exceed 3/32 in. (2.4 mm).
7. Welds shall be free from overlap.

*Courtesy of the American Welding Association.

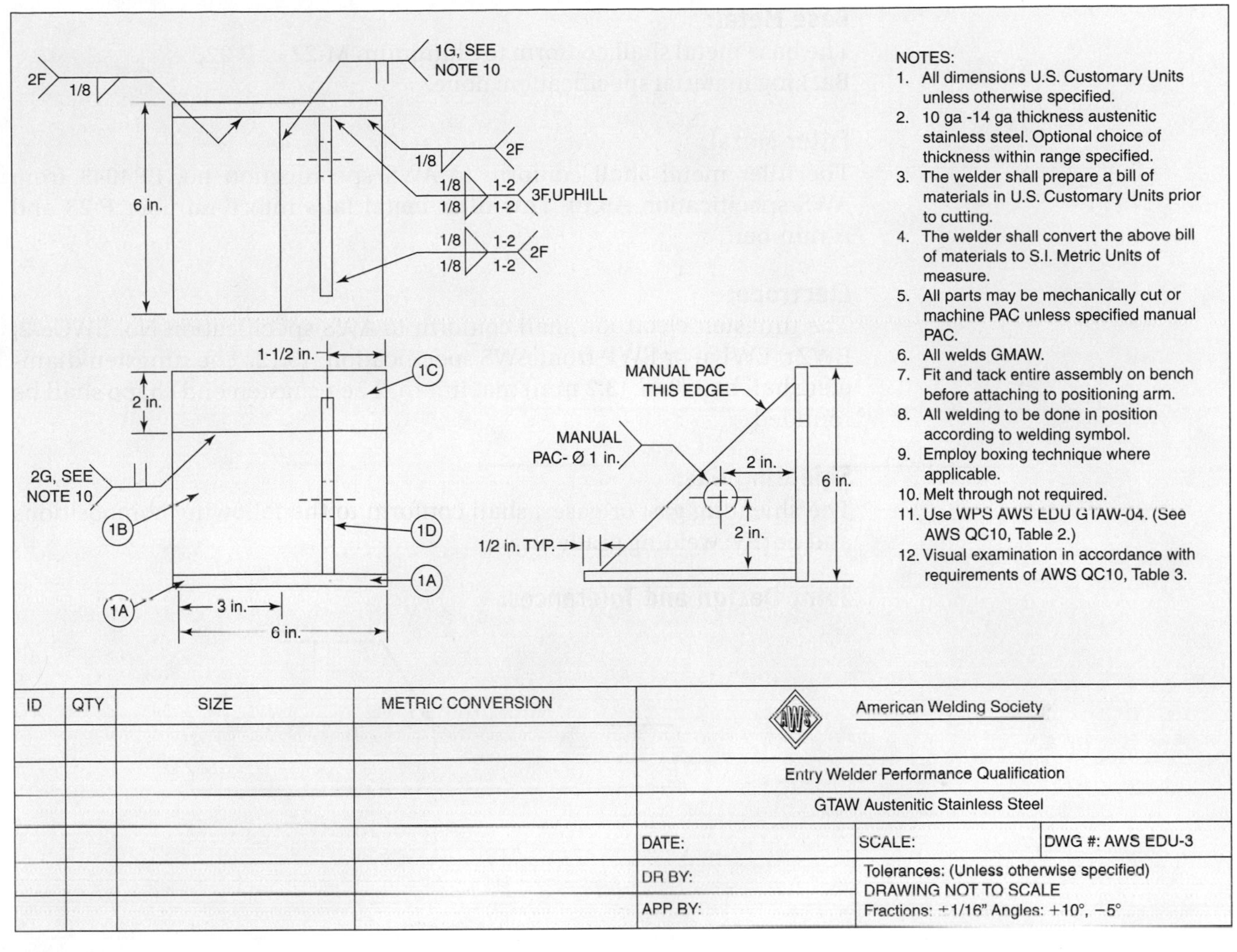

Figure 5.57
GTAW Stainless Steel Workmanship Qualification Test
Courtesy of the American Welding Society

Sketches:

Complete a copy of the "Student Welding Report" listed in Appendix I or provided by your instructor.

PRACTICE 5-38

Gas Tungsten Arc Welding (GTAW) on Aluminum Workmanship Sample

Module 1
Key Indicator 1, 3, 4

Module 2
Key Indicator 1, 2, 3, 4, 6, 7

Module 7
Key Indicator 1, 2
Aluminium
Key Indicator 13, 14, 15, 16, 17

Module 9
Key Indicator 1, 2

Welding Procedure Specification (WPS)

Welding Procedure Specification No.: Practice 5-38. Date:

Title:

Welding GTAW of sheet to sheet.

Scope:

This procedure is applicable for square groove and fillet welds within the range of 18 gauge through 10 gauge.

Welding may be performed in the following positions: 1G and 2F.

Base Metal:
The base metal shall conform to aluminum M-22 or P-22.
Backing material specification: none.

Filler Metal:
The filler metal shall conform to AWS specification no. ER4043 from AWS specification A5.10. This filler metal falls into F-number F-23 and A-number.

Electrode:
The tungsten electrode shall conform to AWS specification No. EWCe-2, EWZr, EWLa, or EWP from AWS specification A5.12. The tungsten diameter shall be 1/8 in. (3.2 mm) maximum. The tungsten end shape shall be rounded.

Shielding Gas:
The shielding gas, or gases, shall conform to the following compositions and purity: welding grade argon.

Joint Design and Tolerances:

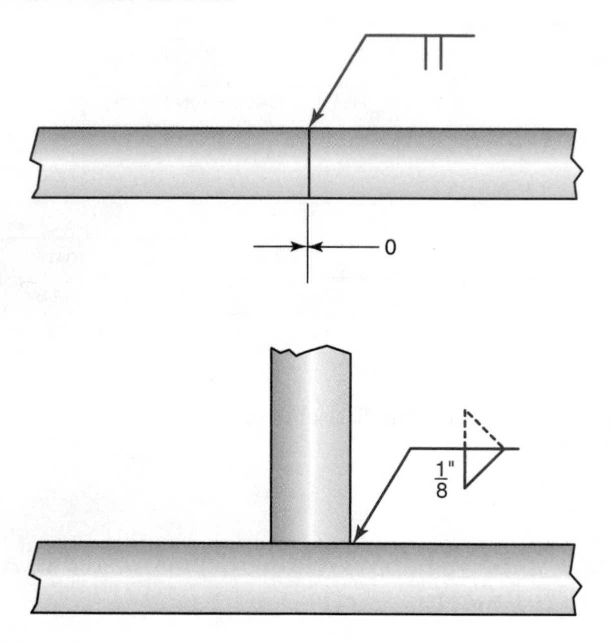

Preparation of Base Metal:
All hydrocarbons and other contaminations, such as cutting fluids, grease, oil, and primers, must be cleaned off all parts and filler metals before welding. This cleaning can be done with any suitable solvents or detergents. The joint face and inside and outside plate surface within 1 in. (25 mm) of the joint must be mechanically or chemically cleaned of oxides. Mechanical cleaning may be done by stainless steel wire brushing, scraping, machining, or filing. Chemical cleaning may be done by using acids, alkalies, solvents, or detergents. Because the oxide layer may reform quickly and affect the weld, welding should be started within 10 minutes of cleaning.

Electrical Characteristics:
The current shall be alternating current high-frequency stabilized (balanced wave preferably). The base metal shall be on the N/A side of the line.

Gas Flow

Preheat:

The parts must be heated to a temperature higher than 50°F (10°C) before any welding is started.

Metal Specifications		Gas Flow			Nozzle Size in. (mm)	Amperage Min. Max.
Thickness	Diameter of ER4043*	Rates cfm (L/min)	Flow Times Preflow	Postflow		
18 ga	3/32 in. (2.4 mm)	20 to 30 (9 to 14)	10 to 15 sec	10 to 25 sec	1/4 to 3/8 (6 to 10)	40 to 60
17 ga	3/32 in. (2.4 mm)	20 to 30 (9 to 14)	10 to 15 sec	10 to 25 sec	1/4 to 3/8 (6 to 10)	50 to 70
16 ga	3/32 in. (2.4 mm)	20 to 30 (9 to 14)	10 to 15 sec	10 to 25 sec	1/4 to 3/8 (6 to 10)	60 to 75
15 ga	3/32 in. (2.4 mm)	20 to 30 (9 to 14)	10 to 15 sec	10 to 25 sec	1/4 to 3/8 (6 to 10)	65 to 85
14 ga	3/32 in. (2.4 mm)	20 to 30 (9 to 14)	10 to 20 sec	10 to 30 sec	3/8 to 5/8 (10 to 16)	75 to 90
13 ga	1/8 in. (3 mm)	25 to 40 (12 to 19)	10 to 20 sec	10 to 30 sec	3/8 to 5/8 (10 to 16)	85 to 100
12 ga	1/8 in. (3 mm)	25 to 40 (12 to 19)	10 to 20 sec	10 to 30 sec	3/8 to 5/8 (10 to 16)	90 to 110
11 ga	1/8 in. (3 mm)	25 to 40 (12 to 19)	10 to 20 sec	10 to 30 sec	3/8 to 5/8 (10 to 16)	100 to 115
10 ga	1/8 in. (3 mm)	25 to 40 (12 to 19)	10 to 20 sec	10 to 30 sec	3/8 to 5/8 (10 to 16)	100 to 125

*Other aluminum AWS A5.10 filler metal may be used if needed.

Backing Gas:

N/A

Safety:

Proper protective clothing and equipment must be used. The area must be free of all hazards that may affect the welder or others in the area. The welding machine, welding leads, work clamp, electrode holder, and other equipment must be in safe working order.

Welding Technique:

The welder's hands or gloves must be clean and oil free to prevent contamination of the metal or filler rods.

TACK WELDS: With the parts securely clamped in place with the correct root gap, the tack welds are to be performed. Holding the electrode so that it is very close to the root face but not touching, slowly increase the current until the arc starts and a molten weld pool is formed. Add filler metal as required to maintain a slightly convex weld face and a flat or slightly concave root face. When it is time to end the tack weld, lower the current slowly so that the molten weld pool can be tapered down in size. When all tack welds are complete, allow the parts to cool as needed before assembling the remaining parts. Repeat the tack welding procedure until the entire part is assembled.

SQUARE GROOVE AND FILLET WELDS: Holding the electrode so that it is very close to the metal surface but not touching, slowly increase the current until the arc starts and a molten weld pool is formed. As the weld progresses, add filler metal as required to maintain a flat or slightly convex weld face. If it is necessary to stop the weld or to reposition yourself or the weld is completed, the current must be lowered slowly so that the molten weld pool can be tapered down in size.

Interpass Temperature:

The plate should not be heated to a temperature higher than 120°F (49°C) during the welding process. After each weld pass is completed, allow it to cool but never to a temperature below 50°F (10°C). The weldment must not be quenched in water.

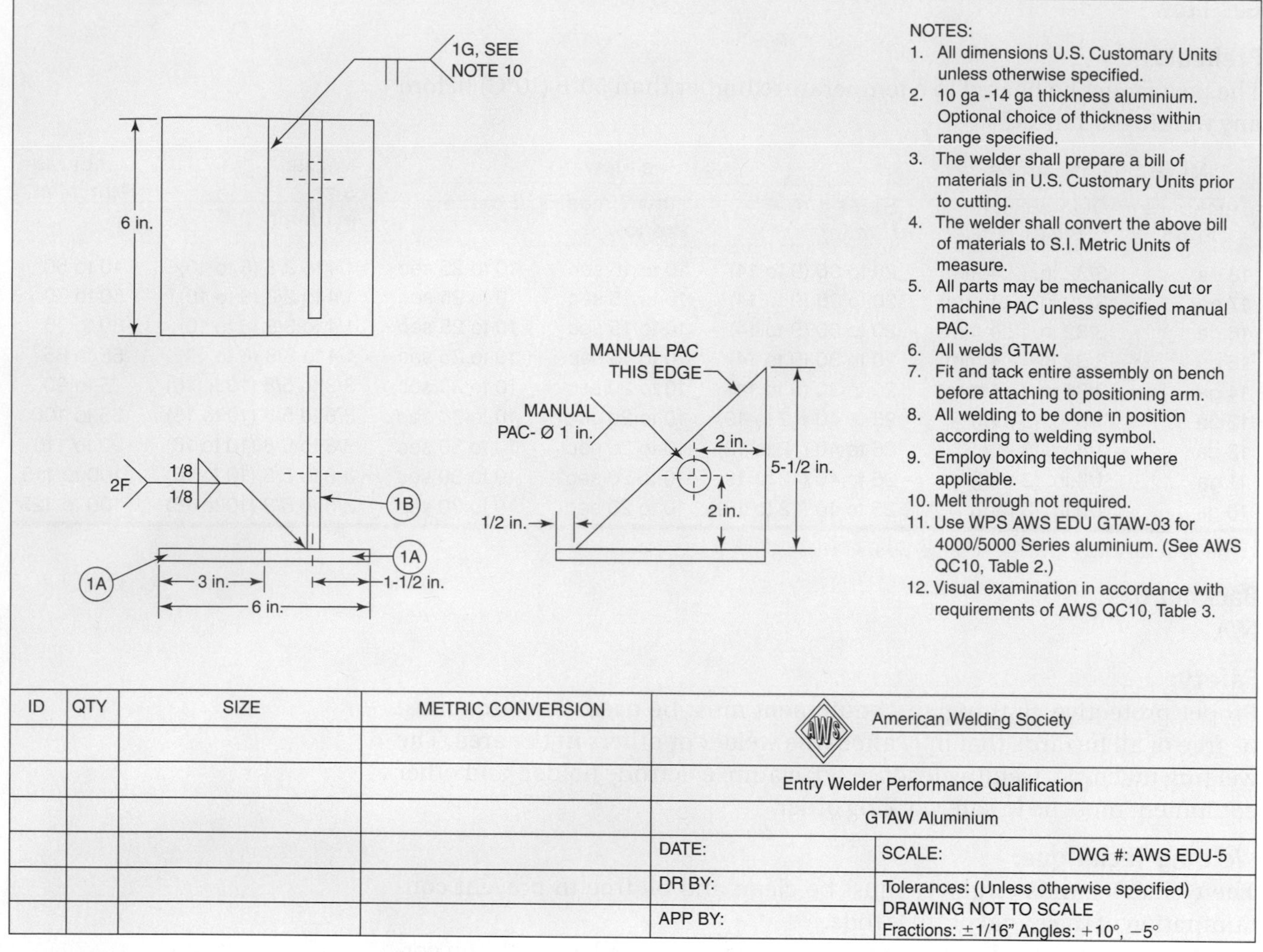

Figure 5.58
GTAW Aluminum Workmanship Qualification Test
Courtesy of the American Welding Society

Cleaning:

Recleaning may be required if the parts or filler metal becomes contaminated or oxidizes to a degree that the weld quality will be affected. Reclean using the same procedure used for the original metal preparation.

Visual Inspection:

Visual inspection criteria for entry welders*:

1. There shall be no cracks, no incomplete fusion.
2. There shall be no incomplete joint penetration in groove welds except as permitted for partial joint penetration groove welds.
3. The Test Supervisor shall examine the weld for acceptable appearance, and shall be satisfied that the welder is skilled in using the process and procedure specified for the text.
4. Undercut shall not exceed the lesser of 10% of the base metal thickness or 1/32 in. (0.8 mm)
5. Where visual examination is the only criterion for acceptance, all weld passes are subject to visual examination, at the discretion of the Test Supervisor.

*Courtesy of the American Welding Association.

6. The frequency of porosity shall not exceed one in each 4 in. (100 mm) of weld length and the maximum diameter shall not exceed 3/32 in. (2.4 mm).
7. Welds shall be free from overlap.

Sketches:
Complete a copy of the "Student Welding Report" listed in Appendix I or provided by your instructor.

SUMMARY

One of the most difficult aspects of learning to produce gas tungsten arc welds is positioning yourself so that you can control the electrode filler metal and see the joint both at the same time. Beginning welders assume they must see the tungsten tip as they make the weld. Experienced welders, however, realize that they need to see only the leading edge of the molten weld pool to know how the weld is being produced. A good view of the leading edge will tell you how the base metal is being melted, or depth of penetration. You can even tell from this small portion whether the filler metal is being added at an appropriate rate. As you learn how to control the weld and develop your skills, it is a good idea to gradually reduce your need for seeing 100% of the molten weld pool. Increasing this part of your skill will be a significant advantage in the field because, unlike welding in the classroom, which is typically done on a comfortable-height table, welding in the field may have to be done out of position.

REVIEW

1. What effect does torch angle have on the shielding gas protective zone?
2. Why must the end of the filler rod be kept in the shielding gas protective zone?
3. What can cause tungsten contamination?
4. What determines the correct current setting for a GTA weld?
5. What is the lowest acceptable amperage setting for GTA welding?
6. List the factors that affect the gas flow setting for GTA welding.
7. When should the minimum gas flow rates be increased?
8. What is the minimum gas flow rate for a nozzle size?
9. What is the maximum gas flow rate for a nozzle size?
10. Which incorrect welding parameters does stainless steel show clearly?
11. Using **Table 5.4,** determine the approximate temperature of metal that has formed a dark blue color.
12. Using **Table 5.3, Table 5.5,** and **Table 5.6,** list the filler metals for the following metals.
    a. 1020 low-carbon steel
    b. 309 stainless steel
13. Why is it possible to control a large aluminum weld bead?
14. What may happen to the end of the aluminum welding rod if it is held too close to the arc?
15. What should be done if someone comes in contact with a cleaning chemical?

16. Using **Table 5.7,** determine the suggested setting for GTA welding of mild steel using a 3/32-in. (2.4-mm) tungsten.
17. What can be done to limit oxide formation on stainless steel?
18. How should the filler metal be added to the molten weld pool?
19. How can the rod be freed if it sticks to the plate?
20. How is an outside corner joint assembled?
21. What must be done with the weld craters when back stepping a weld? Why?
22. How is the lap joint tested for 100% penetration?
23. What can prevent both sides of a stainless steel tee joint from being welded?
24. How is the filler metal added for a 3F weld?
25. What can cause undercutting on a 3F tee joint?
26. What helps hold the weld in place on a 2F lap joint?
27. What helps hold the weld in place on a 4G weld?

APPENDIX I—STUDENT WELDING REPORT

I. STUDENT WELDING REPORT

Student Name: ______________________ Date: ____________

Instructor: ______________________ Class: ____________

Experiment or Practice #: ______________________ Process: ____________

Briefly describe task: ______________________

____

____

____

INSPECTION REPORT			
Inspection	Pass/Fail	Inspector's Name	Date
Safety:			
Equip. Setup:			
Equip. Operation:			
Welding	Pass/Fail	Inspector's Name	Date
Accuracy:			
Appearance:			
Overall Rating:			

Comments:

Student Grade: ____________ Instructor Initials: ____________ Date: ____________

Glossary

amperage A measurement of the rate of flow of electrons; amperage controls the size of the arc.
amperaje Una medida de la proporción de la corriente de electrones; el amperaje controla el tamaño del arco.
amperage range The lower and upper limits of welding power, in amperage, that can be produced by a welding machine or used with an electrode or by a process.
rango de amperaje Los límites máximos y mínimos de poder de soldadura (en amperaje) que puede tener una máquina para soldar o que pueden usarse con un electrodo o a través de un proceso.
anode Material with a lack of electrons; thus, it has a positive charge.
ánodo Un material que carece electrones; por eso tiene una carga positiva.
arc length The length from the tip of the welding electrode to the adjacent surface of the weld pool.
largura del arco La distancia de la punta del electrodo a la superficie que colinda con el charco de la soldadura.
back gouging The removal of weld metal and base metal from the weld root side of a welded joint to facilitate complete fusion and complete joint penetration upon subsequent welding from that side.
gubia trasera Quitar el metal soldado y el metal base del lado de la raíz de una junta soldada para facilitar una fusión completa y penetración completa de la junta soldada subsecuente a soldar de ese lado.
burn-through Burning out of molten metal on the back side of the plate.
metal quemado que pasa al otro lado Metal derretido que se quema en el lado de atrás del plato.
cathode A natural curve material with an excess of electrons, thus having a negative charge.
cátodo Un material de curva natural con un exceso de electrones, por eso tiene una carga negativa.
cellulose-based fluxes Fluxes that use an organic-based cellulose ($C_6H_{10}O_5$) (a material commonly used to make paper) held together with a lime binder. When this flux is exposed to the heat of the arc, it burns and forms a rapidly expanding gaseous cloud of CO_2 that protects the molten weld pool from oxidation. Most of the fluxing material is burned, and little slag is deposited on the weld. E6010 is an example of an electrode that uses this type of flux.
fundentes para electrodos celulósicos Fundentes que usan celulosa de base orgánica ($C_6H_{10}O_5$) (un material normalmente utilizado para fabricar papel), y que se mantienen unidos con un aglomerante de cal. Cuando a este fundente se lo expone al calor del arco, se consume y forma una nube gaseosa de CO_2 que se expande rápidamente y protege de la oxidación al charco de soldadura derretido. La mayor parte del material del fundente se consume, y se deposita poca escoria en la soldadura. El E6010 es un ejemplo de un electrodo que utiliza este tipo de fundente.
chill plate A large piece of metal used in welding to correct overheating.
plato desalentador Una pieza de metal grande que se usa para corregir el sobrecalentamiento.
cleaning action A phenomenon occurring during DCEP in which oxides are removed from the surface to be welded by accelerated ions or electrons. These particles cause surface erosion, also called etching, that assist in the cleaning of the surface to be welded. Cleaning action requires an argon-rich atmosphere.
acción limpiadora Fenómeno que se presenta durante el proceso de DCEP (corriente directa electrodo positivo) en el que los óxidos se eliminan de la superficie para ser soldados por iones acelerados o electrones. Estas partículas provocan que la superficie se erosione (un proceso también conocido como decapado), lo cual auxilia en la limpieza de la superficie que se va a soldar. La acción limpiadora requiere una atmósfera con cantidades abundantes de argón.
collet A cone-shaped sleeve that holds a GTAW electrode in place within a GTA welding torch.
collarín Un manguito cónico que mantiene al electrodo en su lugar dentro del soplete para soldar por arco de tungsteno con gas (en inglés, GTAW).
contamination Undesirable foreign substances found on base metal surfaces, on filler wires, or in gas shielding atmospheres that inhibit welding operations.
contaminación Sustancias extrañas e indeseables en las superficies de metales básicos, alambres de relleno o atmósferas de protección por gas que inhiben las operaciones de soldadura.
cover pass The last layer of weld beads on a multiple pass weld. The final bead should be uniform in width and reinforcement, not excessively wide, and free of any visual defects.
pasada para cubrir La última capa de cordónes soldadura de pasadas múltiples. La pasada final debe ser uniforme en anchura y refuerzo, no excesivamente ancha, y libre de defectos visuales.

duty cycle The percentage of time during an arbitrary test period that a power source or its accessories can be operated at rated output without overheating.
ciclo de trabajo El porcentaje de tiempo durante un período a prueba arbitraria de una fuente de poder y sus accesorios que pueden operarse a la capacidad de carga de salida sin sobrecalentarse.
electrode angle The angle between the electrode and the surface of the metal; also known as the direction of travel (leading angle or trailing angle); leading angle pushes molten metal and slag ahead of the weld; trailing angle pushes the molten metal away from the leading edge of the molten weld pool toward the back, where it solidifies.
ángulo del electrodo El ángulo en medio del electrodo y la superficie del metal; también conocido como la dirección de avance (apuntado hacia adelante o apuntado hacia atras); el ángulo apuntado empuja el metal derretido y la escoria enfrente de la soldadura; y el ángulo apuntado hacia atrás empuja el metal derretido lejos de la orilla delantera del charco del metal derretido hacia atrás, donde se solidifica.
electrons A negatively charged component of an atom. Electrons exist outside of and surrounding the atom's nucleus. Each electron carries one unit of negative charge and has a very small mass as compared with that of a neutron or proton.
electrones Uno de los componentes con carga negativa de un átomo. Los electrones existen en el exterior y alrededor del núcleo del átomo. Cada electrón cuenta con una unidad de carga negativa así como con una masa pequeña en comparación con la de un neutrón o protón.
filler pass One or more weld beads used to fill the groove with weld metal. The bead must be cleaned after each pass to prevent slag inclusions.
pasada para rellenar Uno o más cordones de soldadura usados para llenar la ranura con el metal de soldadura. El cordón debe ser limpiado después de cada pasada para prevenir inclusiones de escoria.
flowmeter A control device used to regulate shield and backing gas flow-in. A flowmeter may control flow on a manifold system or it may be part of a combination flow and pressure regulator device.
flujómetro Un dispositivo de control empleado para regular el flujo del gas de protección y respaldo. Un flujómetro puede controlar el flujo en un sistema colector o bien puede ser parte de un dispositivo con regulación combinada de flujo y presión.
frequency The rate at which alternating electrical current (AC) is measured as it switches back and forth between polarities. Each cycle back and forth is counted as one hertz (Hz).
frecuencia La velocidad registrada por la corriente eléctrica alterna (CA) durante los cambios de polaridad alternados. Cada ciclo de intercambio se cuenta como un hertz (Hz).
gas coverage The protective atmosphere delivered on top of or underneath a weld. Gas coverage may be supplied from the welding torch or from a secondary trailing or backup device.
cobertura de gas La atmósfera protectora aplicada a la parte superior o inferior de una soldadura. La cobertura de gas puede provenir del soplete para soldar o bien de un dispositivo de escape secundario o de respaldo.
guided bend specimen Any bend specimen that will be bend-tested in a fixture that controls the bend radii, such as the AWS bend-test fixture.
probeta de dobléz guiada Cualquier probeta de dobléz en la cual se va a hacer un dobléz guiado en una máquina que controla el radio del dobléz, como la máquina de dobléz guiado del AWS.
hot pass The welding electrode is passed over the root pass at a higher than normal amperage setting and travel rate to reshape an irregular bead and turn out trapped slag. A small amount of metal is deposited during the hot pass so the weld bead is convex, promoting easier cleaning.
pasada caliente El electrodo de soldadura se pasa sobre la pasada de raíz poniendo el amperaje más alto que lo normal y proporción de avance para reformar un cordón irregular y sacar la escoria atrapada. Una cantidad pequeña de metal es depositada durante la pasada caliente para que el cordón soldado sea convexo, promoviendo más fácil la limpieza.
inert gas A gas that normally does not combine chemically with materials.
gas inerte Un gas que normalmente no se combina químicamente con materiales.
interpass temperature In a multiple pass weld, the temperature of the weld area between weld passes.
temperatura de pasada interna En una soldadura de pasadas multiples, la temperatura en la área de la soldadura entre pasadas de soldaduras.
inverter A welding machine that is much smaller than other types of machines of the same amperage range.
inversor Una máquina soldadora más pequeña que otros tipos de máquinas similares con el mismo rango de amperaje.
keyhole A welding technique in which a concentrated heat source penetrates completely through a workpiece, forming a hole at the leading edge of the weld pool. As the heat source progresses, the molten metal fills in behind the hole to form the weld bead.
soldadura con pocillo Una técnica en la cual una fuente de calor concentrado se penetra completamente a través de la pieza de trabajo, formando un agujero en la orilla del frente del charco de la soldadura. Asi como progresa la potencia de calor, el metal derretido rellena detrás del agujero para formar un cordón de soldadura.
lap joint A joint between two overlapping members.
junta de solape Una junta entre dos miembros traslapadas.

magnetic flux lines Parallel lines of force that always go from the north pole to the south pole in a magnet and surround a wire carrying DC current.
líneas magnéticas de flujo Líneas paralelas de fuerza que siempre van del polo norte al polo sur en un magneto, y rodea un alambre que lleva corriente DC.
mineral-based fluxes Fluxes that use inorganic compounds, such as the rutile-based flux (titanium dioxide, TiO_2). These mineral compounds do not contain hydrogen, and electrodes that use these fluxes are often referred to as low hydrogen electrodes. Less smoke is generated with this welding electrode than with cellulose-based fluxes, but a thicker slag layer is deposited on the weld. E7018 is an example of an electrode that uses this type of flux.
fundentes para electrodos de base mineral Fundentes que usan compuestos inorgánicos, como por ejemplo, el fundente a base de rutilo (bióxido de titanio, TiO_2). Estos compuestos minerales no contienen hidrógeno, y a los electrodos que usan estos fundentes se los llama con frecuencia electrodos de bajo hidrógeno. En la soldadura con electrodos se producen menos humos que en la que se realiza con fundentes celulósicos, pero se deposita una capa de escoria más gruesa en la soldadura. El F7018 es un ejemplo de un electrodo que usa este tipo de fundente.
molten weld pool The liquid state of a weld prior to solidification as weld material.
charco de soldadura derretido El estado líquido de una soldadura antes de solidificarse como material de soldadura.
multiple pass weld A weld requiring more than one pass to ensure complete and satisfactory joining of the metal pieces.
soldadura de pasadas múltiples Una soldadura que requiere más de una pasada para asegurar una completa y satisfactoria unión de las piezas de metal.
open circuit voltage The voltage between the output terminals of the power source when no current is flowing to the torch or gun.
voltaje de circuito abierto El voltaje entre los terminales de salida de una fuente de poder cuando la corriente no está corriendo a la antorcha o pistola.
operating voltage Also called closed circuit voltage, it is the voltage measured during welding.
voltaje operativo También conocido como voltaje de circuito cerrado, es el voltaje registrado durante el proceso de soldadura.
output The welding current a particular welding power supply is rated to produce, or the measure of current being produced during welding operations.
potencia de salida La capacidad de la corriente de soldadura de un suministro eléctrico para soldar, o bien la cantidad de corriente producida durante las operaciones de soldadura.
oxide layer An oxide layer is a thin coating that forms on the surface of aluminum alloys. Once the oxide layer has formed, the progress of corrosion is significantly reduced. However, the oxide layer must be removed prior to welding operations because it inhibits penetration. The melting point of aluminum oxide is approximately 3500°F (1926°C), while the melting point of aluminum itself is about 1200°F (650°C).
capa de óxido Una capa de óxido es una película delgada presente en la superficie de las aleaciones de aluminio. Una vez que se forma la capa de óxido el avance de la corrosión se reduce de manera significativa. Sin embargo, la capa de óxido debe eliminarse antes de realizar la soldadura ya que evita la penetración. El punto de fusión del óxido de aluminio es aproximadamente a 3500°F (1926°C), mientras que el punto de fusión del aluminio es a unos 1200°F (650°C).
postflow time The time interval from current shut-off to shielding gas and/or cooling water shut-off.
tiempo de poscorriente El intervalo de tiempo de cuando se cierra la corriente a cuando se cierra el gas de protección y o cuando se cierra el agua para enfriar.
postheating The application of heat to an assembly after welding, brazing, soldering, thermal spraying, or thermal cutting.
poscalentamiento La aplicación de calor a una asamblea después de la soldadura, soldadura fuerte, soldadura blanda, rociado termal, o corte termal.
preflow time The time interval that shield gas is delivered to a welding gun or torch before current is applied.
tiempo de preflujo El intervalo en el que el gas de protección se deposita en un soplete antes de aplicar corriente eléctrica.
preheating The heat applied to the base metal or substrate to attain and maintain preheat temperature.
precalentamiento El calor aplicado al metal base o substrato para obtener y mantener temperatura de precalentamiento.
protective zone The area covered by shield gas during welding operations.
zona protectora El área cubierta con gas de protección durante las operaciones de soldadura.
rectification Arc rectification is a phenomenon that occurs when the surface oxide of a nonferrous metal, such as aluminum, acts as a barrier that obstructs the flow of electrons between the electrode and the work during alternating current GTAW operations. Traditional square wave power supplies are equipped with superimposed high-frequency current, which helps establish and maintain the arc. Inverter power supplies with advanced circuitry switch back and forth so quickly between DCEN and DCEP that rectification is rarely experienced and superimposed high frequency is not needed.
rectificación Un arco de rectificación es un fenómeno que se presenta cuando el óxido superficial de un metal

no ferroso, tal como el aluminio, actúa como una barrera que obstruye el flujo de electrones entre el electrodo y el proyecto durante la realización de la soldadura TIG (en inglés, GTAW) con corriente alterna. Los suministros eléctricos tradicionales de ondas cuadradas están equipados con una corriente de alta frecuencia superpuesta que ayuda a establecer y mantener el arco. El suministro eléctrico de los inversores con circuitos avanzados cambia a una velocidad tal entre DCEN y DCEP que rara vez se presenta la rectificación, por lo que el uso de alta frecuencia superpuesta no es necesaria.

rectifier Allows current to flow in one direction only.

rectificador Permite que la corriente fluya en una sola dirección.

root pass The first weld of a multiple pass weld. The root pass fuses the two pieces together and establishes the depth of weld metal penetration.

pasada de raíz La primera soldadura de una soldadura de pasadas múltiples. La pasada de raíz funde las dos piezas juntas y establece la profundidad de la penetración del metal soldado.

rutile-based fluxes This flux system produces a smooth and stable medium-penetrating arc with easily removeable slag. Rutile-based fluxes are used in several types of electrodes and in all welding positions for SMAW and FCAW.

flujos con base de rutilo Este sistema de flujo produce un arco homogéneo y estable para la penetración de medios con escoria fácil de eliminar. Los flujos con base de rutilo se utilizan en varios tipos de electrodos y en todas las posiciones de soldadura SMAW y FCAW.

spark gap oscillator A component in GTAW power supplies that helps to deliver superimposed high-frequency current that assists in arc initiation and alternating current GTAW welding of aluminum alloys.

oscilador de chispa Un componente en suministros eléctricos para soldar por arco de tungsteno con gas (en inglés, GTAW) que ayuda a administrar una corriente de alta frecuencia superpuesta que auxilia en la iniciación del arco y en el intercambio de corriente en soldaduras GTAW de aleaciones de aluminio.

square butt joint A joint made when two flat pieces of metal face each other with no edge preparation.

junta escuadra de tope Una junta hecha cuando dos piezas planas de metal se enfrentan una a la otra sin preparación de orilla.

step-down transformer A transformer-type welding machine that is quieter, more energy efficient, requires less maintenance, and is less expensive than electric-motor or gas-powered engine welding power supplies.

transformador reductor Máquina para soldar tipo transformador que es más silenciosa, consume menos energía, requiere menos mantenimiento y es menos costosa que los dispositivos para soldar con motor eléctrico o de gas.

stringer bead A type of weld bead made without appreciable weaving motion.

cordón encordador Un tipo de cordón de soldadura sin movimiento del tejido apreciable.

surface tension Surface tension is an attractive property of the surface of a liquid. Surface tension causes the surface portion of liquid to be attracted to another surface, such as that of another portion of liquid (as in connecting bits of water or a drop of mercury that forms a cohesive ball). In welding, surface tension keeps the molten weld bead in the proper shape during overhead joining operations.

tensión superficial La tensión superficial es una propiedad de atracción en la superficie de un líquido. La tensión superficial provoca que la superficie de un líquido se anexe a otra superficie, por ejemplo, la porción de otro líquido (como gotas de agua o una gota de mercurio, las cuales forman una esfera cohesiva). En soldaduras, la tensión superficial mantiene la forma correcta del cordón de soldadura fundido durante la realización de operaciones de acoplamiento elevado.

tee joint A joint between two members located approximately at right angles to each other in the form of a "T."

junta en T Una junta en medio de dos miembros que están localizados aproximadamente a ángulos rectos de uno al otro en la forma de "T."

tungsten The primary element used in making GTAW electrodes.

tungsteno El elemento primario empleado en la creación de electrodos GTAW.

voltage The measurement of electrical pressure.

voltaje La medición de presión eléctrica.

wagon tracks A pattern of trapped slag inclusions in the weld that show up as discontinuities in X rays of the weld.

huellas de carreta Una muestra de inclusiones de escoria atrapadas en la soldadura que enseña que hay discontinuidades en los rayos-x de la soldadura.

wattage A measurement of the amount of power in the arc; the wattage of the arc controls the width and depth of the weld bead.

número de vatios Una medida de la cantidad de poder en el arco; el número de vatios del arco controla lo ancho y hondo del cordón de la soldadura.

weave pattern The movement of the welding electrode as the weld progresses; common weave patterns include circular, square, zigzag, stepped, "C," "J," "T," and figure eight.

muestra de tejido El movimiento del electrodo para soldar a como progresa la soldadura; las muestras de tejidos comunes incluyen circular, de cuadro, zigzag, de pasos, "C," "J," "T" y la figura ocho.

weld groove A channel in the surface of a workpiece or an opening between two joint members that provides space to contain a weld.

soldadura de ranura Un canal en la superficie de una pieza de trabajo o una abertura entre dos miembros de junta que provee espacio para contener una soldadura.

welding cables The work cable and electrode cable of an arc welding circuit. Refer to Figure 1-7 for direct current electrode positive.

cables para soldar Los cables de pieza de trabajo y el portelectrodo de un circuito de soldadura de arco. Refierase al dibujo orriente directa con el electrodo positivo.

welding leads The work lead and electrode lead of an arc welding circuit.

cables para soldar Los cables de pieza de trabajo y el portelectrodo de un circuito de soldadura de arco.

weld specimen A welded component or section of a welded component that is to be inspected.

muestra de soldadura El componente o sección soldado de un componente también soldado que se va a analizar.

Index

Italic page numbers indicate material in tables or figures.